Dieter Armbruster · Karl G. Kempf
Editors

Decision Policies for Production Networks

Springer

Dr. Dieter Armbruster
School of Mathematical
and Statistical Sciences
Arizona State University
Tempe, AZ 85287-1804
USA

Dr. Karl G. Kempf
Decision Engineering Group
Intel Corporation
5000 W. Chandler Boulevard
Chandler, AZ 85226-3699
USA

ISBN 978-0-85729-643-6 e-ISBN 978-0-85729-644-3
DOI 10.1007/978-0-85729-644-3
Springer London Heidelberg New York Dordrecht

British Library Cataloguing in Publication Data
A catalogue record for this book is available from the British Library

Library of Congress Control Number: 2012932906

Printed on acid-free paper

Springer is part of Springer Science+Business Media (www.springer.com)

Acknowledgments

Contributed volumes like this one are only possible through the collaboration of many different people. First among these are the contributing authors. Without them there would be no book and we are deeply grateful to each not only for providing high quality material but also for responding for our requests for repeated revisions. These revisions were suggested by the many reviewers of the material contained in this volume. They gave their time and effort to help improve the chapters and they are gratefully acknowledged.

Thanks are also due to Anthony Doyle of Springer who initially suggested this volume and then facilitated its development. Claire Protherough and Grace Quinn supported the editorial team, nudging when necessary and exhibiting patience and resourcefulness well beyond the call of duty. Thanks to you both. The unsung parties here are of course the production team who converted our various manuscripts, figures, and tables into this handsome volume. Amazing!

Last but not least, we would like to thank our families who have suffered late nights and grumpy mornings over the span of this project. Your understanding and support have been indispensable.

Contents

Contributors

Tarik Aouam holds an M.Sc. in IE from Kansas State University and a Ph.D. in Operations Research from the IE department at Purdue University. He previously worked for United Airlines (USA) as a senior analyst in the Enterprise Optimization Group developing solutions for crew scheduling and supply chain management. Tarik was also chair of the IE department and Director of the Master of Engineering Management Program at Al Hosn University (UAE). He is currently an associate professor in the school of business administration at Al Akhawayn University (Morocco). His research interests include applications of operations research to production planning, supply chain integration, procurement risk management, and telecommunications. He is a member of the Institute of Industrial Engineers (IIE). Contact: t.aouam@aui.ma

Dieter Armbruster received a Ph.D. in physics at the University of Tübingen, Germany in 1984. He was a postdoc at Cornell University in Mathematics and in Theoretical and Applied Mechanics. Since 1990 he is at Arizona State University where he is a Professor in the School of Mathematical and Statistical Sciences at Arizona State University. He is also currently a part time professor in Mechanical Engineering at Eindhoven University of Technology. His research interests are broad based and range from dynamical systems theory and chaos to the dynamics of complex networks and production systems and the simulation and control of semiconductor fabs. Contact: armbruster@asu.edu

Martin W. Braun received a B.S. in Chemical Engineering from SUNY-Buffalo, and M.S.E. and Ph.D. degrees in Chemical Engineering from Arizona State University. He worked on supervisory control and data acquisition projects while employed at International Imaging Materials, Inc. While working at Texas Instruments, Inc. he created and deployed cutting edge run-to-run controllers at the Kilby Development Center, and mentored research in the area of run-to-run control performance monitoring in collaboration with the University of Texas at Austin. Since joining Intel Corporation he has led additional research and production deployment in the areas of run-to-run control, inventory control, as well as in the mid-range planning solver problem space. Martin's interests include system identification, model predictive control, fault detection, signal processing, and portfolio theory. He is a senior member of the Institute of Electrical and Electronic Engineers (IEEE), and a member of the Institute for Operations Research and the Management Sciences (INFORMS). Contact: martin.w.braun@intel.com

Sergey Dashkovskiy has his M.Sc. (1996) and Ph.D. (2002) degrees in Mathematics from the Moscow (Lomonosov) State University, Russia and University of Jena, Germany, respectively. In 2009 he completed his Habiltation in Mathematics at the University of Bremen. Since 2008 he is the head of the research group Mathematical Modeling of Complex Systems at the University of Bremen and he is a principal investigator of several interdisciplinary research projects. From 2002 to 2011 he was a Senior Researcher in Systems and Control Theory at the Center of Industrial Mathematics, University of Bremen in Germany. Since 2011 he is Professor of Mathematics at the University of Applied Sciences Erfurt, Germany. His research interests are in mathematical modeling of systems with complex structures and behavior, including mathematical material science and large-scale dynamical networks. Contact: sergey.dashkovskiy@fh-erfurt.de

Kenneth Fordyce holds a B.S in Mathematics, a M.S in Operations Research and Statistics, a Ph.D. in Administrative and Engineering Systems, and studied computational methods. He has worked for IBM since 1977 in computational decision science where his efforts have covered almost all aspects of planning, scheduling, and dispatch; as well as decision support and statistics. Ken has served as an adjunct professor at Columbia University and Rensselaer Polytechnic Institute. He has been fortunate to have worked for three teams that received Outstanding Technical Contribution awards from IBM, two of which were finalists for the Edelman award of the Institute for Operations Research and the Management Sciences (INFORMS), and one received an innovative application award from AAAI. Contact: fordyce@us.ibm.com or ken12443@aol.com

Michael Görges was awarded a degree in Industrial Engineering from the University of Bremen in 2008. He is currently working as a research scientist at the Collaborative Research Centre 637 'Autonomous Cooperating Logistic Processes: A Paradigm Shift and its Limitations' at University of Bremen, Germany. Contact: goe@biba.uni-bremen.de

Simone Göttlich holds a Diploma in Business Mathematics and a Ph.D. in Applied Mathematics, and did her postdoctoral research at the University of Kaiserslautern. Since February 2011, she is a professor in the School of Business Informatics and Mathematics at the University of Mannheim. Her research interests include the modeling and simulation of transportation networks (e.g. supply chains) as well as the interaction of discrete and continuous optimization problems. Contact: goettlich@uni-mannheim.de

Oleg Gusikhin is a Technical Leader at Ford Vehicle and Enterprise Sciences Research Laboratory. He received his Ph.D. from the St. Petersburg Institute of Informatics and Automation of the Russian Academy of Sciences and an MBA from the Ross Business School at the University of Michigan. For over 15 years he has been working at Ford Motor Company in different functional areas including Information Technology, Advanced Electronics Manufacturing, and Research and Advanced Engineering. During his tenure at Ford, Dr. Gusikhin has been involved in the design and implementation of advanced information technology and intelligent controls for manufacturing and vehicle systems. Dr. Gusikhin is a recipient of two Henry Ford Technology Awards and the 2009 Institute of Industrial Engineers Transactions Best Application Paper Prize in Scheduling and Logistics. He is a Certified Fellow in Production and Inventory Management, chair of Southeastern IEEE Systems, Man and Cybernetics chapter, and a Lecturer in the Industrial and Operations Engineering department at the University of Michigan. Contact: ogusikhi@ford.com

Michael Herty holds a Diploma and a Ph.D. in Applied Mathematics. He is a professor at RWTH Aachen University (Germany). He previously worked at the TU Kaiserslautern (Germany) and has held visiting professor positions in University of Kwa-Zulu Natal (South Africa) and Southeast University (China). He has authored more than 50 research papers and a textbook on supply chain optimization. His primary research is in applied mathematics with an emphasis on modeling, simulation, and optimization of transport processes governed by hyperbolic partial differential equations. His recent studies involve problems with an underlying network structure, such as those appearing in traffic flow, gas transportation, and supply chains. Contact: herty@mathc.rwth-aachen.de

Thomas Jagalski studied Economics and Business Sciences at Humboldt University Berlin, Germany and holds M.Sc. in Economics and in Management. He previously worked as branch manager for 'Texaco/DEA', in Munich, Germany and as editor-in-chief for 'Industrie Management', a German Journal for Engineering. Thomas is currently working as a research scientist at the Collaborative Research Centre 637 "Autonomous Cooperating Logistic Processes: A Paradigm Shift and its Limitations" at University of Bremen, Germany. Contact: jag@biba.uni-bremen.de

Karl G. Kempf holds a B.A. in Physics, a B.S. in Chemistry, a Ph.D. in Applied Mathematics, and engaged in postdoctoral research in Computer Science. He previously worked for Goodyear Tyre and Rubber Company (UK) in their International Racing Division, at Pinewood Movie Studios (UK) creating cinematic special effects, and at McDonnell Douglas Corporation (USA) on factory and spacecraft automation projects. Karl has served as an adjunct professor at the University of Missouri (Computer Science), Arizona State University (CS and Industrial Engineering), and North Carolina State University (IE). He is currently employed by Intel Corporation and is involved in developing decision support tools for use across the company including factory and supply chain design and operation as well as product design and development. He is a member of the National Academy of Engineering (USA), a Fellow of Intel Corporation and a Fellow of the Institute of Electrical and Electronic Engineers (IEEE), and a member of the Institute for Operations Research and the Management Sciences (INFORMS). Contact: karl.g.kempf@intel.com

Erica Klampfl leads a Strategy and Sustainability Analytics Research Group at Ford Research and Advanced Engineering. Her research interests include the application of Operations Research techniques to manufacturing, marketing, logistics, supply chain management, strategic planning, sustainability, and intelligent vehicle functions. She received a Ph.D. in Computational and Applied Mathematics from Rice University in 2001. Dr. Klampfl is an active member of INFORMS and was Chair of the 2011 INFORMS Prize Committee and is Chair of the 2012 INFORMS Analytics Conference. In addition, Dr. Klampfl is a member of the University of Michigan's Industrial and Operations Engineering Advisory Board, an industry advisor for Michigan State University's Industrial Math Program, Ford's representative for the Institute for Mathematics and its Applications (IMA), and has served periodically as Ford's INFORMS Roundtable representative. She was selected to attend the 2005 National Academy of Engineering 11th Symposium on Frontiers of Engineering and is a 2008 INFORMS Daniel H. Wagner Prize finalist: her internal awards over the past few years include the 2010 Henry Ford Technology Award, and the 2010, 2009, and 2008 Ford Technical Achievement Awards. Contact: eklampfl@ford.com

Erjen Lefeber received his M.Sc. in Applied Mathematics from the University of Twente, Enschede, The Netherlands, in 1996. In 2000 he received his Ph.D. from the University of Twente on the subject of tracking control of nonlinear mechanical systems. Since 2000, he has been an Assistant Professor at the Systems Engineering group, currently the Manufacturing Networks group, at the Department of Mechanical Engineering at Eindhoven University of Technology. His current research interests include modeling and control of manufacturing systems. In 2007 he received an Innovational Research Grant (VIDI) for excellent researchers who are among the best of their age group. Contact: A.A.J.Lefeber@tue.nl

R. John Milne received his B.S. and M.Eng. degrees in operations research from Cornell University and his Ph.D. in decision sciences and engineering systems from Rensselaer Polytechnic Institute. His 26-year career at IBM was dominated by the application of operations research to decision problems in supply chain management and resulted in Edelman finalist and Wagner Prize recognition from INFORMS. He presently teaches engineering management at Clarkson University including courses in operations research, operations and supply chain management, and advising students on their senior capstone design projects. His research focuses on the application of OR to supply chain management problems. Contact: jmilne@clarkson.edu

Lars Naujok received his diploma in mathematics at the University of Bremen in 2008. He is working as a Ph.D. student in mathematics at the University of Bremen in the Collaborative Research Centre 637 "Autonomous Cooperating Logistic Processes: A Paradigm Shift and its Limitations" (CRC 637) funded by the German Research Foundation (DFG). His research activities include modeling and analysis of complex networks, more precisely, the stability analysis of interconnected dynamical systems modeled by ordinary or retarded differential equations with applications in logistics. Contact: larsnaujok@math.uni-bremen.de

Dominique Perdaen did an internship, as a student at Eindhoven University of Technology, at Arizona State University in 2005 on the push–pull point strategy in reentrant manufacturing lines. He was awarded a Master of Science degree in Mechanical Engineering from Eindhoven in 2006 based on his work on a data-driven design of a supervisory controller for a wafer scanner at ASML. Since January 2007, Dominique has worked as an image-sheet registration architect at Océ Technologies. Here, he continued his work as lead engineer on the paper feed module of an innovative black and white product. Currently, Dominique works as a Quality Lead at Oce responsible for the quality processes in a multi-disciplinary engineering project. Contact: dominique.perdaen@oce.com

Christian Ringhofer holds a B.A. in computer science and a M.Sc. and a Ph.D. in Applied Mathematics from the Polytechnical University of Vienna. He has held a postdoctoral position at the University of Wisconsin—Madison, and various visiting positions at the University of Hamburg, the Universidad de Buenos Aires, Osaka University and the University of Vienna. Since 1984 he has been on the faculty of Arizona State University. Research interests include: kinetic theory, mathematical modeling for supply networks, partial differential equations, numerical analysis, and quantum mechanics. Contact: ringhofer@asu.edu

Oliver Rose holds the Chair for Modeling and Simulation at the Department of Computer Science of the University of the Federal Armed Forces Munich, Germany. He received an M.S. degree in Applied Mathematics and a Ph.D. degree in Computer Science from Würzburg University, Germany. His research focuses on the operational modeling, analysis and material flow control of complex manufacturing facilities, in particular, semiconductor factories. He is a member of IEEE, INFORMS Simulation Society, ASIM, and GI. His web address is www.simulation-rose.com and his email address is oliver.rose@unibw.de

Bernd Scholz-Reiter holds a degree in Industrial Engineering from the Technical University Berlin (TUB). Following his Ph.D. (TUB, 1990), Bernd Scholz-Reiter served as postdoctoral fellow in the department for Manufacturing Research at IBM T.J. Watson Research Center in Yorktown Heights, U.S.A. From 1994 to 2000 he held the Chair for Industrial Information Systems at the newly founded Brandenburg Technical University at Cottbus, Germany. In 1998 he founded the Fraunhofer Application Center for Logistic Systems Planning and Information Systems in Cottbus, which he headed until 2000. Since November 2000 Bernd Scholz-Reiter is full professor and holds the new Chair for Planning and Control of Production Systems at the University of Bremen, Germany. Since 2002 he also serves as Managing Director of the Bremen Institute for Production and Logistics GmbH (BIBA) at the University of Bremen. Since July 2007, Bernd Scholz-Reiter is Vice President of the German Research Foundation (DFG). Contact: bsr@biba.uni-bremen.de

Jay Schwartz received the B.S.E., M.S.E., and Ph.D. degrees in Chemical Engineering from Arizona State University. He was the recipient of an Intel Corporation Ph.D. Fellowship award, which funded his research into novel methods for modeling and control of large-scale manufacturing networks. As an intern, he applied his research toward the implementation of a predictive control strategy for managing material flows in Intel's supply chain network. After completing the Ph.D., Jay joined Intel and contributed to the development of power management algorithms at multiple scales ranging from individual microprocessors to large datacenters. His research interests include model predictive control, system identification, and numerical optimization. Jay currently works on real-time power estimation and management for next-generation microarchitectures within Intel Corporation's Microprocessor Development Group. Contact: jay.schwartz@intel.com

Reha Uzsoy is Clifton A. Anderson Distinguished Professor in the Edward P. Fitts Department of Industrial and Systems Engineering at North Carolina State University. He holds B.S. degrees in Industrial Engineering and Mathematics and an M.S. in Industrial Engineering from Bogazici University, Istanbul, Turkey. He received his Ph.D. in Industrial and Systems Engineering in 1990 from the University of Florida. His teaching and research interests are in production planning, scheduling, and supply chain management. He is the author of one book, two edited books, and over eighty refereed journal publications. Before coming to the US he worked as a production engineer with Arcelik AS, a major appliance manufacturer in Istanbul, Turkey. He has also worked as a visiting researcher at Intel Corporation and IC Delco. His research has been supported by the National Science Foundation, Intel Corporation, Hitachi Semiconductor, Harris Corporation, Kimberly Clark, Union Pacific, Ascension Health and General Motors. He was named a Fellow of the Institute of Industrial Engineers in 2005, Outstanding Young Industrial Engineer in Education in 1997, and has received awards for both undergraduate and graduate teaching. Contact: ruzsoy@ncsu.edu

Zhugen Zhou is a PhD student at the University of the Federal Armed Forces Munich, Germany. He is a member of the scientific staff of Prof. Dr. Oliver Rose at the Chair of Modeling and Simulation of the Department of Computer Science. He received an M.S. degree in Computational Engineering from Dresden University of Technology. His research interests include advanced dispatching and scheduling concepts for complex production facilities, particularly WIP balance and due date control in wafer fabs. He is involved in a project about developing workload control and balance strategies for Infineon (Dresden, Germany). Contact: zhugen.zhou@unibw.de

An Overview of Decision Policies for Production Networks

Karl G. Kempf

Abstract This chapter provides the reader with an overview of this volume from a number of perspectives. First is an overview of the business problems addressed and the decision policies required stretching from networks of machines in a factory to networks of factories in a company to networks of companies in a supply chain. Next there is a brief overview of each chapter with advice to the reader on useful sequences of study depending on individual goals and tastes. Finally there is an overview of the network of authors who contributed to this book.

1 Introduction

The economic systems that are the focus of this book involve manufacturing and stretch from the suppliers' suppliers to the customers' customers. The network of manufacturing companies involved is usually referred to as a supply chain. Within each company there is often a network of geographically disperse but interconnected factories taking in materials and putting out products. Within each factory there is almost certainly a network of machines executing manufacturing processes.

On the one hand there are many similarities between these levels: variability in supply and demand cause difficulty at all levels, at each level materials flow from suppliers to customers increasing in value while becoming products, and money generally flows from customers to suppliers with cost incurred at every intermediate stop. On the other hand there are many differences between levels, principally in physical, temporal, and financial scales as we will explore. In addition, information flows upstream and downstream and between levels.

K. G. Kempf (✉)
Decision Engineering Group, Intel Corporation,
5000 W. Chandler Boulevard, Chandler, AZ 85226-3699, USA
e-mail: karl.g.kempf@intel.com

D. Armbruster and K. G. Kempf (eds.), *Decision Policies for Production Networks*,
DOI: 10.1007/978-0-85729-644-3_1,

For our purposes we refer collectively to these complex levels as "production networks". For the sustained business success of any production network, all levels must function efficiently and effectively in an integrated manner. A failure at any level radiates to the other levels with negative impact.

Over the past several decades a number of powerful influences have driven the evolution of production networks. One is the demand generated by the end customer who increasingly desires a product customized for a specific need, priced as though it was produced in mass, and delivered overnight to a residential address. This is compounded by the globalization of business. Hence the potential customer base spans multiple cultures, geographies, and economic systems.

Another driver of the changes in production networks is the double effect of the technology treadmill. Expanding technology supports the introduction of new products through innovative materials and functions as well as production and distribution methods. At the same time, advancing technology provides faster, better support for management through all facets of electronic commerce from data availability to integrated decision-making.

A third influence is the ever-increasing level and pace of competition. A broader more demanding customer base stimulates the profit motive to satisfy the needs. A stronger, more versatile technology base raises the confidence that the needs can be satisfied. Production networks compete for the perceived profit using every opportunity to gain an advantage.

In this evolutionary cycle it is difficult to separate cause and effect. Does demand drive technology or technology drive demand? Does competition foster demand, or vice-versa? Are technologies and markets related? The answers are, of course, "Yes," and it is clear that, to be competitive, the managers of any production network must deal with all of these factors on a daily basis. It is equally clear that a primary managerial tool in this effort is an efficient and effective set of decision policies. Such a set can be the basis for a competitive edge through faster and better decision-making. Components of this set ranging from simple heuristics to complex mathematical algorithms are the focus of each chapter in this book.

From a technical perspective, production networks are dynamic systems with many sources of nonlinearity and stochasticity. At every point in a production network there is variability in both supply and demand with demand volatility usually outpacing the responsiveness of the supply. Production networks involve a wide variety of tradeoffs, and balances must be achieved in decision policies to remain profitable including (at least) tactical versus strategic and local versus global. In making production network management decisions, the appropriateness of the metrics chosen as well as the availability, freshness, and accuracy of the data required are critically important.

Since production networks change size and markets over time, the decision policies must easily scale and adapt to these changes. In addition the technical complexities have been growing over time as production networks evolve and, if not held in check, have a tendency to slow down network responsiveness. Over the same time, business complexity in terms of market expectations and competitive pressures has made forecasting much more difficult. Decision policies are intimately related to forecasting since decisions are made to influence some outcomes in the future. Combining these complexities generates a disturbing scenario. From a control theoretic perspective, if your ability to look forward is diminishing at the same time your speed of response is degrading, you will suffer serious consequences. The question is not whether you will lose control, but rather when it will occur and if you will be able to recognize it when it happens. This situation provides a rich set of research and application opportunities for those interested in decision and control—specifically those in academics and practitioners represented by the contributors to this book.

2 Advice to the Reader

Like most edited works, this volume can be approached and studied from many directions depending on the goals of the reader. The chapters are briefly summarized here and are ordered roughly (a) from a focus on networks of machines to networks of factories to networks of companies, (b) in increasing level of mathematical sophistication, and (c) from methods proven in practice to topics of current research with almost all chapters including worked examples.

The conventional approach of reading front to back (Chaps. "Modeling and Control of Manufacturing Systems" through "The Production Planning Problem: Clearing Functions, Variable Leads Times, Delay Equations and Partial Differential Equations") should certainly build a compelling story of developing decision policies to control production networks. The reader who is less mathematically oriented or more practically focused will probably spend more time studying the first six chapters (Chaps. "Modeling and Control of Manufacturing Systems" through "A Control Theoretic Evaluation of Schedule Nervousness Suppression Techniques for Master Production Scheduling"). A more mathematically oriented or more research-focused reader might study the last six chapters (Chaps. "A Control Theoretic Evaluation of Schedule Nervousness Suppression Techniques for Master Production Scheduling" through "The Production Planning Problem: Clearing Functions, Variable Leads Times, Delay Equations and Partial Differential Equations") in more detail.

Readers more interested in networks of companies could read from back to front (Chaps. "Optimal Order and Distribution Strategies in Production Networks"–"WIP-Oriented Dispatching in Complex Manufacturing Facilities") working down the levels ending at networks of machines. Of course each chapter stands on its own as a description of one or more fundamentally important ideas about decision policies for production networks and can be read in any order. It is advised however that non-conventional approaches all start by reading the basics in Chaps. "Modeling and Control of Manufacturing Systems" and "" and end by reading Chap. "The Production Planning Problem: Clearing Functions, Variable Leads Times, Delay Equations and Partial Differential Equations" that synthesizes many of the ideas that are common to many chapters.

3 An Overview of the Chapters

Chapters "Modeling and Control of Manufacturing Systems" and "The Ongoing Challenge for a Responsive Demand Supply Network: The Final Frontier Controlling the Factory" provide an introduction to decision policies for production networks from both an academic and a practical perspective.

Modeling and Control of Manufacturing Systems—Lefeber: This chapter provides the basic framework for understanding modeling and control of manufacturing systems. It should be read as a tutorial that begins by introducing the fundamental concepts. This is followed by descriptions of incrementally more sophisticated models reflecting mass conservation in queuing systems and exploring them via discrete event simulation culminating in decisions models based on model predictive control (MPC). Along the way the very important concepts of effective process times (EPTs), fluid approximations, and clearing functions (CFs) are explained. All of these are important concepts central to many of the later chapters. Five worked example problems are provided illustrating the tutorial nature of this chapter and building the fundamental themes of the book.

The Ongoing Challenge for a Responsive Demand Supply Network: The Final Frontier—Controlling the Factory—Fordyce and Milne: This second tutorial chapter is the counterpoint to Lefeber. The authors have spent years in the factory and supply chain trenches. Staring Murphy (in the United States, Sod in Europe) in the eye on a daily basis results in a very practical view of the basics. In this chapter from two of the leading practitioners in the area we are taken down the hierarchy of demand supply network planning, scheduling, and dispatch. Along the way it becomes clear that there are different physical and temporal scales that are important as well as increasing levels of detail in what is being decided. The set of worked example problems at the end of the chapter generates a practical feel for the problems encountered at the operational level in large industrial corporations at the beginning of the twenty-first century.

Chapters "WIP-Oriented Dispatching in Complex Manufacturing Facilities" and "Controlling a Re-Entrant Manufacturing Line via the Push–Pull Point" focus specifically on networks of machines from the perspective of delivering products on time in the face of supply and demand variability.

WIP-Oriented Dispatching in Complex Manufacturing Facilities—Rose and Zhou: These authors provide a deep dive into the topic of dispatch in networks of machines within a factory. After noting that different dispatch rules have different performance objectives, they contrast due-date-oriented rules that focus on getting work out of the factory on time with work-in-progress-oriented rules that focus on controlling the workload in the factory. The latter approach often includes the release of work into the factory. The contribution of this chapter is the development of a blended workload balance and due-date control approach. Discrete event simulation is used to analyze performance over a suite of metrics.

Controlling a Re-entrant Manufacturing Line Via the Push–Pull Point—Perdaen, Armbruster, Kempf, and Lefeber: This chapter takes a different perspective on dispatch rules. While the previous chapter was concerned with the impact of rules on the variation of factory performance or supply metrics, this chapter considers rule impact on factory response to demand variability. The authors apply two simple dispatch rules that are widely used in practice—push at the beginning of the production line (aka first buffer first served) and pull at the end of the line (aka shortest expected remaining processing time)—and then vary the point in the production flow that push changes to pull—in conjunction with a material release policy at the beginning of the line that maintains a constant workload in the factory (aka CONWIP or only start material to balance work that exits the factory) . When analyzed using a discrete simulation, such policies generate very good results in the most difficult circumstance encountered in practice—high demand with high variance.

Chapters "JEDI: Just-in-Time Execution and Distribution Information Support System for Automotive Stamping Operations" and "A Control Theoretic Theoretic Evaluation of Schedule Nervousness Suppression Techniques for Master Production Scheduling" continue the theme of managing supply and demand variability but highlight the problems with rapidly changing production schedules at the equipment level or material release schedules at the internal supply chain level in support of agility. The techniques presented can be used in an individual factory or for a network of factories within a company.

JEDI: Just-in-Time Execution and Distribution Information Support System for Automotive Stamping Operations—Gusikhin and Klampfl: In a complex network of manufacturing activities, one approach to agility is continuous modification to the factory production schedule. However, this requires collecting many different types of data from a variety of sources. Maintaining the correctness, consistency, and especially timeliness of the data is a daunting problem. The practitioner authors of this chapter address this problem with a system for production personnel that (a) consolidates and organizes relevant data, (b) makes each element easily traceable to its source for validation if required, (c) clearly identifies the decision problem with visibility into supply, demand, and constraints, and (d) supports what-ifs and sensitivity analysis. Including knowledgeable personnel in the process provides quick solutions to routine problems and higher quality input to powerful optimization algorithms for difficult problems.

A Control Theoretic Evaluation of Schedule Nervousness Suppression Techniques for Master Production Scheduling—Braun and Schwartz: Another approach to improve agility of production schedules is the continuous modification of the master production schedule. However this can result in "schedule nervousness" and large swings in production starts leading to excess setup costs, unnecessary staffing changes, disruptive order changes to suppliers, and unstable inventory. The practitioner authors propose three possible remedies: frozen horizon based on traditional control methods, move suppression borrowed from optimization-based control, and an original contribution called schedule change suppression. Since master production schedules are normally generated using linear programing techniques, they evaluate the stability and performance of these decision polices using an empirical stability analysis method.

Chapters "Chance-Constraint Based Heuristics for Production Planning in the Face of Stochastic Demand and Workload-Dependent Lead Times" and "Traffic Flow Models and Service Rules for Complex Production Systems" are research contributions addressing a number of the underlying issues in modeling complex production networks and policies to control them. These issues have been mentioned in previous chapters, but not in nearly the depth or mathematical sophistication contained here.

Chance-Constraint-Based Heuristics for Production Planning in the Face of Stochastic Demand and Workload-Dependent Lead Times—Aouam and Uzsoy: Although mathematical programing has been applied to a wide variety of problems in production networks, at least two basic limitations recur. On the one hand these models fail to capture the nonlinear relationship between production starts, work in progress, and manufacturing throughput times. "Clearing Functions" are proposed as an effective solution to this deficiency. On the other hand uncertain demand is not adequately represented in these models either. This chapter examines several different chance constraint-based models as well as two-stage and multi-stage stochastic programing formulations to address these deficiencies.

Traffic Flow Models and Service Rules for Complex Production Systems—Ringhofer: Traffic flow models are extensions of rate equation and fluid models, and an aggregation of discrete event simulation models. While they allow for a more detailed description of transient phenomena than rate equations, they are not as versatile as discrete event models in including decision policies. This chapter initially introduces the derivation of traffic flow models based either on clearing functions (as described in the previous chapter) or on mean field theories (borrowed from many body physics). It goes on to model arbitrarily complex policies in the context of traffic flow models, and concludes with numerical examples demonstrating the accuracy of the approach against discrete event simulations.

Chapters "Autonomous Decision Policies for Networks of Production Systems" and "Optimal Order and Distribution Strategies in Production Networks" deal with broad network issues including production and transportation processes and the flow of orders, products, and payments. Once again examples and case studies are presented to aid the reader.

Autonomous Decision Policies for Networks of Production Systems—Scholz-Reiter, Dashkowskiy, Görges, Jagalski, and Naujok: This chapter describes integration and coordination between production and transport processes as an essential function of any operating production network. Different central planning and scheduling functions for shop-floor and transport operations are presented. The concepts of autonomous decision policies and autonomous cooperating processes are described, modeled, and analyzed in mathematical terms and simulation approaches. Based on the mathematical modeling approach, criteria for the stability of production networks are derived and subsequently refined by simulations. Two examples for analyzing the stability and the performance of autonomous decision policies in production networks are included.

Optimal Order and Distribution Strategies in Production Networks—Göttlich, Herty, and Ringhofer: This chapter presents an abstract level of a production network integrating the product flow with order and payment flows through the network. It focuses on the dynamics of a production network where each entity (a) receives orders for its output from other entities in the network and a final customer, (b) orders its input from other entities in the network and a raw material supplier, and (c) receives payments for delivered items and pays production costs for each item produced. The authors introduce a coupled system of ordinary differential delay equations where time-dependent distribution and order strategies of individual manufacturers influence the flow of goods and the total revenue. Order and distribution strategies are degrees of freedom which can vary in time. They are determined as solutions to an optimization problem where additionally economic factors such as production and inventory costs and credit limits influence the maximization of profit. A case study for a sample network is included.

Chapter "The Production Planning Problem: Clearing Functions, Variable Leads Times, Delay Equations and Partial Differential Equations" closes the book and addresses common problems encountered in many of the previous chapters.

The Production Planning Problem: Clearing Functions, Variable Leads Times, Delay Equations and Partial Differential Equations—Armbruster: The production planning problem addressing material release into a factory to generate a desired production profile in the future is either explicitly or implicitly a major theme of almost half of the chapters in this book. However, the common problem of determining the output of a production unit in a network (machine, factory, supply chain node) is dealt with at very different levels of sophistication. This chapter connects the approaches used and discusses their applicability and approximation errors. It also addresses the practical questions of choosing an appropriate clearing function model for a production planning problem. Finally it identifies the open questions associated with the production planning problem in general and the clearing function approach in particular.

4 An Overview of the Authors

What may not be obvious is that the authors of this volume also form a network. As you can see from the list of contributors as well as the references at the end of each chapter: (a) some of us work at the same companies or teach at the same universities, (b) the companies fund some of the academic research and our institutions collaborate, and (c) we collaboratively write papers together as well as refer to each others' papers.

There is an underlying reason for this broad and deep collaboration throughout the authors' network. Too often in academia research can focus on an apparently interesting problem that simply is not encountered in practice. Too often in the haste of practice a practical problem is tackled uninformed by research resulting in a weak solution at best or a fundamentally flawed solution at worst. The astute reader will notice that all of the research contributions in this book are directly motivated by actual practical problems (albeit somewhat simplified for tractability) and all of the applied contributions are directly built on strong theory (albeit somewhat relaxed to account for the oddities of the specific application).

Modeling and Control of Manufacturing Systems

Erjen Lefeber

Abstract In this chapter we provide a framework within which concepts from the field of systems and control can be used for controlling manufacturing systems. After introducing some basic notions from manufacturing analysis, we start with the concept of effective process times (EPTs) which can be used for modeling a manufacturing system as a large queuing network. Next, we restrict ourselves to mass production, which enables us to model manufacturing systems by means of a linear system subject to nonlinear constraints (clearing functions). These models serve as a starting point for designing controllers for these manufacturing systems using Model-based Predictive Control (MPC). Finally, the resulting controllers can be implemented on the queuing network model, and ultimately at the real manufacturing system.

1 Preliminaries

In this section we first recall a few basic notions and the main principles from manufacturing system analysis.

1.1 Basic Notions from Manufacturing Analysis

The items produced by a manufacturing system are called *lots*. Also the words product and job are commonly used. Other important notions are throughput, flow time, wip and utilization. These notions are illustrated in Fig. 1 at factory and machine level.

E. Lefeber (✉)
Department of Mechanical Engineering, Eindhoven University of Technology,
P.O. Box 513, 5600 MB, Eindhoven, The Netherlands
e-mail: A.A.J.Lefeber@tue.nl

D. Armbruster and K. G. Kempf (eds.), *Decision Policies for Production Networks*,
DOI: 10.1007/978-0-85729-644-3_2, © Springer-Verlag London 2012

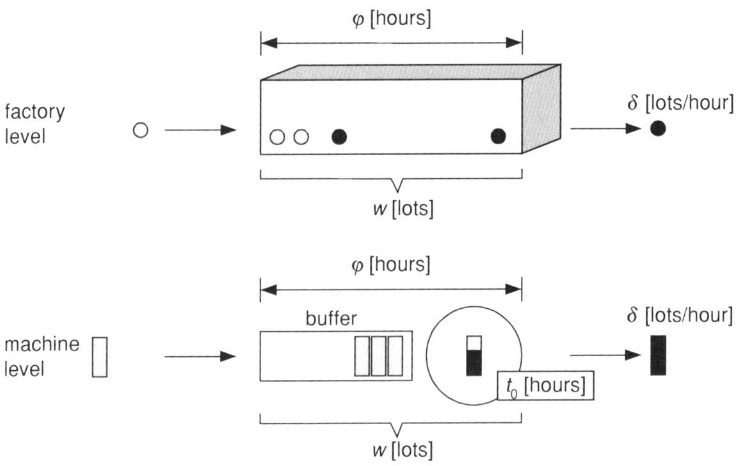

Fig. 1 Basic quantities for manufacturing systems

Raw process time t_0 of a lot denotes the net time a machine needs to process the lot. This process time excludes additions such as setup time, breakdown, or other sources that may increase the time a lot spends in the machine. The raw process time is typically measured in hours or minutes.

Throughput δ denotes the number of lots per unit time that leaves the manufacturing system. At a machine level, this denotes the number of lots that leave a machine per unit time. At a factory level it denotes the number of lots that leave the factory per unit time. The unit of throughput is typically lots/hour.

Flow time φ denotes the time a lot is in the manufacturing system. At a factory level this is the time from the release of the lot into the factory until the finished lot leaves the factory. At a machine level this is the time from entering the machine (or the buffer in front of the machine) until leaving the machine. Flow time is typically measured in days, hours, or minutes. Instead of flow time the words cycle time and throughput time are also commonly used.

Work in process (wip) w denotes the total number of lots in the manufacturing system, i.e., in the factory or in the machine. Wip is measured in lots.

Utilization u denotes the fraction of time that a machine is not idle. A machine is considered idle if it could start processing a new lot. Thus process time as well as downtime, setup-time and preventive maintenance time all contribute to the utilization. Utilization has no dimension and can never exceed 1.0.

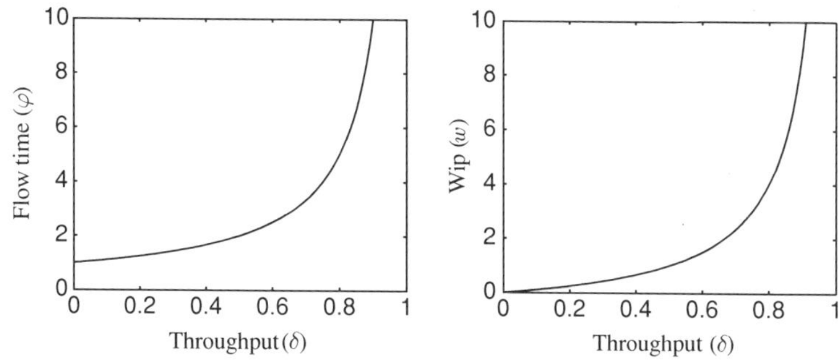

Fig. 2 Basic relations between basic quantities for manufacturing systems

Ideally, a manufacturing system should have both a high throughput and a low flow time or low wip. Unfortunately, these goals are conflicting (cf. Fig. 2) and can not both be met simultaneously. If a high throughput is required, machines should always be busy. As from time to time disturbances like machine failures happen, buffers between two consecutive machines are required to make sure that the second machine can still continue if the first machine fails (or vice versa). Therefore, for a high throughput many lots are needed in the manufacturing system, i.e., wip needs to be high. As a result, if a new lot starts in the system it has a large flow time, since all lots that are currently in the system need to be completed first.

Conversely, the least possible flow time can be achieved if a lot arrives at a completely empty system and never has to wait before processing takes place. As a result, the wip level is small. However, for most of the time machines are not processing, yielding a small throughput.

When trying to control manufacturing systems, a trade-off needs to be made between throughput and flow time, so the nonlinear (steady state) relations depicted in Fig. 2 need to be incorporated in any reasonable model of manufacturing systems. We return to this in Sect. 4.1 when discussing clearing functions.

1.2 Analytical Models for Steady-State Analysis

In order to get some insights in the steady-state performance of a given manufacturing system simple relations can be used. We first deal with mass conservation for determining the mean utilization of workstations and the number of machines required for meeting a required throughput. Furthermore, relations from queueing theory are used to obtain estimates for the mean wip and mean flow time.

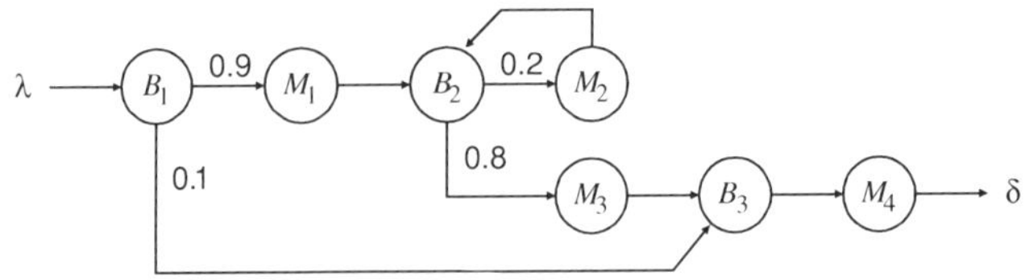

machine	t_0[hrs]
M_1	2.0
M_2	6.0
M_3	1.8
M_4	1.6

Fig. 3 Manufacturing system with rework and bypassing

1.2.1 Mass Conservation (Throughput)

Using mass conservation the mean utilization of workstations can easily be determined.

Example 1 Consider the manufacturing system with rework and bypassing in Fig. 3. The manufacturing system consists of three buffers and four machines. Lots are released at a rate of λ lots/hour. The numbers near the arrows indicate the fraction of the lots that follow that route. For instance, of the lots leaving buffer B_1 90% goes to machine M_1 and 10% goes to buffer B_3. The process time of each machine is listed in the table in Fig. 3.

Let δ_{M_i} and δ_{B_i} denote the throughput of machine M_i $(i = 1, 2, 3, 4)$ and buffer B_i $(i = 1, 2, 3)$, respectively. Using mass conservation we obtain

$$\begin{aligned} \delta_{M_1} &= 0.9\delta_{B_1} & \delta_{B_1} &= \lambda \\ \delta_{M_2} &= 0.2\delta_{B_2} & \delta_{B_2} &= \delta_{M_1} + \delta_{M_2} \\ \delta_{M_3} &= 0.8\delta_{B_2} & \delta_{B_3} &= \delta_{M_3} + 0.1\delta_{B_1} \\ \delta_{M_4} &= \delta_{B_3} & \delta &= \delta_{M_4}. \end{aligned}$$

Solving these linear relations results in:

$$\begin{aligned} \delta_{M_1} &= 0.9\lambda & \delta_{B_1} &= \lambda \\ \delta_{M_2} &= 0.225\lambda & \delta_{B_2} &= 1.125\lambda \\ \delta_{M_3} &= 0.9\lambda & \delta_{B_3} &= \lambda \\ \delta_{M_4} &= \lambda & \delta &= \lambda. \end{aligned}$$

Using the process times of the table in Fig. 3, we obtain for the utilizations:

$$\begin{aligned} u_{M_1} &= 0.9\lambda \cdot 2.0/1 = 1.8\lambda & u_{M_3} &= 0.9\lambda \cdot 1.8/1 = 1.62\lambda \\ u_{M_2} &= 0.225\lambda \cdot 6.0/1 = 1.35\lambda & u_{M_4} &= \lambda \cdot 1.6/1 = 1.6\lambda. \end{aligned}$$

Machine M_1 has the highest utilization, therefore it is the bottleneck and the maximal throughput for this line is $\lambda = 1/1.8 = 0.56$ lots per hour. □

Using mass conservation, utilizations of workstations can be determined straightforwardly. This also provides a way for determining the number of machines required for meeting a given throughput. By modifying the given percentages the effect of rework or a change in product mix can also be studied.

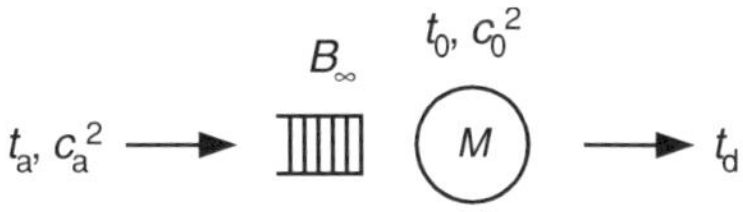

Fig. 4 Single machine workstation

1.2.2 Queueing Relations (Wip, Flow time)

For determining a rough estimate of the corresponding mean flow time and mean wip, basic relations from queueing theory can be used.

Consider a single machine workstation that consists of infinite buffer B_∞ and machine M (see Fig. 4). Lots arrive at the buffer with a stochastic interarrival time. The interarrival time distribution has mean t_a and a standard deviation σ_a which we characterize by the coefficient of variation $c_\mathrm{a} = \sigma_\mathrm{a}/\mu_\mathrm{a}$. The machine has stochastic process times, with mean process time t_0 and coefficient of variation c_0. Finished lots leave the machine with a stochastic interdeparture time, with mean t_d and coefficient of variation c_d. Assuming independent interarrival times and independent process times, the mean waiting time φ_B in buffer B can be approximated for a stable system by means of Kingman's equation [10]:

$$\varphi_B = \frac{c_\mathrm{a}^2 + c_0^2}{2} \cdot \frac{u}{1-u} \cdot t_0 \qquad (1)$$

with the utilization u defined by: $u = t_0/t_\mathrm{a}$. Equation 1 is exact for an $M/G/1$ system, i.e., a single machine workstation with exponentially distributed interarrival times and any distribution for the process time. For other single machine workstations it is an approximation.

For a stable system, we have $t_\mathrm{d} = t_\mathrm{a}$. We can approximate the coefficient of variation c_d by Kuehn's linking equation [11]:

$$c_\mathrm{d}^2 = (1-u^2) \cdot c_\mathrm{a}^2 + u^2 \cdot c_0^2. \qquad (2)$$

This result is exact for an $M/M/1$ system. For other single machine workstations it is an approximation. Having characterized the departure process of a workstation, the arrival process at the next workstation has been characterized as well. As a result, a line of workstations can also be described.

Example 2 (*Three workstations in series*) Consider the three workstation flow line in Fig. 5. For the interarrival time at workstation 0 we have $t_\mathrm{a} = 4.0$ h and $c_\mathrm{a}^2 = 1$. The three workstations are identical with respect to the process times: $t_{0,i} = 3.0$ h for $i = 0, 1, 2$ and $c_{0,i}^2 = 0.5$ for $i = 0, 1, 2$. We want to determine the mean total flow time per lot.

Since $t_\mathrm{a} > t_{0,i}$ for $i = 0, 1, 2$, we have a stable system and $t_{a,i} = t_{d,i} = 4.0$ h for $i = 0, 1, 2$. Subsequently, the utilization for each workstation is $u_i = 3.0/4.0 = 0.75$ for $i = 0, 1, 2$.

Using (1) we calculate the mean flow time for workstation 0

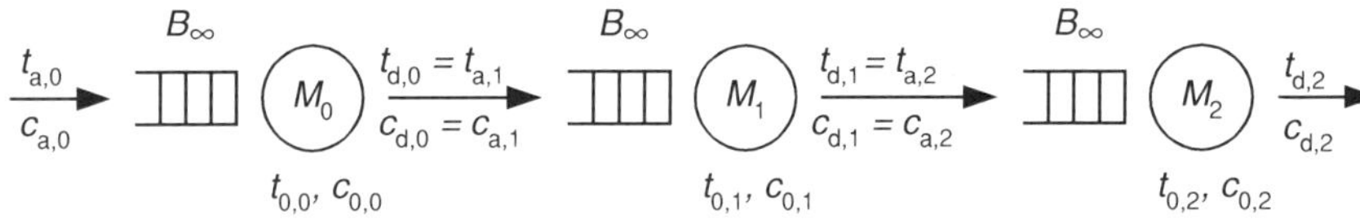

Fig. 5 Three workstation flow line

$$\varphi_0 = \varphi_B + t_0 = \frac{c_a^2 + c_0^2}{2} \cdot \frac{u}{1-u} \cdot t_0 + t_0 = \frac{1+0.5}{2} \cdot \frac{0.75}{1-0.75} \cdot 3.0 + 3.0 = 9.75\,\text{h}.$$

Using (2), we determine the coefficient of variation on the interarrival time $c_{a,1}$ for workstation W_1

$$c_{a,1}^2 = c_{d,0}^2 = (1-u^2) \cdot c_a^2 + u^2 \cdot c_0^2 = (1-0.75^2) \cdot 1 + 0.75^2 \cdot 0.5 = 0.719$$

and the mean flow time for workstation 1

$$\varphi_1 = \frac{0.719+0.5}{2} \cdot \frac{0.75}{1-0.75} \cdot 3.0 + 3.0 = 8.49\,\text{h}.$$

In a similar way, we determine that $c_{a,2}^2 = 0.596$, $\varphi_2 = 7.93$ h. We then calculate the mean total flow time to be

$$\varphi_{tot} = \varphi_0 + \varphi_1 + \varphi_2 = 26.2\,\text{h}.$$

Note that the minimal flow time without variability ($c_a^2 = c_{0,i}^2 = 0$) equals 9.0h. □

Equations 1 and 2 are particular instances of a workstation consisting of a single machine. For workstations consisting of m identical machines in parallel the following approximations can be used [8, 16]:

$$\varphi_B = \frac{c_a^2 + c_0^2}{2} \cdot \frac{u^{\sqrt{2(m+1)}-1}}{m(1-u)} \cdot t_0 \tag{3}$$

$$c_d^2 = (1-u^2) \cdot c_a^2 + u^2 \cdot \frac{c_0^2 + \sqrt{m} - 1}{\sqrt{m}}, \tag{4}$$

where the utilization $u = t_0/(m \cdot t_a)$. Notice that in case $m = 1$ these equations reduce to (1) and (2).

Once the mean flow time has been determined, a third basic relation from queueing theory, Little's law [14], can be used for determining the mean wip level. Little's law states that the mean wip level (number of lots in a manufacturing system) w is equal to the product of the mean throughput δ and the mean flow time φ, provided the system is in steady state:

$$w = \delta \cdot \varphi. \tag{5}$$

An example illustrates how Kingman's equation and Little's law can be used.

Example 3 Consider the system of Example 2 as depicted in Fig. 5. From Example 2 we know that the flow times for the three workstations are respectively

$$\varphi_0 = 9.75 \text{ h}, \varphi_1 = 8.49 \text{ h}, \varphi_2 = 7.93 \text{ h}.$$

Since the steady-state throughput was assumed to be $\delta = 1/t_a = 1/4.0 = 0.25$ lots/hour, we obtain via Little's law

$$\begin{aligned} w_0 &= 0.25 \cdot 9.75 = 2.44 \text{ lots}, \\ w_1 &= 0.25 \cdot 8.49 = 2.12 \text{ lots}, \\ w_2 &= 0.25 \cdot 7.93 = 1.98 \text{ lots}. \end{aligned}$$ □

The above mentioned relations are simple approximations that can be used for getting a rough idea about the possible performance of a manufacturing system. These approximations are fairly accurate for high utilizations but less accurate for lower degrees of utilization. A basic assumption when using these approximations is the independence of the interarrival times, which in general is not the case, e.g., for merging streams of lots. Furthermore, using these equations only steady state behavior can be analyzed. For studying things like ramp-up behavior or for incorporating more details like operator behavior, more sophisticated models are needed, as described next.

1.3 Discrete Event Models

A final observation of relevance for modeling manufacturing systems is the nature of the system signals. In Fig. 6a characteristic graph of the wip at a workstation as a function of time is shown. Wip always takes integer values with arbitrary (nonnegative real) duration. One could consider a manufacturing system to be a system that takes values from a finite set of states and jumps from one state to the other as time evolves. This jump from one state to the other is called an *event*. As we have a countable (discrete) number of states, the name of this class of models is explained.

Manufacturing systems can be modeled as a network of concurrent processes. For example, a buffer is modeled as a process that as long as it can store something is willing to receive new products, and as long as it has something stored is willing to send products. A basic machine is modeled as a process that waits to receive a product; upon receipt it holds the product for a specified amount of time (delay). Upon completion, the machine tries to send the product to the next buffer in the manufacturing line. The machine keeps on doing these three consecutive things. The delay used is often a sample from some distribution.

In particular in the design phase discrete event models are used. These discrete event models usually contain a detailed description of everything that happens in the manufacturing system under consideration, resulting into large models. Since in

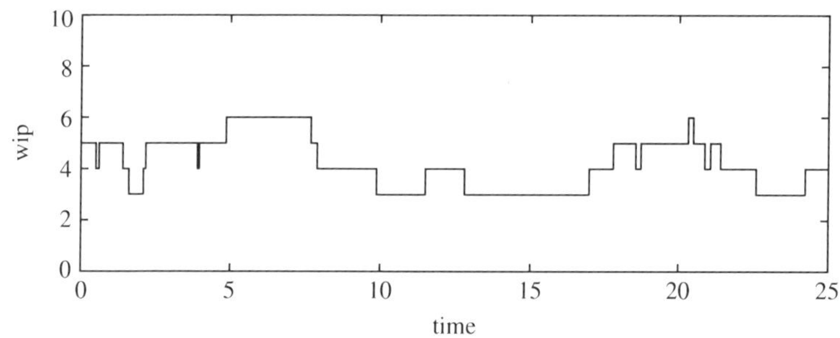

Fig. 6 A characteristic time-behavior of wip at a workstation

practice manufacturing systems are changing continuously, it is very hard to keep these discrete event models up-to-date [4].

Fortunately, for a manufacturing system in operation it is possible to arrive at more simple/less detailed discrete event models by using the concept of Effective Process Times (EPTs) as discussed in the next section.

2 Effective Process Times (EPTs)

For the processing of a lot at a machine, many steps may be required. For example, it could be that an operator needs to get the lot from a storage device, setup a specific tool that is required for processing the lot, put the lot on an available machine, start a specific program for processing the lot, wait until this processing has finished (meanwhile doing something else), inspect the lot to determine if all went well, possibly perform some additional processing (e.g., rework), remove the lot from the machine and put it on another storage device and transport it to the next machine. At all of these steps something might go wrong: the operator might not be available, after setting up the machine the operator finds out that the required recipe cannot be run on this machine, the machine might fail during processing, no storage device is available anymore so the machine cannot be unloaded and is blocked, etc.

Even though one might build a discrete event model including all these details, it is impossible to measure all sources of variability that might occur in a manufacturing system. One might measure some of them and incorporate these in a discrete event model. The number of operators and tools can be modeled explicitly and it is common practice to collect data on mean times to failure and mean times to repair of machines. Also schedules for (preventive) maintenance can be incorporated explicitly in a discrete event model. Nevertheless, still not all sources of variability are included. This is clearly illustrated in Fig. 7, obtained from [9]. The left graph contains actual realizations of flow times of lots leaving a real manufacturing system, whereas the right graph contains the results of a detailed discrete event simulation model including stochasticity. It turns out that in reality flow times are much higher

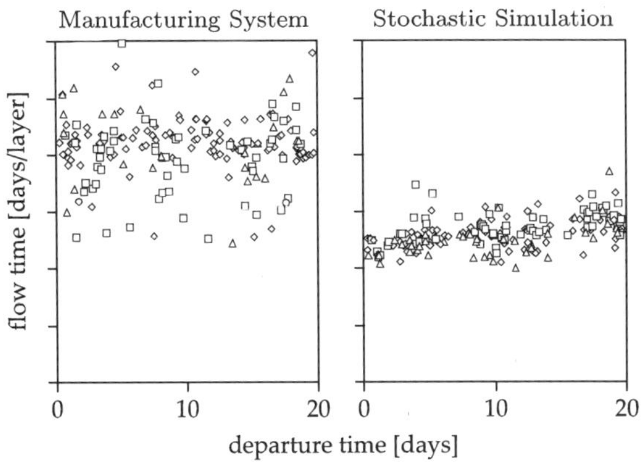

Fig. 7 A comparison

and much more irregular than simulation predicts. So, even if one endeavors to capture all variability present in a manufacturing system, still the outcome predicted by the model is far from reality.

Hopp and Spearman [8] use the term *Effective Process Time* (EPT) as the time seen by lots from a logistical point of view. In order to determine this Effective Process Time, Hopp and Spearman assume that the contribution of the individual sources of variability is known.

Instead of taking the bottom-up view of Hopp and Spearman, a top-down approach can also be taken, as shown by Jacobs et al. [9], where algorithms have been introduced that enable determination of Effective Process Time realizations from a list of events. For these algorithms, the basic idea of the Effective Process Time to include time losses was used as a starting point.

To illustrate this approach, we first deal with a workstation consisting of a single machine, serving one lot type, using a First In First Out (FIFO) policy. Then we deal with the more general case.

2.1 Single Machine, One Lot Type, FIFO Policy

Consider a workstation consisting of a single machine, serving one lot type, using a First In First Out (FIFO) policy. Let the Gantt chart of Fig. 8 depict what happened at this workstation during a certain time interval. At $t = 0$ the first lot arrives at the workstation. After a setup, the processing of the lot starts at $t = 2$ and is completed at $t = 6$. At $t = 4$ the second lot arrives at the workstation. At $t = 6$ this lot could have been started, but apparently no operator was available, so only at $t = 7$ the setup for this lot starts. Eventually, at $t = 8$ the processing of the lot starts and is completed at $t = 12$. The fifth lot arrives at the workstation at $t = 22$, processing starts at $t = 24$, but at $t = 26$ the machine breaks down. It takes until $t = 28$ before the machine has been repaired and the processing of the fifth lot continues. The processing of the fifth lot is completed at $t = 30$.

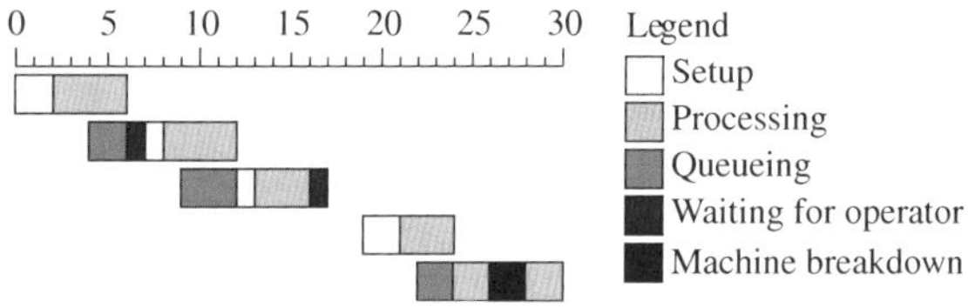

Fig. 8 Gantt chart of 5 lots at a single machine workstation

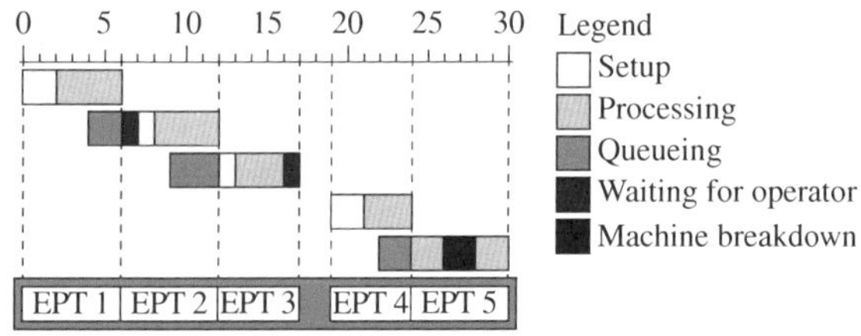

Fig. 9 EPT realizations of 5 lots at a workstation

If we take the point of view of a lot, what does a lot see from a logistical point of view? The first lot arrives at an empty system at $t = 0$ and departs from this system at $t = 6$. From the point of view of this lot, its processing took 6 time-units. The second lot arrives at a non-empty system at $t = 4$. Clearly, this lot needs to wait. However, at $t = 6$, if we forget about the second lot, the system becomes empty again. So from $t = 6$ on the second lot does not need to wait anymore. At $t = 12$ the second lot leaves the system, so from the point of view of this lot, its processing took from $t = 6$ till $t = 12$; the lot does not know whether waiting for an operator and a setup is part of its processing. Similarly, the third lot sees no need for waiting after $t = 12$ and leaves the system at $t = 17$, so it assumes to have been processed from $t = 12$ till $t = 17$. Following this reasoning, the resulting Effective Proces Times for lots are as depicted in Fig. 9. Notice that only arrival and departure events of lots to a workstation are needed for determining the Effective Process Times. Furthermore, none of the contributing disturbances needs to be measured.

In highly automated manufacturing systems, arrival and departure events of lots are being registered, so for these manufacturing systems, Effective Process Time realizations can be determined rather easily. Next, these EPT realizations can be used in a relatively simple discrete event model of the manufacturing system. This discrete event model only contains the architecture of the manufacturing system, buffers and machines. The process times of these machines are samples from their EPT-distribution as measured from real manufacturing data. Machine failures, operators, etc., do not need to be included as this is all included in the EPT-distributions. Furthermore, the algorithms as provided in [9] are *utilization independent*. That is, data collected at a certain throughput rate is also valid for different throughput rates. Furthermore, since EPT-realizations characterize operational time variability, they can be used for performance measuring. For more on this issue, the interested reader is referred to [9].

Recently, the above mentioned EPT-model has been generalized. This generalization is presented next.

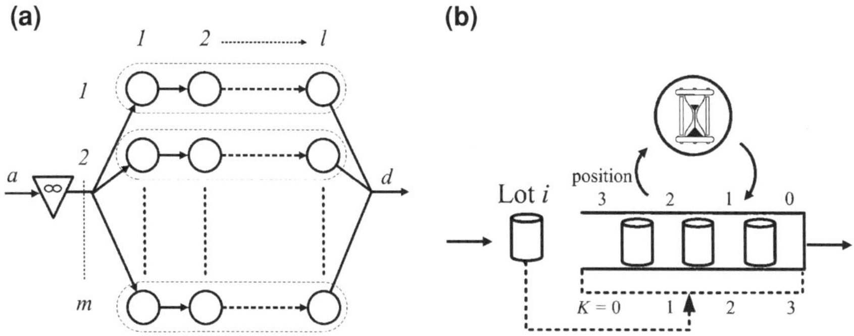

Fig. 10 **a** An example of a workstation. **b** The proposed aggregate model

2.2 Integrated Processing Workstations

Consider an integrated processing workstation consisting of m identical parallel machines, each of which have l sequential integrated processes, cf. Fig. 10. We replace the model of this workstation with a much simpler model, which is not a true physical server anymore, i.e., the structure of the aggregate model differs significantly from the real workstation. Nevertheless, the input/output behavior of the aggregate model closely resembles the input/output behavior of the workstation it models. Lots arrive according to some arrival process to the queue of the aggregate model. Lot i is defined as the ith *arriving* lot in this queue. The queue consists of *all* lots that are currently in the system, including lots that are (supposed to be) in process. Therefore, the queue is *not* a queue as in common queue-server models. Lots are not physically processed, i.e., during "processing" lots stay in the queue. Processing is modeled as a timer that determines when the next lot leaves the queue. When the timer expires, i.e., the "process time" has elapsed, the lot that is currently first in the queue leaves the system. Upon arrival of a new lot i, it is determined how many of the lots already present in the queue w are overtaken by lot i. The number of lots to overtake $K \in \{0, 1, \ldots, w\}$ is sampled from a probability distribution which depends on the number of lots w in the queue just before lot i arrives. The arriving lot is placed on position $w - K$ in the queue, where position 0 corresponds with the head of the queue. The timer starts when either a lot arrives to an empty system, or a lot departs while leaving one or more lots behind. The duration of the "process time" is sampled from a distribution which depends on the number of lots w in the queue just after the timer starts, i.e., including a possibly newly arrived lot. We model the server as a timer to allow newly arriving lots to overtake *all* lots in the system while the timer is running. We need this to model the possibility that a lot which arrives second to a multi-machine workstation leaves first.

Example 4 Consider the Gantt chart in Fig. 11 which depicts what happened at a three machine workstation. At $t = 1$, the first lot arrives at the workstation, service at machine 1 is started, and service is completed at $t = 25$. At $t = 2$, the second lot

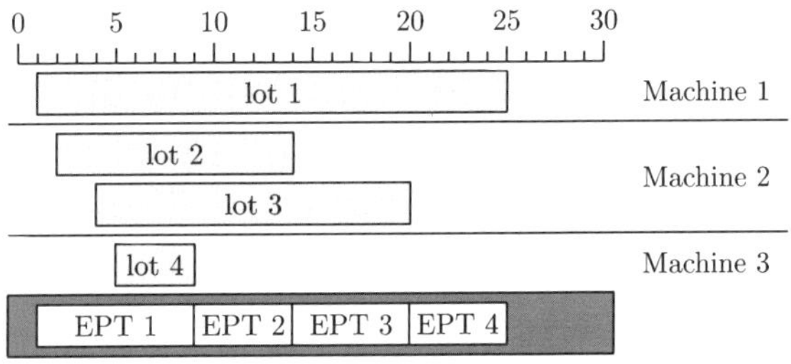

Fig. 11 Gantt chart of 4 lots at a three machine workstation, and the corresponding realization for the aggregate model

arrives at the workstation, service at machine 2 is started, and service is completed at $t = 14$. At $t = 4$, the third lot arrives at the workstation. For some reason it is not served at machine 3, but it waits to be served at machine 2. Its service at machine 2 (effectively) starts at $t = 14$ and is completed at $t = 20$. Finally, the fourth lot arrives at the workstation at $t = 5$, is served at machine 3, and leaves the system at $t = 9$.

In the aggregate model we model the resulting input-output behavior of this system differently. At $t = 1$, the first lot arrives and a timer is set, which expires at $t = 9$. Meanwhile, the second lot arrives at $t = 2$ and is inserted at the head of the queue. Next, the third lot arrives at $t = 4$, and is inserted in the middle of the queue, i.e., behind lot 2, but in front of lot 1. At $t = 5$, the fourth lot arrives which is inserted at the head of the queue, i.e., it overtakes the three lots already in the queue. When the timer expires at $t = 9$, the lot that is at the head of the queue leaves the system, i.e., lot 4 leaves the system. Then the timer is set again to expire at $t = 14$. Again, the head of the queue leaves the system, which is lot 2. The timer is set again to expire at $t = 20$, and lot 3 leaves the system. Next, the timer is set to ring at $t = 25$ and finally lot 1 leaves the system.

For more details about this aggregate model for integrated processing workstations, including implementation issues and algorithms for deriving distributions from real manufacturing data, the interested reader is referred to [19]. In that paper an extensive simulation study and an industry case study demonstrate that the aggregate model can accurately predict the cycle time distribution of integrated processing workstations in semiconductor manufacturing.

Most importantly, EPTs can be determined from real manufacturing data and yield relatively simple discrete event models of the manufacturing system under consideration. These relatively simple discrete event models serve as a starting point for controlling manufacturing systems.

3 Control Framework

In the previous section, the concept of Effective Process Times has been introduced as a means to arrive at relatively simple discrete event models for manufacturing systems, using measurements from the real manufacturing system under

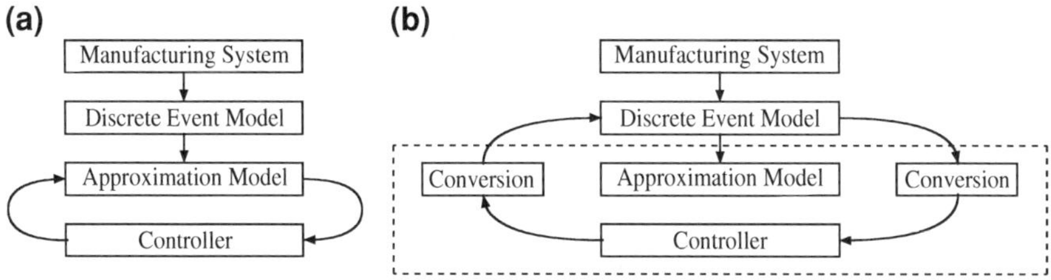

Fig. 12 **a** Control framework (I). **b** Control framework (II)

consideration. This is the first step in a control framework. The resulting discrete event models are large queueing networks which capture the dynamics reasonably well. These relatively simple discrete event models are not only a starting point for analyzing the dynamics of a manufacturing system, but can also be used as a starting point for controller design. If one is able to control the dynamics of the discrete event model of the manufacturing system, the resulting controller can also be used for controlling the real manufacturing system.

Even though control theory exists for controlling discrete event systems, unfortunately none of it is appropriate for controlling discrete event models of real-life manufacturing systems. This is mainly due to the large number of states of a manufacturing system. Therefore, a different approach is needed.

If we concentrate on mass production, the distinction between lots is not really necessary and lots can be viewed in a more continuous way. Instead of the discrete event model we might consider an approximation model. This is the second step in the control framework. Next, we can use standard control theory for deriving a controller for the approximation model. These first three steps in the control framework are illustrated in Fig. 12a. We elaborate on this second and third step in the next two sections. For now it is sufficient to know that time is discretized into periods (e.g., shifts) and that the resulting controller provides production targets per shift for each machine. So for now we assume that the derived controller behaves as desired on the approximation model. As a fourth step this controller could be connected to the discrete event model. This cannot be done directly, since the derived controller is not a discrete event controller. The control actions still need to be transformed into events. It might very well be that the optimal control action is to produce 2.75 lots during the next shift. One still needs to decide how many lots to really start (2 or 3), and also when to start them. This is the left conversion block in Fig. 12b. From this figure, it can also be seen that a conversion is needed from discrete event model to controller. In the remainder of this chapter we assume to sample the discrete event model once every shift. Other strategies might be followed. For example, if at the beginning of a shift a machine breaks down it might not be such a good idea to wait until the end of the shift before setting new production targets. Designing proper conversion blocks is the fourth step in the control framework.

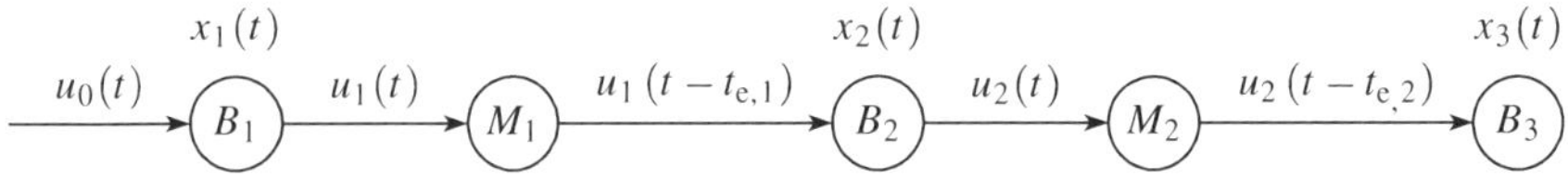

Fig. 13 A simple manufacturing system

After the fourth step, i.e., properly designing the two conversion blocks, a suitable discrete event controller for the discrete event model is obtained, as illustrated in Fig. 12b (dashed).

Eventually, as a fifth and final step, the designed controller can be disconnected from the discrete event model, and attached to the manufacturing system.

4 An Approximation Model

The analytical approximations models of Sect. 1.2 are only concerned with steady state, no dynamic behavior is included. This disadvantage is overcome by discrete event models as discussed in Sect. 2, where each lot is modeled separately and stochastically. In Sect. 2 we derived how less detailed discrete event models can be built by abstracting from all kinds of disturbances like machine failure, setups, operator behavior, etc. By aggregating all disturbances into one Effective Process Time, a complex manufacturing system can be modeled as a relatively simple queueing network. Furthermore, the data required for this model can easily be measured from manufacturing data.

Even though this approach considerably reduces the complexity of discrete event models for manufacturing systems, this aggregate model is still unsuitable for manufacturing planning and control. Therefore, in this section we introduce a next level of aggregation, by abstracting from events. Using the abstraction presented in Sect. 2 we can view a workstation as a node in a queueing network. In this section we assume that such a node processes a deterministic continuous stream of fluid. That is, we consider this queue as a so-called fluid queue.

For example, consider a simple manufacturing system consisting of two machines in series, as displayed in Fig. 13. Let $t_{e,i}$ denote the Effective Process Time of the ith machine for $i \in \{1, 2\}$. Furthermore, let $u_0(t)$ denote the rate at which lots arrive at the system at time t, $u_i(t)$ the rate at machine M_i starts lots at time t, $x_i(t)$ the number of lots in buffer B_i at time t ($i \in \{1, 2\}$) and $x_3(t)$ the cumulative number of lots produced by the manufacturing system at time t.

The rate of change of the buffer contents is given by the difference between the rates at which lots enter and leave the buffer, taking into account the time-delay due to processing:

$$\begin{aligned}\dot{x}_1(t) &= u_0(t) - u_1(t),\\ \dot{x}_2(t) &= u_1\left(t - t_{e,1}\right) - u_2(t),\\ \dot{x}_3(t) &= u_2\left(t - t_{e,2}\right).\end{aligned} \tag{6}$$

In practice, manufacturing systems are often controlled by means of setting production targets per shift. That is, time is divided into shifts for example, 8 or 12 h. For this period of 8 or 12 h it is determined how many lots should be started on each machine. The control problem then reduces to determining these production targets per shift.

To that end, we sample the continuous time system (6) using a zero-order-hold sampling, cf. [2]. Assuming that the longest Effective Process Time is less than the duration of a shift, the resulting zero-order-hold sampling of the system in (6) becomes

$$\begin{bmatrix}\bar{x}_1(k+1)\\ \bar{x}_2(k+1)\\ \bar{x}_3(k+1)\\ \hline \bar{x}_4(k+1)\\ \bar{x}_5(k+1)\end{bmatrix} = \left[\begin{array}{ccc|cc}1&0&0&0&0\\0&1&0&\frac{t_{e,1}}{h}&0\\0&0&1&0&\frac{t_{e,2}}{h}\\ \hline 0&0&0&0&0\\0&0&0&0&0\end{array}\right]\begin{bmatrix}\bar{x}_1(k)\\ \bar{x}_2(k)\\ \bar{x}_3(k)\\ \hline \bar{x}_4(k)\\ \bar{x}_5(k)\end{bmatrix} + \begin{bmatrix}1&-1&0\\0&\frac{h-t_{e,1}}{h}&-1\\0&0&\frac{h-t_{e,2}}{h}\\ \hline 0&1&0\\0&0&1\end{bmatrix}\begin{bmatrix}\bar{u}_0(k)\\ \bar{u}_1(k)\\ \bar{u}_2(k)\end{bmatrix} \tag{7}$$

where $\bar{u}_0(k)$ denotes the number of lots arriving at the system during shift k, $\bar{u}_i(k)$ the number of lots started at machine M_i during shift k, $\bar{x}_i(k)$ the number of lots in buffer B_i at the beginning of shift k ($i \in \{1, 2\}$), and $\bar{x}_3(k)$ the cumulative number of lots produced by the manufacturing system at the beginning of shift k. Furthermore, h denotes the sample period, e.g., 8 or 12 h. The auxiliary variables $\bar{x}_4(k)$ and $\bar{x}_5(k)$ are required to remember the starts during the previous shift, in order to incorporate the lots for which processing is started in shift k on machine M_1 and M_2 respectively but completed in shift $k+1$. If the longest Effective Process Time exceeds the duration of a shift, but not exceed the duration of two shifts, similarly auxiliary variables $\bar{x}_6(k)$, and $\bar{x}_7(k)$ are required.

The model (6) and its discrete time equivalent (7) are also subject to constraints. We present the constraints for the model (7). For the model (6), similar constraints hold.

The first constraint is a non-negativity constraint: buffer contents can never be negative. Also production targets cannot become negative. Expressed mathematically we have the following constraints:

$$\bar{x}_i(k) \geq 0 \qquad i \in \{1, 2, 3, 4, 5\} \qquad \forall k \tag{8a}$$

$$\bar{u}_j(k) \geq 0 \quad j \in \{1, 2, 3\} \quad \forall k \tag{8b}$$

Furthermore, machines can produce at most at maximal capacity. That is, the total time spent on serving the required number of lots during a shift cannot exceed the duration of the shift:

$$t_{e,j} \cdot \bar{u}_j(k) \leq h \quad j \in \{1, 2, 3\} \quad \forall k \tag{8c}$$

where h again denotes the sample period or shift duration.

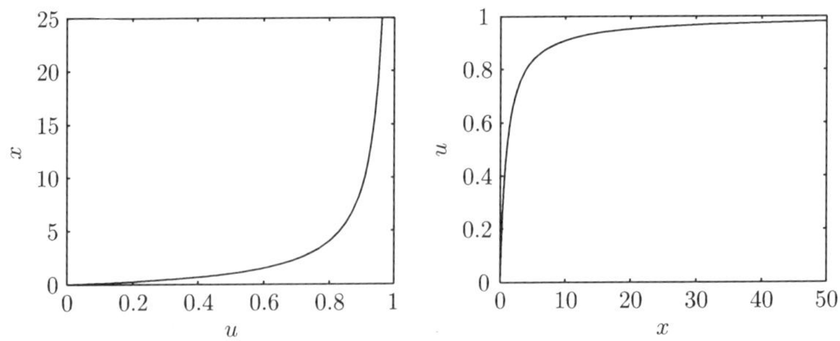

Fig. 14 Effective clearing function of (9) with $c_a = c_e = m = 1$

4.1 Clearing Functions

The model (7) with constraints (8) describes the dynamics of a manufacturing system well. By incorporating delays due to processing, the minimal flow time is also taken into account. Furthermore, steady-state corresponds with the mass conservation results presented in Sect. 1.2.1.

Nevertheless, one property of manufacturing systems is not yet taken into account in the model (7), (8). And that is the queueing relations (3).

In order not to lose the steady state queueing relation between throughput and queue length, we include this relation as a system constraint.

Consider a workstation that consists of m identical servers in parallel that all have a mean Effective Process Times t_e and coefficient of variation c_e. Furthermore, assume that the coefficient of variation of the interarrival times is c_a and that the utilization of this workstation is $\rho < 1$. Then we know from (3), (5) that in steady state the mean number of lots in this workstation is approximately given by

$$x = \frac{c_a^2 + c_e^2}{2} \cdot \frac{\rho^{\sqrt{2(m+1)}-1}}{m(1-\rho)} + \rho. \tag{9}$$

In Fig. 14 this relation has been depicted graphically. In the left-hand side of this figure one can clearly see that for an increasing utilization, the number of lots in this workstation increases nonlinearly. By swapping axes, this relation can be understood differently. Depending on the number of lots in the workstation, a certain utilization can be achieved, or a certain throughput. This has been depicted in the right-hand side of Fig. 14. This relation is also known as the *clearing function* as introduced by [7].

For the purpose of production planning, this effective clearing function provides an upper bound for the utilization of the workstation depending on the number of lots in this workstation. Therefore, for the model (7), in addition to the constraints (8) we also have (using $\rho = \bar{u} \cdot t_e/(h \cdot m)$ and $m = 1$):

$$\begin{aligned}
\frac{c_{\mathrm{a},1}^2 + c_{\mathrm{e},1}^2}{2} \cdot \frac{u_1(k)^2}{\frac{h}{t_{\mathrm{e},1}}\left(\frac{h}{t_{\mathrm{e},1}} - u_1(k)\right)} + \frac{t_{\mathrm{e},1}}{h} u_1(k) \leq \bar{x}_1(k) \quad \forall k \\
\frac{c_{\mathrm{a},2}^2 + c_{\mathrm{e},2}^2}{2} \cdot \frac{u_2(k)^2}{\frac{h}{t_{\mathrm{e},2}}\left(\frac{h}{t_{\mathrm{e},2}} - u_2(k)\right)} + \frac{t_{\mathrm{e},2}}{h} u_2(k) \leq \bar{x}_2(k) \quad \forall k.
\end{aligned} \tag{10}$$

The clearing function model for production planning then consists of the model (7) together with the constraints (8) and (10). When we want to use this clearing function model for production planning, we need the parameters c_{e} and c_{a}. In Sect. 2 we explained how Effective Process Times can be determined for each workstation, which provides us with the parameter c_{e} for each workstation. Additionally, for each workstation the interarrival times of lots can also be determined from arrival events, which provides us with the parameter c_{a} for each workstation. Therefore, both parameters can easily be determined from manufacturing data.

We conclude this section with some remarks about the additional constraints (10). The first remark is that these constraints are convex in the input u, so optimization problems become "simple" convex optimization problems. A second remark is that from a practical point of view, one can easily approximate each convex constraint by means of several linear constraints. A third remark is that the constraints (10) only hold for steady state, whereas our system is never in steady state. A more accurate planning result is obtained by conditioning the expected throughput on the current work in the buffer, resulting in so-called *transient clearing functions*. For the latter subject, the interested reader is referred to [15].

5 Controller Design

In the previous section we derived a fluid model as an approximation for the discrete event model derived earlier. The next step in the control framework presented in Sect. 3 is to control the approximation model using standard techniques from control theory.

Typically two control problems can be distinguished: the *trajectory generation problem* and the *reference tracking control problem*. The solution of the first problem serves as an input for the second problem.

To illustrate the difference between these two problems, consider the problem of automatically flying an airplane from A to B by means of an autopilot. Then also two problems are solved separately. The first problem is to determine a trajectory for the airplane to fly which brings it from A to B. The resulting flight plan is a solution to the trajectory generation problem. The second problem is the design of the autopilot itself. Given an arbitrary feasible reference trajectory for this airplane, how to make sure that it is tracked as well as possible, despite all kinds of disturbances. The latter is the reference tracking control problem. We follow a similar approach for the control of manufacturing systems.

5.1 Trajectory Generation Problem

The trajectory generation problem is the problem of finding a feasible reference trajectory for the system, also known as production planning. So for the example considered previously, the problem is to find a trajectory $\big(x_r(k), u_r(k)\big)$ which satisfies (7) as well as the constraints (8) and (10). Clearly, many trajectories exist that meet these requirements. Typically, "the best" trajectory is looked for. Therefore, the trajectory generation or production planning problem is often formulated as an optimization problem.

Example 5 Consider the system described by (7) together with the constraints (8) and (10). Assume that $c_{\mathrm{a,i}} = c_{\mathrm{e,i}} = 1$, $t_{\mathrm{e,i}} = 1$ $(i = 1, 2)$, $h = 2$, and that the cumulative demand is given by $x_{r,3}(k) = k$. If one would like to satisfy this cumulative demand while having a minimal number of jobs in the system, the trajectory generation problem can be formulated as the following optimization problem:

$$\begin{aligned} &\min_{u_r(k), x_r(k)} \sum_{k=1}^{N} x_1(k) + x_2(k) && \\ &\text{subject to } x_{r,3} = k && k = 1, \dots, N \\ &\qquad\qquad (7), (8), (10) && k = 1, \dots, N \end{aligned}$$

The solution to this problem is given by

$$\begin{array}{lll} x_{r,1}(k) = 1 & u_{r,0}(k) = 1 & k = 1, \dots, N \\ x_{r,2}(k) = 1 & u_{r,1}(k) = 1 & k = 1, \dots, N \\ x_{r,3}(k) = k & u_{r,2}(k) = 1 & k = 1, \dots, N \\ x_{r,4}(k) = 1 & & k = 1, \dots, N \\ x_{r,5}(k) = 1 & & k = 1, \dots, N. \end{array} \tag{11}$$

5.2 Reference Tracking: Model-Based Predictive Control (MPC)

For the reference tracking control problem, we assume that an *arbitrary* feasible reference trajectory is given. So for the example considered before we assume that a reference trajectory $\big(x_r(k), u_r(k)\big)$ is given which satisfies (7) together with the constraints (8) and (10). This could for example be the trajectory (11), but any other feasible reference trajectory can be used as a starting point as well. The goal in the reference tracking control problem is to find an input $u(k)$ which guarantees that the system tracks this reference input, while meeting the constraints (8) and (10).

In order to solve the reference tracking control problem, the tracking error dynamics is considered. For the remainder of this section we assume that the system dynamics is described by

$$x(k+1) = Ax(k) + Bu(k)$$

subject to the linear constraints

$$Ex(k) + Fu(k) \leq g.$$

Without loss of generality this can be extended to nonlinear dynamics with nonlinear constraints.

In addition, a feasible reference trajectory $\big(x_r(k), u_r(k)\big)$ is given, i.e., a trajectory which satisfies

$$x_r(k+1) = Ax_r(k) + Bu_r(k)$$

and

$$Ex_r(k) + Fu_r(k) \leq g.$$

Next, one can define the tracking error $\tilde{x}(k) = x(k) - x_r(k)$, and the input correction $\tilde{u}(k) = u(k) - u_r(k)$. Then the tracking error dynamics becomes

$$\tilde{x}(k+1) = A\tilde{x}(k) + B\tilde{u}(k) \tag{12a}$$

subject to the constraints

$$E\big(\tilde{x}(k) + x_r(k)\big) + F\big(\tilde{u}(k) + u_r(k)\big) \leq g$$

or

$$E\tilde{x}(k) + F\tilde{u}(k) \leq g - Ex_r(k) - Fu_r(k) \tag{12b}$$

Using these error coordinates, the reference tracking control problem can be formulated as to find an input correction $\tilde{u}(k)$ which steers the error dynamics (12a) toward 0, while satisfying the constraints (12b).

Since we have a system with constraints, the most suitable technique from standard control theory is Model-based Predictive Control (MPC).

The basic idea of MPC is to use the model of the system (12a) to predict the state evolution as a function of future inputs. Furthermore, a cost function is used which penalizes the predicted future deviations from the reference trajectory. This cost function is then minimized over the future inputs, subject to the constraints (12b). This optimization takes place over a so-called prediction horizon p, i.e., the first p inputs are determined in this optimization problem. The resulting control action then consists of the first of these inputs. One time period later, the entire procedure is repeated. Therefore, MPC is also called a receding horizon stategy. This is illustrated in Fig. 15.

Assume that at time k, the tracking error $\tilde{x}(k) = \tilde{x}(k|k)$ is measured. So we have the tracking error $\tilde{x}$ at time k given that we are currently at time k. Using a horizon of length p, we can define the input corrections for the times $k, k+1, \ldots, k+p-1$ given that we are currently at time k: $\tilde{u}(k|k), \tilde{u}(k+1|k), \ldots, \tilde{u}(k+p-1|k)$.

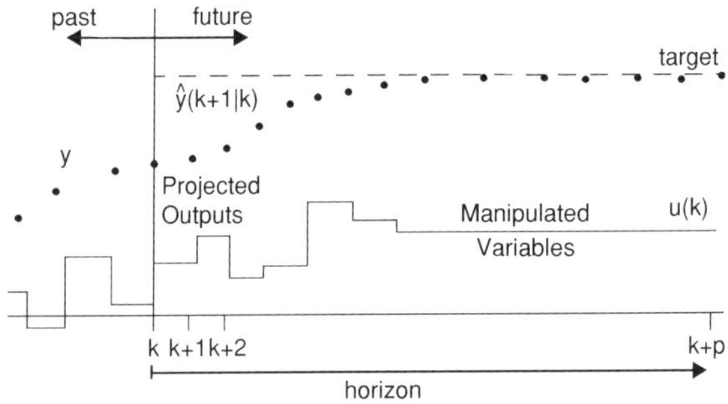

Fig. 15 The ingredients of MPC

By means of the model (12a) we are able to predict the resulting tracking errors as a function of these future input corrections:

$$\begin{bmatrix} \tilde{x}(k+1|k) \\ \tilde{x}(k+2|k) \\ \vdots \\ \tilde{x}(k+p|k) \end{bmatrix} = \begin{bmatrix} A \\ A^2 \\ \vdots \\ A^p \end{bmatrix} x(k|k) + \begin{bmatrix} B & 0 & \dots & 0 \\ AB & \ddots & \ddots & \vdots \\ \vdots & \ddots & \ddots & 0 \\ A^{p-1}B & \dots & AB & B \end{bmatrix} \begin{bmatrix} \tilde{u}(k|k) \\ \tilde{u}(k+1|k) \\ \vdots \\ \tilde{u}(k+p-1|k) \end{bmatrix} \tag{13}$$

Next we define a cost function for having a non-zero tracking error. One of the properties of our controlled system is that if we happen to be on the reference, we should stay on the reference. In particular this implies that the cost function should be such that costs are 0 if and only if the system stay in $(\tilde{x}, \tilde{u}) = (0, 0)$.

In control theory often a quadratic cost function is used:

$$\min_{u(k|k),\dots u(k+p-1|k)} \sum_{i=1}^{p} x(k+i|k)^T Q x(k+i|k) + u(k+i-1|k)^T R u(k+i-1|k) \tag{14}$$

with $Q = Q^T \geq 0$ and $R = R^T > 0$. But also other cost functions can be used, e.g., linear cost functions. What is most important is that costs are 0 if and only if the system stays in $(\tilde{x}, \tilde{u}) = (0, 0)$. Clearly the minimization should take place subject to the constraints (12b). Using a quadratic cost function as in (14) results in a QP (quadratic program) to be solved each time instant, whereas a linear cost function results in an LP (linear program), see e.g., [18].

The result from solving the above-mentioned optimization problem is a vector of future input corrections $\tilde{u}(k|k), \tilde{u}(k+1|k), \dots, \tilde{u}(k+p-1|k)$. At time k the input $\tilde{u}(k|k)$ is applied. Subsequently, at time $k+1$ the whole procedure starts all over again.

We conclude this section with some remarks. First, the stability of the MPC approach is not guaranteed. At least not in the way as presented here. In order to achieve guaranteed stability, one should take the horizon $p = \infty$. This is not desirable from a practical point of view. A second way of achieving stability is by adding the terminal constraint that after the horizon, the system should be on the reference, i.e., one could add the constraint that $\tilde{x}(k+p) = 0$. Notice that in order to have a feasible optimization problem, again one should take p large enough.

For more information about MPC, the interested reader is referred to [5].

6 Concluding Remarks

In this chapter we provided a framework within which concepts from the field of systems and control can be used for controlling manufacturing systems. We presented the concept of Effective Process Times (EPTs) which can be used for modeling a manufacturing system as a large queuing network. Restricting ourselves to mass production enabled us to model manufacturing systems by means of a linear system subject to nonlinear constraints (clearing functions). These models then served as a starting point for designing controllers for these manufacturing systems using Model-based Predictive Control (MPC). Thoughout this chapter we provided examples to illustrate the most important ideas and concepts. We also provided additional references for the interested reader.

We presented MPC as a possible approach from control theory for controlling manufacturing systems. But many more suitable approaches can be used, ranging from classical control theory using z-transforms and transfer functions, dynamic programming and optimal control, to robust control and approximate dynamic programming. A good overview of these kinds of approaches for the dynamic modeling and control of supply chains has been provided in the review paper [17].

But also the approximation model presented in Sect. 4 is only one of the possible choices for modeling manufacturing systems. An overview on aggregate models for manufacturing systems has been given in [13]. In the model presented here a fluid approximation has been presented where the number of jobs was modeled continuously, but the position in the factory was modeled discretely. Using a less detailed model, we can even abstract from workstations and model manufacturing flow as a real fluid using continuum models [1, 3, 6]. Optimal control of PDE models for manufacturing systems has been presented in [12].

From the above it is clear that the modeling and control of manufacturing systems has been, and still is, an open and active research area. In this chapter we provided some of the basic models and standard control approaches, illustrated by examples so that they can be applied straighforwardly.

Acknowledgements Erjen Lefeber is supported by the Netherlands Organization for Scientific Research (NWO-VIDI grant 639.072.072).

References

1. Armbruster D, Marthaler DE, Ringhofer C, Kempf K, Jo TC (2006) A continuum model for a re-entrant factory. Operations Research 54(5):933–950
2. Åström KJ, Wittenmark B (1990) Computer-controlled systems: theory and design, 2nd edn. Prentice-Hall, Englewood Cliffs
3. Daganzo CF (2003) A Theory of Supply Chains. Springer, New York
4. Fischbein S, Yellig E (2011) Why is it so hard to build and validate discrete event simulation models of manufacturing facilities. In: Kempf KG, Uzsoy R, Keskinocak P (eds) Planning production and inventories in the extended enterprise: a state of the art handbook, volume 2, Springer international series in operations research and management science, vol 152, chap 12. Springer, New York pp 271–288
5. Garcia CE, Prett DM, Morari M (1989) Model predictive control: theory and practice—a survey. Automatica 25(3):335–348
6. Göttlich S, Herty M, Klar A (2005) Network models for supply chains. Commun Math Sci 3(4):545–559
7. Graves SC (1986) A tactical planning model for a job shop. Oper Res 34(4):522–533
8. Hopp WJ, Spearman ML (2000) Fact Physics, 2nd edn. McGraw-Hill, New York
9. Jacobs JH, Etman LFP, Campen EJJv, Rooda JE (2003) Characterization of the operational time variability using effective processing times. IEEE Trans Semicond Manuf 16(3):511–520
10. Kingman JFC (1961) The single server queue in heavy traffic. Proc Camb Philos Soc 57:902–904
11. Kuehn PJ (1979) Approximate analysis of general queueing networks by decomposition. IEEE Trans Commun 27:113–126
12. La Marca M, Armbruster D, Herty M, Ringhofer C (2010) Control of continuum models of production systems. IEEE Trans Autom Control 55(11):2511–2526
13. Lefeber E, Armbruster D (2011) Aggregate modeling of manufacturing systems. In: Kempf KG, Uzsoy R, Keskinocak P (eds) Planning production and inventories in the extended enterprise: a state of the art handbook, volume 1, Springer international series in operations research and management science, vol 151, chap 17, pp 509–536 Springer, New York.
14. Little JDC (1961) A proof of the queueing formula $l = \lambda w$. Oper Res 9:383–387
15. Missbauer H (2009) Models of the transient behaviour of production units to optimize the aggregate material flow. Int J Prod Econ 118:387–397
16. Sakasegawa H (1977) An approximation formula $L_q \approx \alpha\rho^{\beta}/(1-\rho)$. Ann Inst Stat Mech 29:67–75
17. Sarimveis H, Patrinos P, Tarantilis CD, Kiranoudis CT (2008) Dynamic modeling and control of supply chain systems: A review. Comput Oper Res 35(11):3530–3561
18. Vargas-Villamil FD, Rivera DE, Kempf (2003) A hierarchical approach to production control of reentrant semiconductor manufacturing lines. IEEE Trans Control Syst Technol 11(4):578–587
19. Veeger CPL, Etman LFP, Lefeber E, Adan IJBF, Herk Jv, Rooda JE (2011) Predicting cycle time distributions for integrated processing workstations: an aggregate modeling approach. IEEE Trans Semicond Manuf 24(2):223–236

The Ongoing Challenge for a Responsive Demand Supply Network: The Final Frontier—Controlling the Factory

Kenneth Fordyce and R. John Milne

Abstract Over the past 20 years organizations have put significant energy into making smarter decisions in their enterprise wide central planning and "available to promise" processes to improve responsiveness (more effective use of assets and more intelligent responses to customer needs and emerging opportunities). However, firms have put only limited energies into factory floor decisions and capacity planning and almost none into generating a tighter coupling between the factory and central planning. The bulk of the work to make "smarter factory decisions" has focused on two simple metrics: increasing output and reducing cycle time—often without accommodating the need to run lots at different velocities and without recognizing how the operating curve (trade-off between lead time and tool utilization—Appendix 3) links them. In fact, many of the recent Lean initiatives have focused on eliminating variability to induce simplicity to achieve improved output or cycle time without concern for the impact on responsiveness or capacity. The purpose of this paper is to (a) make clear the critical, and often overlooked, role of factory responsiveness with respect to central planning; (b) explain how traditional factory planning and the current application of Lean can severely impact the firm's responsiveness; (c) elaborate on touch points between central and factory planning demonstrating simple tactical methods that can improve responsiveness and protect the factory from churn; (d) explain why smarter dispatch scheduling is critical to successful responsiveness; and (e) outline the basics of smarter dispatch scheduling. Although the focus of this

Tighter Coupling with Central Planning, Smarter Near Term Tactical Planning, and More Intelligent Dispatch.

K. Fordyce (✉)
IBM Systems and Technology Group, 2455 South Rd, Poughkeepsie, NY 12601, USA
e-mail: fordyce@us.ibm.com

R. J. Milne
Clarkson University School of Business, P.O. Box 5790, Potsdam, NY 13699-5740, USA
e-mail: jmilne@clarkson.edu

D. Armbruster and K. G. Kempf (eds.), *Decision Policies for Production Networks*,
DOI: 10.1007/978-0-85729-644-3_3,

paper is the factory, many of the core concepts apply to a wide range of industries from restaurants to health care delivery.

1 Positioning the Factory Within an Enterprise Wide Demand-Supply Network

Organizations, from healthcare facilities to manufacturing giants to small restaurants, can be viewed as an ongoing sequence of loosely coupled activities where current and future assets are matched with current and future demand across the demand-supply network.

These planning, scheduling, and dispatch decisions across a firm's demand-supply network are best viewed as a series of information flows and decision points organized in a decision hierarchy or tiers and further classified by the type of supply chain activity creating a grid for classification. The row dimension is the decision tier and the column is the responsible unit (Fig. 1). Observe the decisions in each tier limit and the options in the tiers below it.

The time frame for the first decision tier, *strategic planning*, is typically driven by the lead time required for business planning, resource acquisition, new product development and introduction, and to produce a product. Depending on the actual lead times for these activities, decision makers are concerned with a set of problems that are 3 months to 7 years into the future even with the same industry. For example, acquiring and validating a new tool may only take 3 months and re-orientating the product line takes 1 year. In both cases these decisions are removed from the production and delivery of current product. Issues in this tier include, but are not limited to, what markets the firm will be in, general availability of equipment and skills, major process changes, risk assessment of changes in demand for existing products, required or expected incremental improvements in the production or delivery process, and the lead times for adding additional equipment and skills.

The second tier, *tactical planning*, deals with aggregate level plans, estimates, and commitments. The time frame can range from 1 week to 6 months and is typically based on production lead times and the pace of change for demand and factory performance. Estimates are made of yields and cycle times (lead times), the likely profile of demand, productivity and reliability of equipment, etc. Decisions are made about scheduling releases into the manufacturing line or staffing levels. Delivery dates are estimated for orders or response times for various classes of patients are estimated. Deployment of equipment and staffing is adjusted. The order release plan is generated or regenerated, and (customer-requested) reschedules are negotiated.

The third tier, *operational scheduling*, deals with the daily execution and achievement of a weekly, biweekly, or monthly plan. Shipments are made, patients receive treatments, customers are waited on, serviceability levels are measured, and recovery actions are taken. Optimal capacity consumption and product output are computed. The time frame is again dependent on the production lead time and the rate of change in the factory.

Demand-Supply (DS)Network Planning, Scheduling, and Dispatch (PSD) Activity Areas and Decision Tiers			
		Enterprise Wide. global view - central planning	**Enterprise Subunits (manufacturing, distribution, retail) factory planning**
Decision Tiers	**Tier 1 Strategic**	**Enterprise wide Central Plan once or twice a year for 2–5-year horizon** at aggregate level with forecasted demand focused on business scenarios. Net result strategic direction established and financial commitments made	**Capacity Analysis** typically at tool family level and overall manpower to support forecasted demand, creation of production flow and capacity information for central plan, determining new production processes to introduce and estimated learning curve
	Tier 2 Tactical	**Enterprise wide central planning weekly/biweeky/monthly** > create demand statement (current orders, forecasts) > capture capacity, WIP, BOM, business policy > central planning engine to match assets with demand > estimate supply line linked to demand, early warning, production requirements, chase situations	**Capacity (tools and manpower) analysis** to gauge impact of changing product mix, identify challenges, review and modify deployment decisions and manufacturing engineering requirements, and **create capacity constraint information for central planning and WIP status**. Monitor tool level performance and take appropriate actions. Establish rules and metrics to set global lot importance - example, how many priority classes, algorithm to set lot importance within a class, limits on number of exppedites.
	Tier 3 Operational "daily"	**Enterprise Wide central planning reduced focus / what if** > what if commit on large orders > what if on major asset change > status of key WIP and actions to take if needed > cross factory signals	**Provide information to central plan and daily factory adjustments** > establish target outs, due dates on lots > maintenance priorities > short term changes in deployment > review key lot status and change priority (up or down) based on progress (eithermanually or dynamically) > one time changes in lot importance guidance > establish mfg lot vs development lot preference > revised projected outs for enterprise planning
	Tier 3.5 sub daily guidance	Change in Priorities, updated supply projections based on updated WIP or capacity status; change in customer reserved supply	**As needed Updates to Guidance to support response decisions** > regular updates to lot status based its progress, entering a time process window, status of short term manufacturing targets, WIP position and tool status > regular updates to tool status based on manufacturing engineering requirements, tool events, etc
	Tier 4 - Response	Available to Promise or Automated Order Commit process, cross factory signals	**Dispatch Scheduling & Tool Response** > assign sequence of lots to a tool > change status of a lot (for example on or off hold) > monitor signals from tools and respond as needed

Fig. 1 Decision grid for demand-supply networks

Tier "3.5" *straddles operational and real time response*. For example, a monitoring system might observe a lot has entered a "process time window" and its "urgency" to be assigned has "increased." A process time window is a sequence of activities that must be accomplished within a certain time limit or the lot might need to be scrapped due to some type of contamination. A non-factory example would be the "triage" system that occurs regularly in an emergency room where a patient is placed in one of four or five categories based on urgency. Although this decision does not directly assign the patient to a healthcare provider, it has a strong influence over the type and urgency of the assignment.

The fourth tier, *real-time response system*, addresses the problems of the next hour to a few weeks by responding to conditions as they emerge in relevant time. Within the demand-supply network, relevant time response is often found in two areas: manufacturing dispatch (assign lots to tools) and order commitment (available to promise, or ATP). For the emergency room setting it would the initial assignment of the patient to a health care professional and then a sequence of assignments based on the initial review (for example go to X-ray, immediately call in the senior resident, and run a blood test).

Within manufacturing, the decisions made across the tiers are typically handled by groups with one of two responsibilities: maintaining an enterprise-wide global view of the demand-supply network and ensuring that subunits (such as manufacturing location, vendor, and warehouse) are operating efficiently. Ideally all planning would be centralized; in practice complexity precludes this. Capacity planning is a good example. At the enterprise level, capacity is modeled at some level of aggregation, typically viewing a key tool set as a single capacity point. At the factory level, each tool, or potentially each chamber within a tool, is modeled.

2 Challenges and Opportunities

Over the past 20 years organizations have put significant energy into making smarter decisions in their enterprise-wide central planning and "available to promise" process to improve responsiveness (more effective use of assets and more intelligent responses to customer needs and emerging opportunities) [9]. However, firms have put limited energy into factory floor decisions and capacity planning, and almost none into a tighter coupling between the factory and central planning. Much of the recent work to improve factory performance has attempted to implement Lean planning [6] concepts of (a) elimination of variability, (b) establishing uniform flow (every part every interval), (c) supermarket-like goods flow (kanbans), and (d) elimination of due dates and on time delivery metrics. Clearly, every factory will run "better" with steady output and predictable lead times—however, the real world always injects variability that sets the price of implementing such methods as reduced responsiveness and/or excess capacity.

The net result is that many factories still operate with the mind set: "establish a set of starts for the month; set a fixed schedule with target outs; and measure actual outs versus target outs." For this approach to work, demand must be accurately forecasted over an extended period of time and uniformly spread across time; all lots must travel at the same speed; tool sets should operate with clockwork precision (never suffering "surprises"); and the flow of parts in the line (even with stable capacity) must never create "piles" or "gaps" due to the variations (for example batch versus single lot tools) intrinsic in the manufacturing process. In today's world, accurate detailed forecasts of demand remain an illusion; even the best factories which have "tool set surprises" (breakdowns and quality excursions), product mix introduces variability in speeds, and the competitive nature of the market precludes carrying excess capacity and insists on responsiveness. Those demand-supply networks that can get their factories more engaged in responsiveness while recognizing the importance of "tools" and "output" will flourish. They will eliminate the variability that matters—a failure to deliver a part on its committed date and the inability to capture a market opportunity that could be handled with "intelligent" factory decisions.

Accomplishing this goal requires "retooling" the approaches for interaction with central planning, near-term tactical planning, and dispatch scheduling to be more adaptive without a loss in productivity. Bob Bixby observed—the optimization time horizon is ever shrinking. The purpose of this paper is to (a) make clear the critical, and often overlooked, role of the factory within central planning; (b) explain how traditional factory planning and the current application of Lean methodology can severely impact the firm's responsiveness; (c) elaborate on touch points between central and factory planning demonstrating simple tactical methods can improve responsiveness and protect the factory from churn; (d) explain why smarter dispatch scheduling is critical to successful responsiveness; and (e) outline the basics of smarter dispatch scheduling.

3 Basics of Enterprise-Wide End-to-End Central Planning

To understand how factory floor decisions can limit responsiveness in central planning, we need to review the key elements of central planning which are given in the list below and in Fig. 2 [9, 31].

1. Create a demand statement
2. Gather and collect key supply information from the factory
    - 2.1. Project the completion of WIP to a decision point (often completion of the part).
    - 2.2. Statement of capacity consumption rates and capacity available.
    - 2.3. Statement of lead time or cycle time to complete a new start.
3. Create a model that captures key enterprise relationships of the demand-supply network (Central Planning Engine—CPE).
4. Create an enterprise-wide central plan by matching current and future assets with current and future demand using the CPE to create a future projected state of the enterprise and the ability to soft peg the current position of the enterprise to the projected future position. Information from the CPE model includes
    - 4.1. Projected supply linked with exit demand
    - 4.2. Identification of at risk customer orders either to a commit date or request date
    - 4.3. Synchronization signals across the enterprise
    - 4.4. Capacity utilization levels

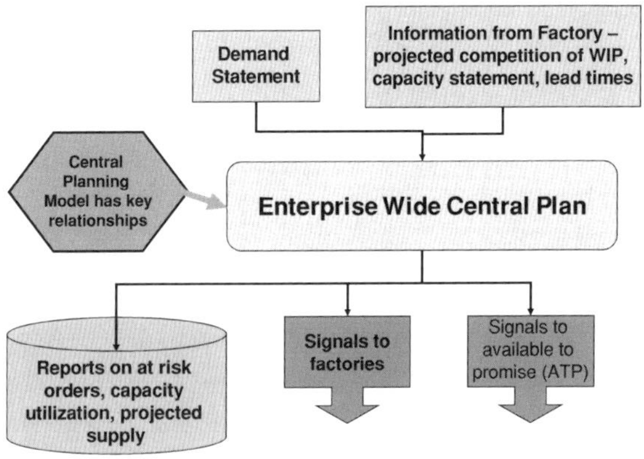

Fig. 2 Basic steps in central planning

4.5. Ability to trace each production and distribution activity that supports meeting a demand.

5. Typically this is an iterative process where each iteration is done with different assumptions and conditions, for example

   5.1. Different assumptions about capacity available
   5.2. Different business policies for protective stock
   5.3. Different commit dates and/or demand priorities for orders

6. Execute the plan, that is,

   6.1. Send signals to each manufacturing facility and distribution center
   6.2. Send projected supply to available to promise (ATP)

4 Basics of Factory Planning and Dispatch

For central planning organizations, the plan (developing the plan, making the customer commits, and monitoring conformance to the plan) is their primary end product. For the factory, planning and dispatch are always secondary in importance to successfully making the parts. Factories are foremost concerned about making a product that works (yields), second keeping their tools operational, and third keeping their output levels high (either operational outs or exit outs).

Dispatch refers to assigning a lot to a tool and requires balancing effective tool utilization with stable delivery (either to the commit date on the lot or to the number of exit lots per day or week). Factories see dispatch as important since it generates the assignment of lots to tools and therefore impacts output. Typically there are different camps with substantially different views on how to make this decision. For example, manufacturing is looking to maximize output while the business team is as concerned about the lots for key clients. Both groups are always suspicious of the applications that do automated dispatch combining rules and models—either

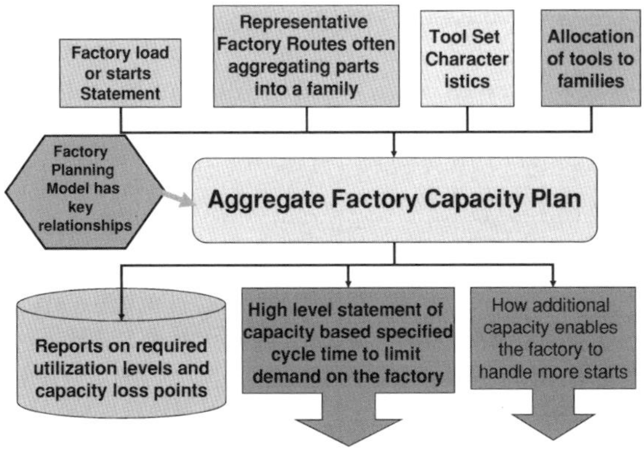

Fig. 3 Aggregate tool planning steps

thinking the automated methods are too simple and they know best how to balance complex trade-offs or "complaining" the automated decisions are too complicated to understand.

Aggregate tool planning [1, 34, 38] is typically focused on assessing the ability of the factory to satisfy certain demand (demand is stated as manufacturing starts a.k.a. planned manufacturing releases) and creating the capacity inputs required by central planning. The basic steps of aggregate factory planning are given below and in Fig. 3:

1. Capturing representative product routes—sequence of operations, raw process time for each operation, and tool set consumed for each key operation.
2. Capturing a specific factory load—typically given in starts.
3. Gathering data on tool set characteristics: number of tools, tool availability, which operations the tool set handles, overlapping tool sets (shared operations between toolsets called a cascade [1], and its operating curve (Appendix 3, [2, 5, 12, 14, 15]) which establishes the trade-off between cycle time and tool utilization/capacity available.
4. Allocation of tool sets to product parts or families either as user input or based on history
5. A model that captures key relationships—often a spreadsheet based single iteration model
6. Executing the model to determine how this load impacts toolsets

   6.1. Required utilization levels
   6.2. Capacity loss points (planned maintenance, high raw process time, etc.)

7. Providing information for central planning

   7.1. High level statement of capacity based on starts or a few key tool sets fixed for the specified cycle time to limit demand on the factory
   7.2. How additional capacity enables the factory to handle more starts

Near term tool planning (deployment) refers to determining which operations each tool will be qualified to handle over the short term. Typically a tool can service many operations, but a factory will limit the number of operations it is "allowed" to service to reduce workload on manufacturing engineering and make dispatching simpler by reducing options.

Observe the lack of influence of an order book on factory planning and decisions.

5 Current Interaction Between Factory and Enterprise: Factory Limits Responsiveness—Opportunities Abound

Steps 3 and 4 in Sect. 3 on the central engine planning process (CPE) are often viewed as the planning "hub" and the focus of making a firm more responsive through "smarter" engines and better (quality and timeliness) data [9]. However, in Step 2 the factory sets the boundaries of "responsiveness" [29]. The CPE relies on the factory to provide

- Estimated completion date for each lot in the line (either to completion or staging point)
- Statement of capacity available and required for each manufacturing start (typically at an aggregate level)
- Estimated lead or cycle time to complete a start fixed for some time interval

Additionally,

- The central planning process cannot change the due date on the lot or the lot's priority without extensive manual negotiations with manufacturing
- Central planning has no control (and typically no knowledge) of the lot importance metric used by the factory or how it balances utilization and delivery.
- Each piece of information supplied by the factory to the central planning process is "fixed"—stripped of all of the information that enables trade-offs to be made. For example, the following possibilities are invisible to central planning:
  - Slowing one lot down to enable another lot to go faster
  - Trade-offs between cycle time and capacity available based on the operating curve
  - Redeploying tools to handle a different mix of manufacturing processes or products

Additionally, limiting "change or variation" within the "factory black box" to improve responsiveness fits the factory culture and is reinforced by Lean principles. Factories "dislike" change. Factories "like" steady rates of production referred to as smooth flow. This has been reinforced with Lean initiatives that view variation as evil and desperately attempt to create smooth demand and production flow with aggregation and kanbans or "super markets" that essentially serve as inventory replenishment decision points to absorb variability and generate smooth (every part every interval) production flow in the factory [6]. They try justifying this by claiming all variability can be eliminated and higher productivity will occur. The reality is this view

- Requires excess capacity to facilitate the "smoothing."
- Is completely divorced from client needs, variability in production flows, and tool availability.
- Has no inherent ability to allocate scarce capacity or project a supply line.
- Fails to account for the operating curve.

Despite the substantial forces to limit change, constant pressure from emerging market opportunities to manufacturing quality excursions to inaccuracies in planning (deviations between the plan and the actual) drive an ongoing sequence "off line one of a kind" negotiations between the central planner and the factory planner to make adjustments that rely on quasi-manual decision support tools with limited function. For example

1. A client may need three lots 4 days earlier than committed and this can be accommodated by placing these lots on expedite.
2. The demand for product A requires 30 units of capacity from Tool Set A1 on average each day. Tool Set A1 only has 25 units of capacity available. Tool Set A2, which is not listed as a capacity option for product A, can service product A, but it runs slower. A review of capacity utilization for Tool Set A2 indicates it will be underutilized. A decision is made to qualify Tool Set A2 to handle product A.
3. A client has had a steady order for 10 units daily of product W with a cycle time of 15 days where the constraining tool set is Tool Set W3. The business has been able to achieve an on time delivery rate of 97%. The client would like to increase its standing order from 10 to 12 units. Central planning initially rejects this opportunity since the stated maximum daily capacity in their model for Tool Set W3 is 10. However, when the two planners look at the details of the tool set and its operating curve, the business decides it can commit to 12 per day if the cycle time is increased to 16 days (or if the OTD commit percentage is lowered).
4. Assume a client has placed an order for five lots of "part A" per day with a cycle time of 10 days. On average there are 50 lots of "part A" in WIP and the factory completes five lots per day. The due date posted on each lot is the start date plus 10 days. For example, lots started on day 6 are due on day 16. Due dates on the lots can only be changed manually by a factory planner. The factory has an abnormal set of tool outages and goes 3 days without delivering any lots—it is past due 15 lots ($= 3 \times 5$). It has continued to start five lots per day. At the start of the fourth day the number of lots in the line is 65 ($=50$ normal $+$ 15 past due). On the evening of the third day the client and Central Planning meet about a recovery strategy. The client determines demand has been soft for this part and agrees to "forgive" five lots and have the remaining ten lots "caught" up at a pace of one per day (in addition to the regular five per day). Therefore the new order book for this client is six lots per day for the next 10 days and then returns to five per day. Without changes to the due date on the lots in WIP, the factory continues to see 15 lots past due and will drive to "catch up" as quickly as possible. The factory

may decide to delay lots for a second client to catch up all 15 past due lots for the first client in 5 days. Therefore the factory planner has to manually change the due dates on the lots to insure the factory floor has the correct guidance.

These examples make it clear that when central planning can make effective use of the flexibility within the factory that is hidden from its traditional view—good things can happen. The opportunity for improved responsiveness simply needs to "widen and straighten" this trail with appropriate planning and dispatch tools, processes, and protocols. Each one would be considered muda (wasteful) by Lean which would say to eliminate them, not to build tools to make doing this more intelligence and efficient. We contend a goal of any firm is to eliminate unnecessary complexity, but ignoring the complexity that remains is like tackling snow storms in the north with bald tires.

Additionally, such tools and processes can keep bad things from happening. For example, if many lots are being "expedited" already there is no room for an additional expedited lot. Tool W3 may be needed for engineering lots not in the central planning data or it may have a history of time consuming qualifications making it too large a risk. Just as the factory prefers "steady" and conservative, central planning often fall preys to an overly optimistic mindset that is fine with constant churn. Tools for improving factory/enterprise coordination fall into three groups

1. Direct interaction with central planning tools (for example WIP projection, expedite decisions, demand pegging, and specialized capacity planning models for flexibility in manufacturing)—Sect. 6
2. Tactical decision models (tool deployment, allocation of cycle time, or tool capacity referred to as operational outs or moves)—Sect. 7
3. Dispatch scheduling (assigning lots to a tool)—Sect. 8

The following sections outline methods in each of these three areas that can provide additional flexibility to the factory without destroying factory output and cycle times. The differences in the length of the sections reflect the amount of detail required to convey the core issues, rather than the relative importance of each area.

6 Dynamic Interaction Between Central Planning and Factory Planning

As previously described in Sects. 3 and 4, the contact points between the factory and central planning include:

- Accurate projection of when lots already in the factory (WIP) will arrive at stock
- Setting due date for lots.
- Changing the committed date or speed for lots.
- Capacity and cycle time information that influence planning manufacturing start decisions and customer commits.

We will explore examples in each of these four areas.

6.1 Smarter WIP Projections by Considering Capacity

Typically each part moves sequentially through a set of manufacturing steps (route) that can be characterized by a raw process time (RPT) at each step, a cycle time multiplier (CTM) that adjusts for the average wait time, total cycle time (TCT) which is RPT × CTM, and the tool set that handles this manufacturing step or activity. A sample route is provided in Table 1.

A factory planner typically uses one of two methods to project when a lot will finish

- Use the commit date for the lot
- Add the remaining cycle time for the lot to the current date. For example if lot 101 was at step 04, we would project its completion date to be NOW + 91 h (= 40 + 41 + 10).

The following methods have proven effective in improving the quality of this projection:

1. *Status of the lot at the current step*: Instead of solely using the TCT to estimate the time a lot will spend at its current step, directly examine the number of lots that are expected to be processed ahead of this lot at this step and adjust for their processing times.
2. *Different CTM estimators*: Typically the CTM is based on a planned value. The quality of this estimate can sometimes be improved using recent manufacturing history to create an estimated CTM for the next 7–14 days.
3. *WIP Projector on a Parcheesi Game Board*: In this method [9] we project the movement of each lot step-by-step according to its cycle time, but incorporate capacity constraints by limiting the number of moves (i.e. number of operation completions) (or time) per day allowed at certain tool sets and allocating these moves based on lot importance. For example, assume LOT201 and LOT202 are at Step 01 and LOT301 and LOT302 are at Step 05. All four lots are serviced by the lion tool set for their present steps. LOT 201 is an expedited lot, LOT 302 is three days behind schedule, and the other two lots are on time. Additionally, the lion tool set has a daily capacity limit of two lots per day. Only LOT201 and LOT302 would move to the next manufacturing step on the game board today (LOT201 to Step 02 and LOT302 to Step 06). The other two lots would have a chance to move tomorrow based on how their priorities rate relative to the competition from other lots.
4. *Queuing Network Equations*: The most complicated, but also the most accurate and flexible is to represent the route in its entirety as a system of queuing network equations. This has been used successfully in some situations [5, 37, 38].

The caveat in each option is to avoid "thrashing" the estimate by over reacting to the normal day-to-day variations in manufacturing flow. When a manufacturing line has sufficient buffer capacity and is appropriately managed (e.g. not too many

Table 1 Route or process steps to manufacture a part

Manufacturing step	Raw process time (RPT) for this step	Cycle time multiplier (CTM) applied to RPT	Total cycle time (TCT) for this step = RPT* multiplier	Tool set
Step 01	10.0	3.3	33.0	Lion
Step 02	15.0	3.8	57.0	Tiger
Step 03	8.0	4.0	32.0	Apple
Step 04	20.0	2.0	40.0	Furnance
Step 05	10.0	4.1	41.0	Lion
Step 06	5.0	2.0	10.0	Squirrel
Total / average	68.0	3.1	213.0	

jobs being expedited, commitments are reasonable), most lots will complete at or near their original commit date by reallocating capacity from lots that are ahead of schedule to lots that are behind. The goals of the projection mechanisms are: to identify when attempts to reallocate capacity so all due dates are met is not likely to succeed; find lots that have fallen too far behind to finish by their commit date; and issue an early warning when it is clear that the assumptions in the planning model are at variance with reality—resulting in many lots finishing late (or early). The goal is not to overreact to normal fluctuations, but catch systemic issues—especially since factories are notorious for "convincing" themselves they will catch up next week once tool availability stabilizes.

6.2 Dynamically Resetting Due Dates

In Sect. 5 we described a situation where the factory fell behind in meeting its commitment to a client, the central planning organization worked with the client to reset the demand, and then the factory planner needed to manually recalculate and reset the due dates on the lots. This is one example of many where the actual "need" date for a lot is different from the due date posted to the lot when the lot starts. A second example occurs when a lot (LOT51) started on day five is placed on hold for 2 days and another lot (LOT61) started on day six "passes" it. If we assume a cycle time of 10 days, then LOT51 has an initial due date of 15, but is behind (further from finishing than) LOT61 with a due date of day 16. This is called "leapfrogging."

Typically when the demand driving that starts on a factory is a complex combination of fixed orders, loose orders, build to forecast, line stock replenishment, etc, and the manufacturing process is long and complex the eventual need date will often be different than the initial due date. A tool to dynamically reset the "due date" on the lot to match real need improves responsiveness. The algorithm works essentially as follows [9, 25].

1. The factory must maintain a "demand" statement or order book on what the business currently expects it to produce (part, quantity, and date). This is the hardest part.
2. The lots of the same part are sorted according to raw process time remaining (low to high).
3. The demand with the nearest (earliest) due date is assigned to the lot with the least remaining raw process time remaining and the pattern continues.
4. Some adjustments need to be made if the lots have different quantities, some lots are manufacturing expedites, etc.

This is a standard MRP algorithm that can and does enable the factory to appropriately focus energy on lots and facilitates early warning when a demand will not be met on time. This is the same MRP that is constantly maligned by Lean advocates. It would appear that having quality need dates would be an asset in any factory concerned about on time delivery.

6.3 Committing Some Lots to Run a little Faster: Collateral Impact

A common, yet manual, practice is for central planning to negotiate with factory planning to "speed" up certain lots to meet a customer request, overcome a manufacturing delay, or compensate for a planning failure. Typically, the analysis is limited and ad hoc with no comprehensive process as seen in central planning or tool planning. There are rules of thumb such as:

- The number of "expedite" lots cannot exceed some fixed number *N* or a maximum percentage of total lots.
- The fastest an expedite lot can run is some CTM less than that of the normal lots. If normal lots have a CTM of five, expedite lots might have a CTM of three.

A closer look makes it clear the core of this decision is a reallocation of either wait time or factory moves [23] that enables some lots to run faster by having others run slower over some subset of the manufacturing line over some time duration. For example, assume the factory has five lots (LOT01 ... LOT05) in the last stage of production; each lot requires four moves (manufacturing actions) to complete; the maximum number of total moves (capacity) per day is five; and the most moves a lot can have in 1 day is two. If capacity is allocated "fairly," then each lot gets one move per day and each lot finishes in 4 days. Now assume the business decides LOT01 and LOT02 must finish in 2 days, then each needs two moves per day for 2 days, and therefore on each of these days two of the other three lots sit "idle" during these 2 days to enable this expedite.

Appendix 1 develops this allocation concept in more detail focusing on wait time allocation instead of moves. In each case lots that look essentially the same are required to run at different speeds on the factory floor. The planned speedup is worthless without successful factory execution. This places a substantial burden on dispatch and precludes the use of simple methods (and Lean favorites) such as first in first out (FIFO) and elapsed time. Again we see that Lean and responsiveness are not in sync.

This topic is part of an area called *General Plan Repair Process*. In this process central and factory planners identify actions to take that will enable orders that are currently flagged as "late" to be met on time. Fordyce et al. [9] reviews this challenge from the central planning perspective—only through increased intelligence on both sides of the fence can responsiveness be improved.

6.4 Smarter Central Planning Through Better Modeling of Factory Capacity

As we outlined before, the central planning process requires as critical inputs from the factory: capacity (consumption rates and availability) and cycle times. Since the 1980s manufacturing resource planning (MRP) and material balance equations (MBE) in optimization formulations have been the two dominant methods used in central planning [30]. In these methods the factory representation is "static" and linear. The cycle times and capacity information are fixed across some time period and handled with linear relationships. For detailed information about central planning, the reader is referred to Refs. [9, 13, 20, 28, 31, 35, 36].

Historically intricacies of factory tool planning (availability, deployment decisions, cascading, setup times, batching, et al.) and the dynamic interaction between equipment utilization (effective capacity) and cycle time through the operating curve have for the most part been ignored. This will not be sustainable in the future as the burden on responsiveness resulting in under utilization or delivering products late is increasingly unacceptable.

In Sect. 5 we described a situation where the client needed 30 units per day and the initial central planning analysis determined the maximum the factory could make was 25. Appendix 2 elaborates on the method to improve responsiveness by capturing alternative deployments of tools to manufacturing operations.

In the same section we outlined that we could trade longer cycle time for more tool capacity (and hence output) based on the operating curve. Appendix 3 contains a simple example that makes it clear the assumption in typical central planning processes that cycle time and capacity are independent is not correct—the two are clearly coupled. We can view this as classical planning meets its uncertainty principle. It is a rich ground for improved responsiveness and a headache for classical planners. Since Lean advocates believe that all variation can be eradicated, it has no awareness of an operating curve, and no methods to capture this opportunity. It is like attempting to ignore special and general relativity and still produce GPS locations [26].

For additional information about work that pushes beyond traditional methods for handling capacity in central planning and factories see: [1, 7, 19, 22, 27, 38].

7 Tactical Decisions in the Factory: Only the Shadow Knows

There are series of ongoing tactical decisions in factories that fall well below interaction with central planning and are not part of dispatch—but have a strong influence on dispatch by constraining the available options to assign lots to tools. We refer to these as the "shadow" decisions—powerful, but difficult to find and capable of substantially restricting responsiveness.

One area is manufacturing engineering requirements (MER). Manufacturing Engineering's (ME) first concern is producing quality products (keeping high yields) and in their zeal can create collateral damage. For example, assume tool A01 is being qualified to run a new process called "yellow tiger"; ME might put in place two rules:

- Only 25% of the total widgets produced over a 24 h period can run on tool A01 (the other 75% has to run on other tools in the tool set) in case tool A01 has quality issues.
- Most of the other processes that can run on tool A01 are soft coded as not available to tool A01 to insure enough widgets for the "yellow tiger" process visit this tool.

On the surface, this sounds logical. In practice, especially when the factory is busy, most simple dispatch decisions systems (automated or human) will initially drive "yellow tiger" widgets to tool A01 and place other widgets on the other tools in the toolset. However, quickly the ME "police" will shutoff assigning these widgets to tool A01 since the 25% limit is met. Typically, there is no method to increase the importance of running "yellow tiger" widgets on the other tools or dynamically alter either rule. Some simple tactical models and dynamic guidance (defined in the next section) will catch this imbalance before it becomes an issue that can, in the heat of the "battle," take days to find without the appropriate diagnostic tools. As Gary Sullivan [33] observed—it is usually better to blow out the lighted match before it gets to the gasoline unless you are measured by putting out fires as opposed to preventing them!

A second area is called deployment decisions. Here the tools that make up a group of similar tools (toolset) are allocated to the operations covered by this toolset. Again this limits the dispatch options. Appendix 4 describes an approach that helps us gauge near term the effectiveness of the deployment decisions for the WIP currently waiting to be processed. A third area is the deployment of manufacturing operators. Again, nothing in the Lean literature tackles these tough questions that live in the shadows.

8 Fundamentals of Dispatch Scheduling for Better Factory Performance

As we observed in the prior sections, for the factory to be responsive, simple dispatch applications are insufficient to ensure planned actions are executed on the floor. In addition, simple dispatch cannot handle the ever increasing complexity and variability factories face on a daily basis—from manufacturing equipment whose throughput is very sensitive to batch sizes and the sequence lots are placed on the tool; diversity in the product mix and quantity which eliminates the ability to run a fixed quantity per day and still meet client expectations; ever tighter boundaries on quality control; competitive pressures that require factories to run at higher utilization rates without an increase in cycle time, more specialty and design parts, etc. Smarter dispatch is required to offset increases in variability and keep the operating curve from shifting in an unfavorable direction.

Essentially, the factory is constantly balancing effective tool utilization with stable delivery against a complex demand statement. This drives the requirement for intelligent dispatch scheduling applications to optimally achieve these goals and limit the quantity of variability the factory introduces into the system. This leads to simple applications for dispatch scheduling being replaced with applications that make the "complex" manageable. Some of the essential components of dispatch scheduling are given in the next subsection. For a comprehensives review of this topic the reader is referred to: [3, 8, 11, 16–18, 32, 33].

8.1 Basics of Dispatch Scheduling

The key inputs to dispatch scheduling can be broken down as follows:

1. Tool—Lot affinity (usually linked by the operation)
    - 1.1. What lots can run on this tool? What tools can handle this lot?
    - 1.2. What are preferred tools? What are preferred lots?
    - 1.3. Manufacturing engineering requirements
        - 1.3.1 Count limits (avoid too many lots on certain tools)
        - 1.3.2 Time limits (tool requires re-qualification after a specified amount of activity)
        - 1.3.3 Process time windows (lot must finish a sequence of steps within a time limit)
        - 1.3.4 Special customer specifications
2. Global importance of the lot to the supply chain or business—priority
3. Pacing lot movement: fluctuation smoothing, flow balance cycle time allocation, delta schedule, critical ratio

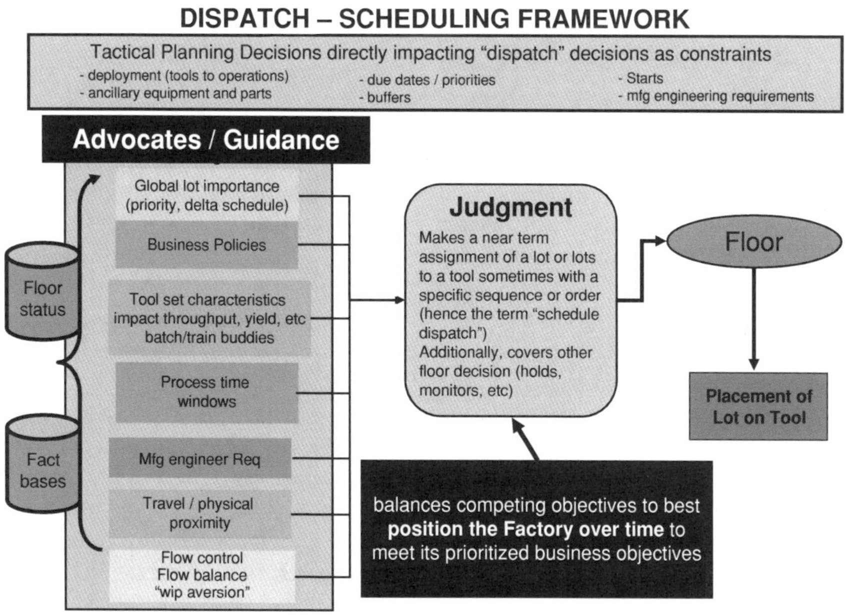

Fig. 4 Dispatch scheduling framework

4. Local tool characteristics and performance
   4.1. Batching and operational trains
   4.2. Setup times dependent on previous job run at the tool
   4.3. Parallelization opportunities
   4.4. Differences in raw process time
   4.5. Multiple chambers within a tool
   4.6. There may be ancillary equipment required at an operation in addition to the core toolset and labor
5. Upstream and down stream requirements
   5.1. Sending wafers to tools with limited WIP in queue in front of them
   5.2. Avoiding tools with large piles of WIP in queue
   5.3. Balancing across repeated levels which use the same tool set

The core dispatch decision-making activities can be divided into two primary components: guidance and judgment (Fig. 4). Appendix 5 has additional details on both components.

9 Conclusion: Slow Steady Progress in Extending the Borders of Bounded Rationality

Herbert Simon [29] observed, "As humans, we have 'bounded rationality' and break complex systems into small manageable pieces." The challenge for organizations is to integrate information and decision technology to push boundaries out and improve performance. Nick Donofrio [4], retired IBM Senior Vice President, observed, "Access to computational capability will enable us to model things that would never have believed before." The challenge reaches beyond coding algorithms, linking to data, and turning it on. Each decision-science team must execute its role as "intelligent evolutionist" to ensure the organization adopts complex decision technology in a sustained incremental fashion. Each management must be willing to push their organization beyond its comfort zone.

Little [21] observed: "Manufacturing systems are characterized by large, interactive complexes of people and equipment in specific spatial and organizational structures. Because we often know the sub units already, the special challenge and opportunity is to understand interactions and system effects. There are certainly patterns and regularity here. It seems likely that researchers will find useful empirical models of many phenomena in these systems. Such models may not often have the cleanliness and precision of Newton's laws, but they can generate important knowledge for designers and managers to use in problem solving."

Improving responsiveness in the factory is one of the most difficult challenges in the near-term horizon, but clearly one of the most important. For many firms substantial gains in end-to-end supply chain responsiveness is limited by the modeling tools and approaches in factories for matching assets with demand and flowing production and the false illusion from Lean advocates that variability and complexity can be eliminated.

Appendix 1: Committing Some Lots to Run Faster—Collateral Impact

Deciding Which Lot are Candidates to Speed Up

Prior to any allocation decisions, the planners must first decide candidate lots to speed up. This requires two critical pieces of information:

- An assessment of whether the lot is currently behind or ahead of schedule
- The exit demand supported by the lot

The second requirement places a burden on the central planning process to be able to link each lot to a specific exit demand(s) and trace all intermediary manufacturing steps. Since this linkage can and will change, creation of the linkage must be dynamic.

This is called *demand pegging* or *coverage analysis* [9] and responsibility for this foundation of factory responsiveness belongs to central planning.

Model 1: Expediting a Set of Lots from Release into the Line

Assume the factory makes two parts (A and B) with the following routes (Tables 2 and 3).

Each lot for Part A (Table 2) goes through six manufacturing steps and takes 20.7 time units to complete. Of this 20.7 units, 6.8 represent actual processing time and 13.9 is wait time. Part A spends 67.1% (= 13.9/20.7) of its time waiting and 32.9%(= 6.8/20.7) of its time being processed. Each lot for Part B (Table 3) goes through four steps (different than Part A) and takes 15.8 time units. Of this 15.8 units, 4.0 represent actual processing time and 11.8 is wait time. Part B spends 74.7% (= 11.8/15.8) of its time waiting and 25.3% (= 4.0/15.8) of its time being processed.

A move is defined as the completion of one manufacturing step. Part A accomplishes six moves in 20.7 units of time. Therefore it averages 0.290(= 6/20.7) moves per unit time in the factory. Part B lots average 0.253 (= 4/15.8) moves per unit time. The flow information for the average lot for each part is summarized in Table 4.

Table 4 also contains information on Part B* which is exactly the same as a Part B lot, but travels faster ("fast track" or expedite lots). Each B* lot goes through the same steps as a regular Part B lot and incurs the same RPT. The difference is average wait time is smaller generating a smaller total cycle time (CT) and CTM. In this example B* has a CTM of 2.50 (= 10/4). Since it moves faster than regular Bs, its move per unit time is higher 0.400 (= 4/10) compared to 0.253. We will refer to this set of parts with these cycle times as Case 1.

Of particular importance in Table 4 is the last column—average wait time per unit time. It is the amount of wait time on average a lot sees per unit time. A lot is in one of two states—waiting to be processed or being processed (a move). Therefore for each part "actual process time (RPT) per unit time" (+) "wait time per unit time" equals 1. In Table 4 we observe the sum of the last two columns for each part is 1.

Now assume the factory starts 10 Part A lots, 20 Part B lots, and 0 Part B* lots per unit time. Then it has a total of 207(= 10 × 20.7 = starts per day × total cycle time) Part A lots, 316 (= 20 × 15.8) Part B lots, and 0 (= 0 × 10) Part B* lots in the line. The total accumulated wait time per unit time for Part A lots is 139 (= 207 × 0.671 = total lots in the line for this part times average wait time per lot for this part per unit time) and 236 (= 316 × 0.747) for Part B lots. Wait time for Part B* is 0, since no Part B* lots were started. This information is summarized in Table 5.

Assume this load on the factory (10 starts per unit time of Part A at CTM = 3.04 and 20 starts per unit time of Part B at CTM = 3.95) leaves little unused capacity—putting the factory on the steep part of the operating curve (Appendix 3 [14, 23]). Therefore any attempt to make some lots run faster requires other lots to run slower. The slower lots have to absorb wait time from the faster lots.

Table 2 Route or process steps to manufacture a Part A

Manufacturing step	Raw process time (RPT) for this step	Cycle time multiplier (CTM) applied to RPT	Total cycle time (TCT) fot this step = RPT* multiplier
Step 01 (for Part A)*	1.0	3.3	3.3
Step 02	1.5	3.8	5.7
Step 03	0.8	4.0	3.2
Step 04	2.0	2.0	4.0
Step 05	1.0	3.5	3.5
Step 06	0.5	2.0	1.0
Total / average	6.8	3.04	20.7

* Steps for Part A and Part B are not the same

Table 3 Route or process steps to manufacture a Part B

Manufacturing step	Raw process time (RPT) for this step	Cycle time multiplier (CTM) applied to RPT	Total cycle time (TCT) fot this step = RPT* multiplier
Step 01 (for Part B)*	0.8	3.4	2.7
Step 02	1.4	4.5	6.3
Step 03	1.2	4.0	4.8
Step 04	0.6	3.3	2.0
Total / average	4.0	3.95	15.80

* Steps for Part A and Part B are not the same

One way to capture this constraint is to assume the total wait time for all lots in the line per unit time is fixed ($375 = 139 + 236$). When we sum the wait time for all lots in the line, it must be equal to this value. If I reduce wait time on one lot by two units, another lot has to gain wait time by two units. With this constraint we can use a model to gauge the impact of a decision to have some lots run faster. For example:

Assume Central Planning decides it needs five out of the 20 Part B lots started each day to complete production in ten time units and makes them "fast track" Part B* lots (Table 6). This is called Case 2. The total wait time burden per unit time for the Part B* lots is 30. Five starts per day for 10 days puts 50 lots in the factory. Each has an average wait time burden per unit time of 0.600 which results in a total wait time burden for the 50 Part B* lots of 30 ($= 50 \times 0.6000$). The remaining 15 B lots have a wait time burden of 177 ($= 15 \times 15.8 \times 0.747$). The total wait burden on B (regular and fast track) is 207 ($= 30 + 177$) compared to the prior burden of 236 units. We are now "short" 29 time burden units (=236–207)—some lots have to gain waiting time.

Cycle time balance and wait time conservation require other lots in the line to absorb the 29 units of wait time. There are many possible solutions; one is the normal CTM for regular Part B lots increases from 3.95 to 4.44, which increases

Table 4 Flow information for average lot for each part Case 1

Part	Total RPT	Total CT	Total wait time	Number of steps	Ave RPT per step	Average CTM	Average wait time per step	Moves per time unit	Actual process time (RPT) per unit time	Wait time per unit time
Part A	6.8	20.7	13.9	6.0	1.13	3.04	2.31	0.290	0.329	0.671
Part B	4.0	15.8	11.8	4.0	1.00	3.95	2.95	0.253	0.253	0.747
Part B*	4.0	10.0	6.0	4.0	1.00	2.50	1.50	0.400	0.400	0.600

Table 5 Flow information on all lots in the line Case 1

Part	Starts per time unit	Total lots in line	Total moves per time unit	Total process time (RPT) per unit time	Total wait per unit time
Part A	10	207	60	68	139
Part B	20	316	80	80	236
Part B*	0	0	0	0	0
Total	30	523	140	148	375

Table 6 Flow information on all lots in the line Case 2

Part	Starts per time unit	Total lots in line	Total moves per time unit	Total process time (RPT) per unit time	Total wait per unit time
Part A	10	207	60	68	139
Part B	15	237	60	60	177
Part B*	5	50	20	20	30
Total	30	494	140	148	346

the normal cycle time to 17.8 from 15.8. This is called Case 3 and the details are in Tables 7 and 8.

From this base the collateral impact can be estimated through trial and error with a spreadsheet model or with an optimization formulation. It should be noted, this approach makes some simplifying assumptions that are not a problem when considering the impact of minor adjustments; major ones would require more elaborate models. *The goal is to get the central planner and factory planner to develop a process and formally recognize waiting time constraints and trade-offs.*

Model 2: Expediting Lots Close to the End of the Manufacturing Line

Often the decision to expedite does not occur for all of the lots of a certain part or group in the manufacturing line, but only for selected lots in the final third of their route. At this juncture planners have a reasonable sense of the likelihood these lots will complete on time and their importance to clients. Planners will look to selectively pick key lots to push faster (expedite) for four reasons: (a) attempt to finish on time; (b) attempt to finish early to meet a customer request; (c) meet quarter end revenue targets; or (d) build some buffer to insure the lots finishes on time. The following example illustrates a simple methodology to organize this decision process.

Table 7 Flow information on individuals Case 3

Part	Total RPT	Total CT	Total wait time	Number of steps	Ave RPT per step	Average CTM	Average wait time per step	Moves per time unit	Actual process time (RPT) per unit time	Wait time per unit time
Part A	6.8	20.7	13.9	6.0	1.13	3.04	2.31	0.290	0.329	0.671
Part B	4.0	17.8	13.8	4.0	1.00	4.44	3.44	0.225	0.225	0.775
Part B*	4.0	10.0	6.0	4.0	1.00	2.50	1.50	0.400	0.400	0.600

Table 8 Flow information on all lots in the line Case 3

Part	Starts per time unit	Total lots in line	Total moves per time unit	Total process time per unit time	Total wait per Total wait per
Part A	10	207	60	68	139
Part B	15	266	60	60	206
Part B*	5	50	20	20	30
Total	30	523	140	148	375

In this example (Table 9), the factory has six lots with only about 80 h of raw process time remaining (REMRPT) until completion. At this juncture the planners review these lots, assess whether are they are ahead or behind schedule to finish on their commit date, and decide if certain lots should be "sped" up—which means other lots have to slow down.

The "Due Time" column is the number of hours left between now and when the lot is committed to be finished. D2_CTM (drive to cycle time multiplier) is how fast (given as a multiple of REMRPT) the lot must travel to finish on time. For lot A01, its D2_CTM is 3.08 ($=240/78$) where 240 is "Due Time" and 78 is REMRPT. FA_CTM is the average factory cycle time multiplier or the average factory speed based on the factory load and its point on the operating curve. For this example its value is 3.5. "Factory Time" is the time the lot will remain in the factory if it travels at FA_CTM. For lot A01, this is $273 = 3.5 \times 78$. "Delta Schedule" is an assessment of whether a lot is ahead or behind schedule. It is "Due Time" minus "Factory Time." Positive values indicate the lot is ahead of schedule and negative values behind schedule. For lot A01, the delta schedule is $-33(=240-273)$.

The next to last column in Table 9 is "wait time burden." Conceptually the expected completion date for the lot depends on the "wait time burden" allocated to the lot plus its remaining raw process time (REMRPT). The lot's estimated exit (last column) is REMPRT + Wait Burden. In Table 9, the wait burden initially assigned to each lot is based on the lot traveling at speed of FA_CTM, where "Wait Burden" = "Factory Time"—REMRPT. For lot A01, "wait burden" is $195(=273-78)$. However, the *wait time burden assigned to each lot is a decision controlled* by the planners based on business needs and subject to some constraints—three are:

(1) System balance requires the sum of the wait time burden across all lots match the current factory operating curve performance point which requires a fixed amount of total wait time
(2) There is a minimum wait time burden no lot can avoid
(3) There are limits in diversity of wait time burden across lots

How do we calculate the total required wait time? Since the average factory velocity stated at CTM is 3.5 and the RPT remaining for this group of lots is 478, this establishes a fixed amount of total wait time that must be burdened across all lots. This total wait time burden is 1195 ($=(3.5 \times 478) - 478 = 2.5 \times 478$). Observe this is the total for the "Wait Burden" column.

Table 10 provides an example of a simple "what if" spreadsheet-based tool, where a planner can try out various wait burden allocations to lots and gauge the impact of these decisions on the estimated completion time for each lot. The seventh column (wait time allocation, shaded) is the decision variable. As planners try different values, columns 8–10 (estimated finish, delta schedule, and required velocity) are automatically updated. The total for wait time allocation must be greater than or equal to 1195.

It is straightforward to enhance this modeling method to provide more automated support for "what if" analysis and incorporate constraints (2) and (3). Additionally, the model can be adapted to an optimization application.

Table 9 Example for allocating wait for lots almost completed

Lot id	REMRPT remaining RPT until lot is finished	"Due time" time remaining between now and the lot's committed due date	D2_CTM speed required for lot to finish on time	FA_CTM average CTM for lots in the factory	"Factory time" remaining cycle time for each lot if it ravels at FACCTM for REMRPT	"Delta schedule" estimate of late (negative) or ahead (positive) for each lot comparing "due time" with "factory time"	"Wait burden" is the total wait time allocated to lot initial burden if lot runs at FA_CTM speed	"Estimated exit" is REMPT + Wait burden is the estimate of when the lot finishes
A01	78	240	3.08	3.5	273	−33	195	273
A02	80	248	3.10	3.5	280	−32	200	280
A03	82	280	3.41	3.5	287	−7	205	287
A04	75	220	2.93	3.5	263	−43	188	263
A05	85	270	3.18	3.5	298	−28	213	298
A06	78	231	2.96	3.5	273	−42	195	273
Total	478	1489			1673	−184	1195	

Table 10 Example for allocating wait for lots that generates equal lateness

Lot id	REMRPT remaining RPT until lot is finished	"Due time" time remaining between now and the lot's committed due date	D2_CTM speed required for lot to finish on time	FA_CTM average CTM for lots in the factory	"Factory time" remaining cycle time for each lot if it ravels at FACCTM for REMRPT	**Wait time allocation decision by planners to allocate wait time based on business needs**	"Estimated exit" is REMPT + Wait burden is the estimate of when the lot finishes **Planner Decision**	"Delta schedule" due time– estimated exit	RQ_CTM required lot velocity to finsh at estimated time
A01	78	240	3.08	3.5	273	**192.7**	270.7	– 30(7	3.5
A02	80	248	3.10	3.5	280	**198.7**	278.7	– 30(7	3.5
A03	82	280	3.41	3.5	287	**228.7**	310.7	– 30(7	3.8
A04	75	220	2.93	3.5	263	**175.7**	250.7	– 30(7	3.3
A05	85	270	3.18	3.5	298	**215.7**	300.7	– 30(7	3.5
A06	78	231	2.96	3.5	273	**183.7**	261.7	– 30(7	3.4
Total	478	1489			1673	**1195**		– 184	

The Planned Speedup is Worthless Without Execution

Both previous examples require lots that look essentially the same to run at different speeds on the factory floor. In the first example most of the Part B lots will run at a velocity of 4.44 (the CTM), while a few will need to run substantially faster at 2.50. In the second example, we observe six lots now must run at their required velocity (RQ_CTM) instead of a standard factor velocity to meet the new business objectives.

This requirement places a substantial burden on factory floor execution, specifically dispatch scheduling—assigning lots to tools. Simple methods such as FIFO (first in first out) and elapsed time will not work (the lot that has been waiting at the tool the longest or the lot whose wait time exceeds a certain threshold goes next), since each inherently assumes equal wait time for all lots waiting to be processed at a tool set.

Why? Let us look at the case referenced in Table 10. To achieve the planner's goal of equal lateness, lot A03 needs to travel at a speed of 3.8 (CTM) and lot A04 at 3.3. A04 must travel faster than A03, which means it needs to absorb less wait time than A04.

Assume for a moment A03 and A04 are both waiting to be processed at tool LION, A03 has been waiting at the tool for 185 h and A04 has been waiting at the tool for 176 h. If jobs are processed FIFO, then based on its longer elapsed time—A03 would be selected for processing first. However, if we look at the allocation of wait time burden, we see A04 at 176 is past its burden point of 175.7, while A03 at 180 is well below its burden point of 228.7.

Simple dispatch rules such as FIFO and elapsed time worked when factories made only a few products in large quantities with steady demand—a rare environment today. Therefore we observe factory responsiveness requires not just smarter planning, but the ability to execute which requires smarter dispatch.

Appendix 2: Revisiting Capacity Allocation: a Rabbit Out of the Hat

The core elements of resource allocation in central planning engines (CPE) are: (a) linking a manufacturing activity to one or more resources; (b) establishing a consumption rate for each unit of production by that manufacturing activity for the selected resource; (c) providing the total available capacity for the resource; and (d) connecting manufacturing releases (starts) to resource consumption with a linear relationship. In Table 11, we see operations 101 and 151 can be handled by Tool A or B. Operations 201, 202, 301, and 302 can be handled only by Tool A. Table 12 tells us the available capacity for Tool A and Tool B is 1152 working minutes per time unit (for example per day).

Assume, we have a uniform start rate of one lot per day and each lot goes through each operation (101, 151, 201, 202, 301, and 302) once. In steady state, each operation

Table 11 Operation resource linkage

Manufacturing activity	Resource	Consumption rate
Operation 101	Tool A	10
Operation 101	Tool B	10
Operation 151	Tool A	10
Operation 151	Tool B	10
Operation 201	Tool A	15
Operation 202	Tool A	15
Operation 301	Tool A	15
Operation 302	Tool A	15

Table 12 Available capacity

Resource	Consumption rate
Tool A	1152
Tool B	1152

would need to process one lot per day. The optimal way to allocate operations to tools is to assign operations 101 and 151 to Tool B and the remaining four operations (201, 202, 301, and 302) to Tool A. This creates a load on Tool B of 20 (= 10+10) minutes of processing and 60 (= 15 + 15 + 15 + 15) minutes on Tool A. Since Tool A has 1152 units available, the maximum number of lots per day is 19.2 (= 1152/60).

If the demand rises to 30 lots, the CPE would indicate this is not feasible and most likely would push some production out in time showing the pieces being delivered late.

However, the CPE lacks access to the tactical deployment detailed information and therefore has no method to identify better solutions. That is a solution that enables the enterprise to deliver lots on time. If the delivery delay is large and the customer is important, the central planner will contact the factory planner and a review of the detailed deployment decision will occur to "mine for capacity." That is look for opportunities to reallocate capacity to satisfy the demand on time. In many industries a tool can potentially handle many different operations, but at a given point in time is only actively deployed (linked) to a small subset of these operations. This "reduced" deployment occurs for a number of reasons including:

- It is physically impossible for the tool to be "actively" ready for more than a small number of operations. If we want the tool to handle an operation different than those currently selected, the tool has to be brought down for a while, "reconfigured," and brought back up.
- There are manufacturing performance advantages to limit the number of operations a tool is currently deployed to handle.
- The manufacturing team often uses deployment decision to attempt to "balance" tool load by estimating future workload.
- The manufacturing team prefers to deploy its fastest tools to certain operations to keep total cycle time low.
- Habit or prior practice.

Table 13 Operation resource linkage

Manufacturing activity	Resource	Consumption rate
Operation 101	Tool A	10
Operation 101	Tool B	10
Operation 151	Tool A	10
Operation 151	Tool B	10
Operation 201	Tool A	15
Operation 201	Tool C	30
Operation 202	Tool A	15
Operation 301	Tool A	15
Operation 301	Tool C	30
Operation 302	Tool A	15

Table 14 Available capacity

Resource	Consumption rate
Tool A	1152
Tool B	1152
Tool C	2000

The reality is the tactical deployment decision made by Manufacturing when reflected in the capacity information sent to Central Planning understates the flexibility of Manufacturing to produce parts to meet an increase in demand. This flexibility can only be uncovered through manual intervention when central planning presses factory planning.

In our example, it might be Tool C can, after being retooled, be switched from working on "gadgets" to "widgets"—specifically it could be re-configured to handle operations 201 and 301. This change is reflected in Tables 13 and 14.

Additionally, the model in the central planning engine can be enhanced to accommodate:

- Differences in how effectively different tools can process work at a specified set of operations. Typically these differences are speed (Tool A is faster than Tool B), cost (Tool B's unit cost is lower than Tool A), or yield (a manufacturing error). For example, this might result in a view that operation 301 will be assigned to Tool C only as a last resort.
- Minimum tool usage (if operation 301 is assigned to Tool C then it needs least two units of work per day to remain qualified)
- Limitations of tool-operation pairings (Tool C can do work from Operation 301 or 201, but not both)

We might be tempted to state the old adage—garbage in garbage out—but that would be wrong and fail to understand the environment that generated the "limited" but accurate capacity information. In fact, the capacity information provided by Manufacturing to the central supply chain model as reflected in Tables 11 and 12 was accurate, but limited. It was limited to the current near-term production requirements and the near-term ability to use the tools. Tool C can not be used for operations 201

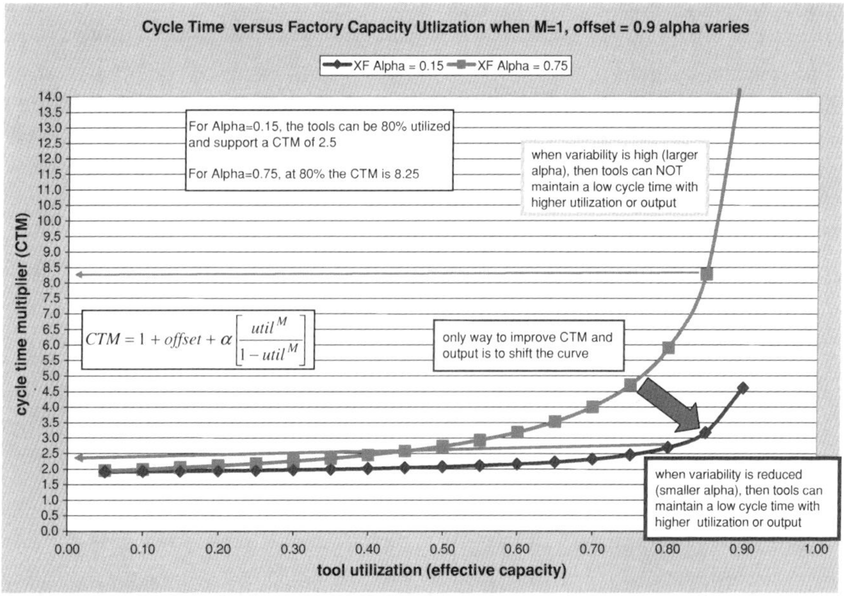

Fig. 5 Cycle time versus factory capacity for Morrison and Martin equation (2006)

and 301 until it is "retooled" and qualified. This may take a day or may take a week—but Manufacturing cannot afford to "retool" on a daily basis.

Appendix 3: Recognizing the Trade-Off Between Capacity and Cycle Time

Review of the Operating Curve

When variability exists either in arrivals or in services there is a trade-off between server (tools, people) utilization and the lead time or cycle time to complete an activity or service. The higher the server utilization, the longer the cycle time. Since higher planned utilization translates to more effective output the trade-off can be reframed as effective capacity available or ouput versus cycle time. The curve that describes this trade-off is called the **Operating Curve** (OC). Lead time or Cycle time is often measured as a cycle time multiplier (CTM), where CTM equals total elapsed cycle time divided by raw process time (RPT). Typically the curve is almost flat for low utilization levels, then spikes sharply upward from the steep part of the curve Fig. 5.

There are a number of equations that can generate this curve, but the one we will use is $\text{CTM} = 1 + \text{offset} + \alpha \left[\frac{\text{util}^M}{1-\text{util}^M}\right]$ [24].

- **CTM** is the cycle time multiplier of RPT—measure of cycle time as a function of RPT.
- **util** is the fraction of utilization in the entity—facility, tool set, checkout clerks, etc.
- **Offset** represents several aspects of the process that generate wait time and cannot be eliminated. For example: travel time, hold hours, and post-processing hours relative to total RPT. A common value for offset may be about 0.9. When offset is 0.9 this sets the minimum CTM at 1.9.
- **M** is the number of identical parallel machines or servers. Typically this value ranges from 1 to 4 (even when the number of tools or servers exceeds 4) work best.
- $\boldsymbol{\alpha}$ represents the amount of variation in the system and controls how long the curve stays flat. The lower the value of $\boldsymbol{\alpha}$ the less variation and longer the curve stays flat. Common values for α range between 0.35 and 0.65 [10].

Solving for Util, we have $\text{Util} = \left(\frac{\text{CTM}-(\text{offset}+1)}{\text{CTM}-(\text{offset}+1)+\alpha}\right)^{\frac{1}{m}}$

Linking Capacity and Cycle Time

Assume that the product XYZ is processed five times by tool set AAA during its production route. Each time a widget goes through tool set AAA it is referred to as a "pass." In this case product XYZ has five passes on tool set AAA. Additionally, assume 100% process yields and that the average RPT for each XYZ widget on tool set AAA is two units. In steady state, this makes for a total RPT required per day of 10 (= 5 × 2 units per widget of XYZ on tool set AAA. Assume we start one widget per day and we have ten units of capacity, what would the cycle time be? The CTM equation makes it clear the cycle time would be infinite since the capacity required matches capacity available making tool utilization 100%.

The business states it wants to run this product with a CTM of 4.0. This requires some portion of time that the tool set is available to produce, but does not have WIP. How do we incorporate that into the planning process?

We calculate a burden or uplift factor (ULF) per widget based on the target CTM and the specific characteristics of the Operating Curve for this tool set. Assume the Operating Curve for this tool set has offset = 1, alpha = 0.5, and m = 1. Using the above equation to solve for UTIL, the required utilization to achieve the CTM target of 4.0 is 0.80 (80%). For each unit of raw capacity required, we need 1.25 units available to meet the CTM target. The value 1.25 is the uplift factor (ULF) and determined by:

$ULF = 1/tool\_utilization\_meet\_cycle\_time\_target\_from\_opcurve.$

If we have 250 units of capacity available per day, how many widgets can we start per day at committed cycle time? The answer is 20 (= 250/(1.25 × 10)).

If the business wanted to achieve a cycle time of 3.5, how many widgets could it start per day? Using the same equation for UTIL, the utilization required to achieve a

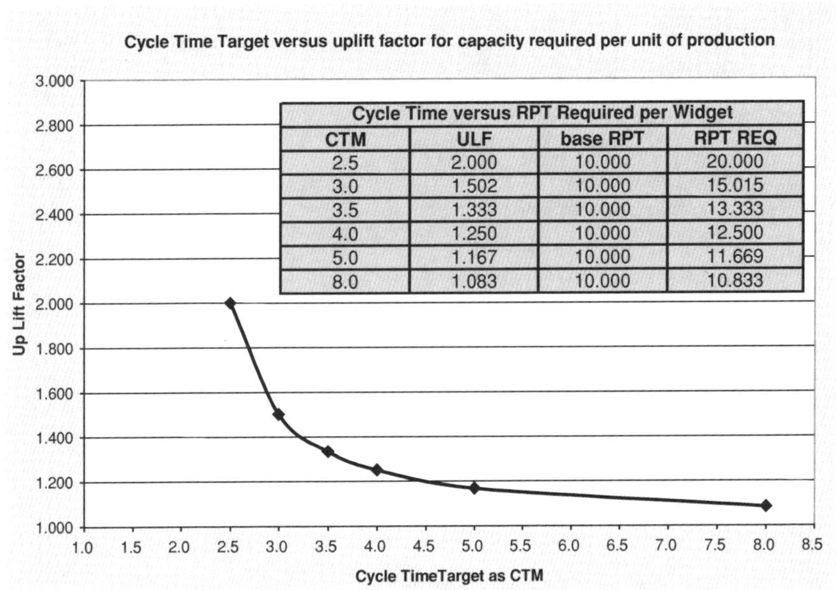

Fig. 6 Uplift in capacity consumption (required) based on cycle time commit

cycle time of 3.5 is 0.75 (75%). The ULF is 1.333 (= 1/0.75). The maximum number of widgets per day at this cycle time commit is 18.75 (= 250/(1.333 × 10)). Shorter cycle time equates to reduced widget starts.

Alternatively, instead of decreasing the capacity available, we can uplift the capacity required per unit of production. For the 4.0 cycle time commit, we uplift the capacity consumption rate of ten units per day per widget to 12.5 (= 10 × 1.25). For the 3.5 commit, the ten units is uplifted to 13.33 (= 10 × 1.33). Figure 6 provides the uplifted capacity consumption rates to input to the central planning process based on the cycle time the business wishes to achieve.

This example makes it clear that the assumption in a typical central planning process that cycle time and capacity are independent is not correct—the two are clearly coupled. We can view this as classical planning meets its uncertainty principle [10]. It is a rich ground for improved responsiveness and a headache for classical planners. Since Lean advocates believe that all variation can be eradicated, they have no awareness of an operating curve and no methods to capture this opportunity. It is like attempting to ignore special and general relativity and still produce GPS locations [26].

Table 15 Deployment decision

	Tool A	Tool B	Tool C	No tools covering oper
Operation 1	1	1	1	3
Operation 2	1	1	0	2
Operation 3	0	1	0	1
Operation 4	0	0	1	1
Operation 5	1	1	1	3
Operation 6	1	0	0	1
Operation 7	1	0	0	1
Number opers tool covers	5	4	3	

Table 16 RPT, lots, available capacity

	Tool A	Tool B	Tool C	# lots
Operation 1	4	20	5	40
Operation 2	15	20	9999999	30
Operation 3	9999999	15	9999999	10
Operation 4	9999999	9999999	20	60
Operation 5	5	5	5	10
Operation 6	8	9999999	9999999	200
Operation 7	10	9999999	9999999	200
Capac avl	720	1152	1296	

Appendix 4: How Well Does the WIP Match the Tool Deployment?

In this example we have three tools (A, B, and C) and seven operations (operation 1–7) to handle the current WIP at this tool set. Table 15 shows the current deployment decisions made by the factory planner. In the table, a 1 means the tool (column) is able to service the operation (row). For now these decisions are fixed.

Table 16 provides all of the key pieces of information for the model. The value in operation / tool cell is the raw process time (RPT) that the tool requires to process one lot at this operation. For example, the RPT for Tool A to process one lot at operation 1 is 4 time units. A value of 9,999,999 indicates this operation/tool combination is not currently active, corresponding to a 0 in Table 15. The last column (# lots) provides the number of lots requiring service of that operation at this point in time. For example, there are 40 lots at operation 1 waiting to get on either Tool A, B, or C. The last row ("capac avl") provides the number of time units of capacity available for that tool over the service period. Tool A has 720 units available.

Table 17 has the basic "what if" model. The value in each operation/tool cell is the *business decision*- the number of lots the tool is assigned to handle for this operation over some time period. For example, Tool B will handle 30 of the 40 lots that require

Table 17 Allocation decision and results

	Tool A	Tool B	Tool C	Lots served	Goal	# lots	Delta
Operation 1	0.0	30.0	10.0	40.0	=	40	0.0
Operation 2	3.0	0.0	0.0	3.0	=	30	−27.0
Operation 3	0.0	10.0	0.0	10.0	=	10	0.0
Operation 4	0.0	0.0	60.0	60.0	=	60	0.0
Operation 5	0.0	10.0	0.0	10.0	=	10	0.0
Operation 6	90.0	0.0	0.0	90.0	=	200	−110.0
Operation 7	0.0	0.0	0.0	0.0	=	200	−200.0
Cap used	765.0	800.0	1250.0		Total unmet dmd		−337.0
Constraint	<	<	<				
Cap avl	720.0	1152.0	1296.0				
Delta	−45.0	352.0	46.0				

service at operation 1, ten lots for operation 4, and ten lots for operation 5. These 21 values (cells) (7 operations by 3 tools) represent allocation decisions.

The results of these decisions are found in column 5 (lots served) and row 10 (cap used). The "lots served" column is the total number of lots served for this operation across all tools. For example 40 lots at operation 1 will be served—0 on Tool A, 30 on Tool B, and 10 on Tool C ($40 = 0 + 30 + 10$). The row "cap used" tells us how much capacity is used for each tool. This is the sum of the product of the allocation times RPT (Table 16). For tool A, the cap used is 765 $(=(0\times4)+(3\times15)+(0\times9999999)+(0\times9999999)+(0\times5)+(90\times8)+(0\times10))$.

The last component of the model is comparing the results of the business (allocation) decision made with the goals of the factory. The factory has two goals: service as many lots as possible and use all available capacity.

The "lots served" goal comparison information is in columns 6 (goal type), 7 (target), and 8 (delta). Our goal is to service all lots waiting (=). The target is the number of lots that are currently waiting (last column in Table 16). The result is posted in the "delta" column which is simply "lots served" minus "target." For example, for operation 2 the value is -27, since our target was 30 and the actual number of lots served was 3 ($-27 = 3 - 30$). This tells us the current allocation decision leaves 27 lots waiting at operation 2 "still waiting." The last value in the delta column is the total unmet demand based on the current allocation decision. The value -337 is simply the column sum.

The "capacity goal" information is in the last three rows (goal type, cap maximum, and delta cap goal). It is a constraint—do not make allocation decisions that exceed available capacity. The target is the capacity available for each tool (last row in Table 16). The result is posted in the last "delta" row which is simply "cap used" minus "cap maximum." For example, for Tool B the value is 352, since the actual capacity used was 800 and the maximum capacity was 1152 ($352 = 1152 - 800$). This tells us the current allocation decision leaves 352 units of capacity at Tool B idle.

Table 18 Revised allocation decision and results

	Tool A	Tool B	Tool C	Lots served	Goal	# lots	Delta
Operation 1	0.0	37.0	3.0	40.0	=	40	0.0
Operation 2	0.0	10.0	0.0	10.0	=	30	−20.0
Operation 3	0.0	10.0	0.0	10.0	=	10	0.0
Operation 4	0.0	0.0	60.0	60.0	=	60	0.0
Operation 5	0.0	10.0	0.0	10.0	=	10	0.0
Operation 6	90.0	0.0	0.0	90.0	=	200	−110.0
Operation 7	0.0	0.0	0.0	0.0	=	200	−200.0
Cap used	720.0	1140.0	1215.0		Total unmet dmd		−330.0
Constraint	<	<	<				
Cap avl	720.0	1152.0	1296.0				
Delta	0.0	12.0	81.0				

This simple model enables planners to manually assess the quality of their tentative deployment decisions and estimate the maximum number of lots that can be serviced with this deployment. Typically a planner will try different allocation decisions (leaving the deployment decision unchanged) to determine how to best allocate WIP to tools to meet prioritized demand and then send guidelines to Manufacturing. Table 18 shows an improved allocation plan eliminating overusing capacity on tool A and reducing overall unmet demand from 337 to 330.

When capacity is highly utilized and tensions are high, most planners will welcome an upgrade to a small optimization model where the decision variables are the allocation values, the constraints are not exceeding the capacity available, and the objective is to minimize unmet prioritized demand. The extensions to handle integer values, demand priorities, multiple periods, and partial deployments (which occur during a phase in) are straightforward.

To enable the planner to model the impact of deployment decisions, we need to couple the decisions made in Tables 15 and 17. He or she can change the deployment decision (Table 15); then revisit the allocation decision (Table 17); then assess the impact on the WIP waiting to be serviced. Again, optimization methods can be used to reduce the workload on the analyst and handle demand priorities and multiple time periods [1, 34].

The Planned Deployment Decision is Worthless Without Execution

As with changing lot velocities, finding an allocation of WIP to tools is only the first part of success for the factory. The second part is execution. This requirement places a substantial burden on factory floor execution, specifically dispatch schedule decision making—assigning lots to tools. Simple methods will not work.

In this example (Table 18), the plan requires all of the lots for operation 5 to run on Tool B even though all three tools are equally proficient (Table 16) at processing operation 5. This decision was made because lots at operation 6 can only run on Tool A and lots at operation 4 can only run on Tool C. However, on the factory floor doing this type of analysis at best would be very difficult. Second, from the floor's point of view, the RPT for lots at operation 5 is the same (5) for all three tools. Therefore a casual analysis would make the floor indifferent to which tool handled lots at operation 5—with dire consequences to the factory! What dire consequences? If the operator places the lots for operation 5 on tool A, then there is no tool available to run lots for operation 6. The result would be lower utilization of Tool B and additional delays for the lots at operation 6.

Appendix 5: Dispatch Scheduling Details on Guidance and Judgment

Referring to Fig. 4, *Guidance* or advocate logic is the set of computational activities (which may be a computer program or manual) to create information posted to some location (often a table structure) that the assignment logic accesses or to trigger an assignment module to execute. The most common example is a calculation to determine whether a lot is ahead or behind its planned pace. Another example is the updating of a fact base that may contain operation—tool preference based on static information (such as difference in raw process times between tools executing the same manufacturing action) or dynamic information (the amount of time will take to set the tool up to handle this manufacturing action). Other types of guidance include flow balance (avoid starving a tool set), manufacturing requirements (avoid running all lots of a certain type on a single tool, but distribute them across three tools), and process control time windows (lot must complete the next three steps within 5 h or it will need to be scrapped due to contamination).

Judgment or assignment is the set of computational activities that when completed, result in a change of state or action on the manufacturing floor. The judgment logic must balance competing requirements such as meeting on time delivery, demand priorities, improving throughput with batches and trains, current WIP position, and tool status.

The real goal of any judgment application is make a sequence of decisions over time that in aggregate improve the future position of the factory relative to its role in the total supply chain or demand supply network. The decisions are based on impact on the future state of the factory, not based on prior events. The sum total of the prior events has resulted in the current state of the factory. All other measures are attempts to create an interim goal that can be measured and decisions made against that is a reasonable approximation of the ultimate goal. Additionally, under some circumstances, these goals can be at odds with each other and the overall good of the demand-supply network.

Simple Judgment typically

- Does not consider the assignment of lots to other tools in the tool group
- Does not consider the assignment of lots over time
- Does not consider upstream or downstream conditions (WIP level and Tool status and near term throughput rates)
- Uses simple rules of thumb for complex trade-offs
- Written with decision tree one iteration logic
- Generates a single decision, through a series of filters and if-then conditions
- Gives a reasonable (though myopic) decision
- Relies on manual intervention for process time windows
- Typically have to be rewritten for different WIP levels (static adjustment)

Advanced Judgment typically

- Looks across the tool set and upstream and downstream
- Handles all process time windows
- Establishes an anticipated sequence of assignments at all tools in a tool group over time for lots at the tool or which will arrive soon
- Measures the quality of a proposed solution and anticipates impact on factory performance
- Uses an iterative search process in judgment logic
- Dynamically adjusts for WIP levels and other business conditions

References

1. Bermon S, Hood S (1999) Capacity optimization planning system (CAPS). Interfaces 29(5): 31–50
2. Bitran G, Tirupati D (1989) Tradeoff curves, targeting and balancing in manufacturing networks. Oper Res 37(4):547–555
3. Bixby R, Burda R, Miller D (2006) Short-interval detailed production scheduling in 300 mm semiconductor manufacturing using mixed integer and constraint programming. *Semiconductorfabtech*, 32nd edn, http://www.fabtech.org, pp 34–40
4. Buchholz J (2005) Interview with Nick Donofrio. IBM on the spot series, posted on http://www.ibm.com. Accessed 9 Aug 2005
5. Chen H, Harrison M, Mandelbaum A, Ackere A, Wein L (1988) Empirical evaluation of a queuing network model for semiconductor wafer fabrication. Oper Res 36(2):202–215
6. Dennis P (2007) Lean production simplified. Productivity Press, New York
7. Denton B, Forrest J, Milne RJ (2006) Methods for solving a mixed integer program for semiconductor supply chain optimization at IBM. Interfaces 36(5):386–399
8. Fordyce K, Bixby R, Burda R (2008) Technology that upsets the social order—a paradigm shift in assigning lots to tools in a wafer fabricator—the transition from rules to optimization. In: Proceedings of the 2008 winter simulation conference
9. Fordyce K, Wang C-T, Chang C, Degbotse A, Denton B, Lyon P, Milne RJ, Orzell R, Rice R, Waite J (2011a) In: Kempf, Keskinocak, Uzsoy (ed). The ongoing challenge—creating an enterprise-wide detailed supply chain plan for semiconductor and package operations. Planning production and inventories in the extended enterprise: a state of the art handbook, Vol 2 (Chapter 14)

10. Fordyce K, Fournier J, Milne RJ (2011b) Basics of the operating curve—classical planning meets its uncertainty principle. Working paper, fordyce@us.ibm.com, jmilne@clarkson.edu
11. Fox B, Kempf K (1985) Complexity uncertainty, and opportunistic scheduling. In: Proceedings of the IEEE second conference on artificial intelligence applications: the engineering of knowledge based systems, Miami, FL, pp 487–492
12. Gross D, Harris C (1998) Fundamentals of queueing theory, 3rd edn. Wiley, New York
13. Hackman ST, Leachman RC (1989) A general framework for modeling production. Manag Sci 35(4):478–495
14. Hopp W, Spearman M (2008) Factory physics, 3rd edn. McGraw-Hill Irwin, New York
15. Horn G, Podgorski W (1998) A focus on cycle time-vs.-tool utilization "paradox" with material. In: Advanced semiconductor manufacturing conference and workshop proceedings, pp 405–412
16. Kempf K, Pape D, Smith S, Fox B (1991) Issues in the design of AI based schedulers. AI Mag 11(5):37–45
17. Kempf K (1989) Manufacturing scheduling: intelligently combining existing methods. In: Working notes of AAAI AI in manufacturing symposium. Fox M (ed.), AAAI, Burgess Drive Menlo Park
18. Kempf K (1994) Intelligent scheduling semiconductor wafer fabrication. In: Mark Fox, Monte Zweben (eds) Intelligent scheduling. Morgan Kaufman Publishers, pp 473–516 (Chapter 18)
19. Kempf K (2004) Control-oriented approaches to supply chain management in semiconductor manufacturing. In: Proceedings of the 2004 American control conference, Boston, MA, pp 4563–4576
20. Leachman R, Benson R, Liu C, Raar D (1996) IMPReSS: an automated production planning and delivery-quotation system at Harris corporation–semiconductor sector. Interfaces 26(1):6–37
21. Little J (1992) Tautologies, models and theories: can we find laws of manufacturing? IEEE Trans 24(3):7–13
22. Liu J, Yang F, Wan H, Fowler J (2010) Capacity planning through queuing analysis and simulation-based statistical methods: a case study for semiconductor wafer FABS. web.ics.purdue.edu/ hwan/docs/IJPR
23. Morrison J, Dews E, LaFreniere J (2006) Fluctuation smoothing production control at IBM's 200mm wafer fabricator: extensions, application and the multi-flow production index (MFPx). In: Proceedings of the 2006 IEEE/SEMI Advanced Semiconductor Manufacturing Conference, Boston, MA
24. Morrison J, Martin D (2006) Cycle time approximations for the G/G/m queue subject to server failures and cycle time offsets with applications. In: ASMC 2006 Proceedings, p 322
25. Orlicky J (1975) Material requirements planning: the new way of life in production and inventory management. McGraw-Hill, New York
26. Pogge R (2009) Real world relativity, http://www.astronomy.ohio-state.edu/~pogge/Ast162/Unit5/gps.html
27. Shirodkar S, Kempf K (2006) Supply chain collaboration through shared capacity models. Interfaces 36(5):420–432
28. Shobrys D (2003) History of APS. Supply chain consultants (www.supplychain.com), Wilmington, DE 19808, USA
29. Simon HA (1957) Administrative behavior, 2nd edn. The Free Press, New York
30. Singh H (2009) Supply chain planning in the process industry. Supply Chain Consultants, Wilmington
31. Singh H (2009) Practical guide for improving sales and operations planning. Supply Chain Consultants, Wilmington
32. Sullivan G (1990) IBM Burlington's logistics management system (LMS). Interfaces 20(1): 43–61
33. Sullivan G (1994) Logistics management system (LMS): integrating decision technologies for dispatch scheduling in semiconductor manufacturing. Intelligent scheduling. Morgan Kaufman Publishers, San Francisco pp 473–516

34. Sullivan G (1995) A dynamically generated rapid response fast capacity planning model for semiconductor fabrication facilities. the impact of emerging technologies on computer science and operations research, Kluwer Academic Publishers, Boston (presented at Winter 1994 computers and operations research conference)
35. Uzsoy R, Lee C, Martin-Vega LA (1992) A review of production planning and scheduling modules in the semiconductor industry, Part 1: system characteristics, performance evaluation, and production planning. IIE Trans Sched Logist 24(4):47–60
36. Uzsoy R, Lee C, Martin-Vega LA (1994) A review of production planning and scheduling modules in the semiconductor industry, Part 2: shop floor control. IIE Trans Sched Logist 26(5):44–55
37. Zisgen H, Ments I, Wheeler B, Hanschle T (2008) A queuing network based system to model capacity and cycle time for semiconductor fabrication. In: Proceedings of the 2008 winter simulation conference
38. Zisgen H, Brown S, Hanschke T, Meents I, Wheeler B (2010) Queuing model improves IBM's semiconductor capacity and lead-time management. Interfaces 40(5):397–407

WIP-Oriented Dispatching in Complex Manufacturing Facilities

Oliver Rose and Zhugen Zhou

Abstract Most of the current dispatching approaches for complex manufacturing facilities like semiconductor fabs are related to due dates. They are variants of classical dispatching rules such as Critical Ratio (CR), Apparent Tardiness Cost (ATC), or Operation Due Date (ODD). Besides that there are a number of operational control policies which target the control of the inventory level of the work centers such as Kanban, Starvation Avoidance, or Minimum Inventory Variability Scheduler (MIVS). While the first set of dispatching rules does not primarily lead to low inventory levels, the latter ones do not always lead to good on-time delivery performance. We are currently developing an approach which combines both ideas, i.e., keeping a low WIP level, avoiding bottleneck starvation, and meeting the due dates. While due dates are usually given by the planning department, adequate WIP levels usually have to be set appropriately by means of pilot studies or educated guessing. As a consequence, an adaptive procedure to determine the adequate inventory levels should be implemented. In our contribution, we provide an overview of current dispatching approaches of both types and discuss their pros and cons. Then, we present our approach in detail and compare its performance with the classical approaches from the literature. Recently, we were able to outperform ODD with respect to WIP levels while having the same on-time delivery performance. The disadvantage is that the optimal target WIP levels (minimum and maximum workload level for the work centers) had to be set experimentally. In our future study, we intend to develop a back-propagation neural network for adaptive parameter setting.

O. Rose (✉) · Z. Zhou
Department of Computer Science,
University of the Federal Armed Forces Munich,
85577 Neubiberg, Germany
e-mail: oliver.rose@unibw.de

D. Armbruster and K. G. Kempf (eds.), *Decision Policies for Production Networks*,
DOI: 10.1007/978-0-85729-644-3_4,

1 Introduction

Nowadays to survive in the global market with increasing and fierce competition, fast, reliable product and service are the key to success for the companies. Inventory is one of the most important performance measures in the factory because it has a major influence on overall manufacturing costs. Excessive inventory requires additional floor space, storage equipment, and handling system to transport and manage, which increases the costs as non-value added. Furthermore, high WIP implies the risk of quality degradation because it is difficult to detect the defects that result in high rework and scraps. More importantly, according to the Little's Law [13], a lower work-in-process (WIP) level leads to a shorter production cycle time given the same throughput, which has significant economic importance to respond to today's quick market change fashion. Besides that, due date commitment is another critical factor, especially for the customer oriented company, to meet customer satisfaction. A missed due date causes not only penalty to the company, but also confidence lost to customer.

During the past 30 years, a number of researchers have investigated the performance of various dispatching rules for complex manufacturing facilities like the ones found in the semiconductor industry (flow shop or job shop). We refer the interested reader to [1, 25] for details. Different dispatching rules have different performance objectives. Some rules target due date control to achieve on-time delivery or at least minimal tardiness. Some rules target WIP balance for operations or work centers which can also lead to cycle time reduction. While the due date-oriented rules do not primarily lead to low inventory levels, the WIP-balance rules do not always guarantee a good on-time delivery performance.

1.1 Single Due Date-Oriented Rules

When the performance objective involves meeting a given due date, due date-oriented dispatching rules are generally employed to minimize the proportion of tardy jobs, mean tardiness of tardy jobs, and the like. They can be categorized into static rules such as Earliest Due Date (EDD) and dynamic rules such as Least Slack Time (LST), Critical Ratio (CR), Operation Due Date (ODD), and Modified Operation Due Date (MOD). The EDD rule aims at meeting the due date, and gives the highest priority to the job which has the earliest due date. Least Slack Time (LST) and Critical Ratio (CR) are variants of EDD. Besides the due date information, LST and CR consider the remaining raw processing-time of a job as well. The LST rule calculates the slack for each job as: *Slack = Due–Now–RemainingRPT*, where Due is the due date of a job, Now is the current time, and RemainingRPT denotes the remaining raw processing time. The job with the smallest slack is favored. LST is an extension to EDD for the reason that it tells us if two jobs have the same due date, the lot with longer remaining raw processing time is more urgent because its due date allows less delay. The CR rule distinguishes job urgency by a ratio between remaining time to the due date and remaining raw processing time, *Critical Ratio = (Due-Now) / RemainingRPT*, instead of computing a difference like LST. A CR value <1 denotes a job which falls behind schedule; a CR value =1 means that a job is on schedule, a CR value >1 represents a job which is ahead of schedule and has slack time left. CR assigns the highest priority to the job with the smallest CR value. Baker and Bertrand [4] presented a simulation study of combining due date assignment rules with due date-oriented dispatching rules. Considering minimizing Mean Tardiness as performance measure, they concluded that compared with SLT and CR, Shortest Processing Time (SPT) is effective with tight target due dates and EDD is superior with loose target due dates. While considering Conditional Mean Tardiness as performance measure, Muhlemann et al. [23] found out that CR outperforms EDD and LST. Rose [27] presented a detailed study of the CR rule, and showed that CR leads to sudden performance degradation when the target due date is too tight. This issue arises because CR only focuses on the final due date and speeds up jobs which are close to due date or already late. In contrast, the fresh jobs run out of their slack time and have to wait in early operations.

The ODD rule [2, 28] succeeds to avoid the above problems of CR. ODD breaks up the slack time into as many segments as the number of operations of a job, which means ODD considers due dates for all intermediate operations, unlike CR which only considers due date of the final processing operation. The ODD value of operation i is defined as: *ODD = Release Time + RPT(i) * FF*, where RPT(i) denotes the RPT for a sequence of processing steps or operations from operation 1 to operation i (including operation i) and FF denotes the target due date flow factor which is the ratio of target cycle time and raw processing time of a job. The ODD rule gives priority to the job with the smallest ODD value. For the final operation of a job the ODD is equal to the classical due date as it is used for CR, because slack times for young jobs assigned by the ODD rule are smaller than in the CR case. Therefore they do not have to let old jobs pass before they are processed. As a consequence, it is not possible with the ODD rule that problems at operations at the end of the processing sequence propagate back to the operations at the beginning. Once the operation due date has been established, the jobs are strictly kept at the right pace to meet their due date through the factory from the early operations on. Thus, the ODD rule is able to minimize the variance of job lateness relative to the due date and typically also leads to a low cycle time variance.

1.2 Composite Due Date-Oriented Rules

The performance of these due date-oriented dispatching rules is mainly affected by how tight or loose the due date is set [9]. Some rules perform better with tight target due dates like SPT, although SPT does not use any due date information, while some rules perform better with loose target due dates such as EDD and ODD. By noticing the complementary strengths of different rules working with different target due dates, Baker and Bertrand [3] presented the composite MOD rule which is a combination of SPT and ODD. It performs like SPT if the target due date is tight and like ODD if the target due date is loose. For each job in the queue of a work center at time t MOD is calculated in the following way: $MOD = \text{Max}\ (ODD, t + PT)$, where ODD is the operation due date of the job at work center, t is current time, and PT is the processing time of the job at the work center. The MOD rule gives priority to the job with the smallest value of MOD. It tends to combine the advantages of SPT and ODD and provides short cycle times and minimizes cycle time variance while working with different target due date simultaneously.

There is another composite rule called Apparent Tardiness Cost heuristic (ATC) [31]. The ATC rule combines the Weighted Shortest Processing Time (WSPT) rule and the LST rule. There are two characteristics of this rule. Firstly, apart from processing time, the ATC rule utilizes a look-ahead strategy and takes waiting time estimates of jobs on downstream work centers into consideration to calculate the slack time of each operation. Secondly, the ATC rule uses an exponential decay function to calculate the weight/processing time to allocate priority to jobs. The simulation results demonstrate that the ATC rule outperforms other due date-oriented dispatching rules with regard to minimization of weighted tardiness penalties. However, there are several user-defined parameters in the ATC rule. The application and accuracy of ATC rule depend considerably on defining appropriate parameters.

Most of these due date-oriented dispatching rules above are local rules. They only focus on the information of jobs which wait in the local work center buffer instead of taking into account information from elsewhere in the shop about, e.g., machines failures, machine utilizations, etc. Furthermore, they only work with due date information and focus only on on-time delivery. Thereby, sometimes they are incapable to handle WIP imbalances because of multiple re-entrant flows, machines breakdowns, etc. Consequently, the shop runs at a high inventory level with considerable cycle times.

1.3 WIP-Oriented Release Rules

In contrast to due date-oriented dispatching rules, WIP-oriented dispatching rules focus on workload control [14] which is a combination of job release approaches and dispatching/scheduling approaches used to control how jobs flow through work centers to achieve WIP balance in the line. WIP-oriented rules are typically global rules which utilize information not only from the local work center where the dispatching decision is made, but also from upstream and downstream work centers. Push and pull rules are two classical job release approaches for workload control. On one hand, the push rule is a make-to-order approach and originated from Material Requirements Planning (MRP) in the early 1970s [33, 29]. The product (job) release is based on shop throughput targets. The weakness of the push approach is that excessive WIP will cause considerable cycle times. On the other hand, the appearance of Japanese manufacturing techniques such as Just-In-Time (JIT) supported the introduction of pull approaches in the early 1980s. With the pull approach product (job) releases are based on the downstream shop status. A downstream work center tries to pull a job from an upstream work center. The pull approach has been proven to lead to less WIP congestion and to easier inventory control than the push approach [29]. Kanban and CONstant Work In Process (CONWIP) are two popular representatives of the pull approach. For the Kanban approach [21, 22], there is a card set between each pair of work centers, and the total system WIP level is limited to the sum of the numbers of cards in all card set. A job is pulled by each work center from the previous work center only if the job receives a card authorization. Kanban controls the WIP at the individual workcenter level. In contrast to Kanban, CONWIP [21, 30] only uses a single global set of cards to control the WIP level of the whole shop. Every job seizes a card when it is released to the system for the first time. If all cards are taken by jobs, a fresh job expecting to enter the system has to wait until a job leaves the system and the corresponding card is released. Kanban pulls jobs between each pair of work centers, while CONWIP pulls jobs only at the beginning of the line. Recently there is a strong interest in CONWIP. Firstly, CONWIP is similar to an input/output control rule. It is easy to understand and robust to control only requiring understanding the relationship between WIP and throughput [14]; Secondary, due to product mix changes, the bottleneck may shift over time. The Kanban approach needs to adjust the number of cards in each card set to avoid bottleneck starvation and make sure throughput. Therefore, the CONWIP approach is easier to manage because there is no tight WIP control between each pair of work centers [20].

Due to the success of Kanban and the appearance of Theory of Constraint (TOC) [17], the bottleneck-oriented pull approach was developed. Wein [32] introduced a Workload Regulation (WR) input approach for job releases to the shop. For WR a target workload of the bottleneck has to be defined. If the actual workload of the bottleneck drops to the target workload, a new job is released into the shop. Wein carried out a design of experiments which combines four job release approaches (Poisson arrival, Constant arrival, CONWIP, and WR) with several dispatching rules. He found out that the effects of specific dispatching rules rely considerably on both the type of job release approach and the number of bottlenecks in the shop. The WR approach is quite intuitive and only requires understanding the relationship between the target workload of the bottleneck and the system throughput. Therefore, it has been already widely adapted in real factory environments. However, setting the appropriate target workload is the key issue of WR. Currently, using a simulation model or a queuing network approximation to estimate the target workload of bottleneck is popular methods [14].

Glassey and Resende [16] presented another well-known bottleneck-oriented job release approach called Starvation Avoidance (SA). They defined a virtual inventory of the bottleneck which is used as a measure to keep a proper inventory level at the bottleneck. The virtual inventory includes the total bottleneck processing time of the next operations of all jobs which reach the bottleneck work center within a given lead time plus the expected time to repair the bottleneck machines which are currently broken down. The lead time is the sum of the processing times of all jobs required to arrive at the bottleneck the first time after their release. Glassey and Resende [16] compared the SA rule with three other job release approaches (Uniform arrival, WR, and CONWIP). They concluded that SA is more effective than the other job release approaches concerning near-capacity throughput while maintaining lower average job delays. However, compared to the WR approach, the SA approach requires more conceptual understanding and considerable implementation effort because it requires global inventory information about the whole shop.

1.4 WIP-Oriented Dispatching Rules

Although some researchers claimed that the job release approach is more important regarding workload control than dispatching [16, 32], there is no doubt that dispatching is still a powerful way to assist or improve the workload control, because dispatching approaches have low computational requirements and an intuitive appeal. In addition, they can be used to avoid machine starvation and they can handle re-entrant flows to effectively balance the line.

A promising WIP-oriented dispatching rule named Minimum Inventory Variability Scheduler (MIVS) was proposed by Li et al. [12]. MIVS considers both upstream and downstream operations, and tries to keep the WIP of each operation close to an average target WIP level. It gives highest priority to an operation which has a high WIP level while its downstream operation has a low WIP level to avoid starvation at downstream operations. In contrast, it gives the lowest priority to an operation which has a low WIP level while its downstream operation has a high WIP level. MIVS succeeds in adapting to the nature of re-entrant flows and in reducing the WIP imbalance through pulling jobs into low WIP operations. The results are reduced WIP variability and reduced cycle times. Collins and Palmeri [17] compared 1-step ahead MIVS with K-steps ahead MIVS and concluded that there is no obvious evidence that 3-step ahead MIVS outperforms 1-step ahead MIVS.

Based on MIVS, Ham and Fowler [19] introduced the Balanced Machine Workload (BMW) dispatching approach. The BMW considers K-machines look ahead and J-machines look back, while considering the WIP balance from the machine viewpoint instead of operation viewpoint like MIVS. Similar to MIVS, Dabbas and Fowler [18] proposed a global Line Balance (LB) algorithm with the objective of minimizing the deviations of actual WIP to target WIP for each operation. Through calculating throughput signals, cumulative signals, and unconstrained quantities, LB determines portions of WIP at all operation stages required to be pushed forward to balance the downstream operations. The main novelty and contribution of this approach is that the authors considered LB as a global dispatching approach combined with several local dispatching rules such as CR, Flow Control (FC) and Throughput (TP) into a single rule, with the objective of optimizing different performance measures simultaneously. Defining an appropriate target WIP level is the key issue of applying MIVS or LB. In general, as shown in previous studies [5, 8, 24], using simulation models or queuing models to estimate the target WIP level is an appropriate way to estimate target WIP levels. Kuo et al. [11] proposed a back-propagation neural network model to determine the target WIP level for the bottleneck and non-bottleneck work centers instead of a queuing model with the purpose to guarantee a maximum throughput of the bottleneck while achieving a minimum WIP level.

Perdaen et al. [26] proposed an interesting dispatching concept combining a push policy (first in first out) and a pull policy (shortest remaining process time) together via a push–pull point (PPP) to control a typical re-entrant manufacturing line, with the objective to reduce the mismatch between the daily output and demand. The novelty of this approach which has not been considered in the literature before is to introduce the PPP to divide the line where push policy is applied in the upstream of PPP and pull policy is employed in the downstream of PPP. Through simulation experiment, they found out that when the PPP control works together with CONWIP release policy, significant improvement was obtained for the high demand with high variance compared with pure pull policy or pure push policy, or CONWIP combined with pure pull policy. The next chapter of this book is dedicated to this approach.

2 Proposed Workload Balance and Due Date Control Approach

Facilitated by the complementary strength of due date-oriented and WIP-oriented rules, we are currently developing an approach which combines WIP balance and due date control. On the one hand, WIP balance leads to cycle time reduction. On the other hand, focusing on due date control achieves better on-time delivery and tardiness performance. We propose the following approach:

(1) Bottleneck workload control

According to TOC, the performance of the whole shop, e.g., its throughput is mainly determined by the bottleneck performance [18]. It is necessary to determine an adequate WIP level for the bottleneck buffer to avoid starvation and to support the whole shop to achieve its maximum throughput while running at the minimum WIP level. However, if the WIP level of bottleneck exceeds the desired WIP level while achieving the maximum throughput of the whole shop, the cycle time is degraded [10]. Jobs will spend a significant queue time in front of the bottleneck work center, which will also cause a WIP imbalance of the line. Similar to the WR rule, we define a minimum workload for the bottleneck work center. If the actual workload of the bottleneck drops to the minimum workload, the bottleneck is fed with jobs to prevent starvation. Besides that, a maximum workload is also taken into account. If the actual workload of the bottleneck is higher than the maximum workload, bottleneck feeding is stopped to avoid extraordinary queue time, especially, when the bottleneck is broken down. In this study, we only consider a single dynamic bottleneck in the shop where the bottleneck is the work center with the highest utilization.

(2) Feeding empty non-bottleneck work centers

Although the bottleneck is the most critical work center which determines the performance of the whole shop, feeding empty non-bottleneck work centers can also smooth the material flow, avoid capacity losses of machines, and improve product cycle times. Therefore, a minimum workload is also defined for the non-bottleneck work centers. If the workload of non bottlenecks drops to this minimum workload level, lots are scheduled to feed it to avoid starvation.

(3) Acceleration of maximum tardiness lots

In general, any WIP balance algorithm tends to push jobs to work centers that are running out of WIP without taking due dates into consideration. In this case, overemphasizing WIP balance has a negative impact on on-time delivery. In fact, sometimes it would be better to push a delayed job to a relative high WIP work center instead of pushing an early job to a relative low WIP work center. Because of customer commitments, keeping the due date is the first priority for customer-oriented companies. Therefore, a compromise is necessary in order to meet due dates and reduce tardiness. Pushing a delayed job despite WIP balance requirements to downstream work centers can give the delayed job a chance to speed up, to save cycle time, and reduce tardiness, although work center capacity might be lost. The acceleration algorithm works as follows:

Step 1 In the queue of the upstream work center, we determine the job which has the maximum tardiness 'MaxTardinessUp'.
Step 2 Then, we identify the target downstream work center where the 'MaxTardinessUp' job will be processed. Next, we find the job which has the maximum tardiness 'MaxTardinessDown' in the queue of the target downstream work center (like in Step 1).
Step 3 If 'MaxTardinessUP' is greater than 'MaxTardinessDown', the job which has 'MaxTardinessUp' is assigned a high priority in the upstream work center.

(4) Acceleration of jobs close to their due date

Acceleration of delayed jobs can only reduce tardiness instead of improving on-time delivery performance. Thus, we propose to speed up the jobs which are close to their due dates. This provides a mechanism for those jobs to catch up with their due date. If there is still a predefined number of hours left for the job to chase after the due date and the CR value of job is less than 1—which means the job has already been close to its due date and possibly fallen behind schedule—this job will obtain a higher priority since there is a high probability that it will be late in the future.

In order to test our approach we extended a simplified version of the global dispatching rule which is in use at Infineon Technologies AG Dresden, a German semiconductor manufacturer, with our ideas. We call the rule Workload Balance & Due Date Control (WB&DDC). In each queue of a work center, jobs are categorized into six classes in descending priorities according to their states. The following six WB&DDC job priority classes were defined:

(1) Jobs waiting more than 48 h in the queue:

- (1.1) Delayed job;
- (1.2) Non-delayed job:
  - (1.2.1) Close to due date;
  - (1.2.2) On schedule.

(2) Acceleration of maximum tardiness jobs.

(3) Feeding empty bottleneck (Bottleneck workload control):

- (3.1) Delayed job;
- (3.2) Non-delayed job:
  - (3.2.1) Close to due date;
  - (3.2.2) On schedule.

(4) Feeding empty non-bottleneck work centers:

- (4.1) Delayed job;
- (4.2) Non-delayed job:
  - (4.2.1) Close to due date;
  - (4.2.2) On schedule.

(5) Jobs for non-empty non-bottleneck work centers:

- (5.1) Delayed job;
- (5.2) Non-delayed job:
  - (5.2.1) Close to due date;
  - (5.2.2) On schedule.

(6) Jobs for overloaded bottleneck (Bottleneck workload control):

- (6.1) Delayed job;
- (6.2) Non-delayed job:
  - (6.2.1) Close to due date;
  - (6.2.2) On schedule.

In the first priority class, jobs spending more than 48 h in the queue waiting for processing have to be processed immediately to reduce the cycle time variability of the operation. The delayed jobs which fulfill the criterion for accelerating of maximum lateness jobs belong to second priority class. This priority class is more critical than the priority class of the bottleneck workload control method and of the feeding empty non-bottleneck method because customer commitment is more important than WIP balance in our approach. Accelerating maximum tardiness jobs is considered as a compromise to WIP balance. Therefore, the upstream work centers would rather push the maximum tardiness job to downstream work centers which may be highly loaded instead of pushing an early job to downstream work centers which may be starved to maintain WIP balance. The maximum tardiness job has to be moved to the next operation to minimize delay. Feeding empty bottlenecks is more urgent than feeding empty non-bottleneck work centers because the bottleneck determines throughput of the whole shop. Jobs which are processed next at the overloaded bottleneck belong to the lowest priority class, since overloaded bottlenecks are more likely to be subject to congestion and breakdown than normal or even high-WIP non-bottleneck work centers. In this case, long queues in front of the bottleneck will result in an irregularity in the process flow. The consequence will be that the average cycle time will increase considerably although the WIP of the whole factory stays approximately on the same level. Except for the second priority class, jobs which belong to the other five priority classes are divided into two sub classes which are the delayed job class and the non-delayed job class. The delayed job class has higher priority than non-delayed job class. Furthermore, the non-delayed job class is also split into two sub classes which separate jobs close to their due dates from jobs on schedule. According to the acceleration of jobs close to their due date method, jobs which are close to their due date are more preferential than jobs on schedule. If jobs belong to the same priority class, the ODD rule is applied as the dispatching rule.

3 Simulation Model

Wafer fabrication facilities (wafer fab) have been intensively studied by academic and industrial researchers for many years. Scheduling a wafer fab is considered as one of the most complicated scheduling problems encountered nowadays. There are several features that differentiate wafer fabs from traditional flow shops or job shops: (1) product mix (2) hundreds of processing steps for each product (3) reentrant flows (4) tool dedication (5) batch processing (6) tool random failure and preventative maintenance, etc. In a wafer fab, the reentrant nature requires that lots at different processing operations have to compete with one another for the same tools, especially for those expensive tools like photolithography. Additionally, there are hundreds of tools in a wafer fab, and they are all subject to random failures and preventative maintenance. In such an environment, WIP imbalance occurs rather often, some tools are starved, while some tools are overloaded. This WIP imbalance phenomenon has a great impact on cycle times and on-time delivery. Hence, customer-oriented companies have to deal with the WIP imbalance carefully to achieve their customer commitment. In this study, we choose wafer fab as the simulation model not only because of its popularity, but also because it will provide an insight to understand the importance of WIP balance and due date control in other complex manufacturing environments.

We use the wafer fab dataset Measurement and Improvement of MAnufacturing Capacities (MIMAC6) to test the proposed WB&DDC approach. MIMAC6 is a typical 200 mm wafer fab model. The following list gives an overview of the main characteristics of MIMAC6 model. For further detail about this model, please refer to Fowler and Robinson [15].

- Product profile:
  - 9 products, 24 wafers of one lot size;
  - Avg. mask layers: 30;
  - Max. static capacity: 2,777 lots released per year (approx. 5,554 wafers per month).
- Process flow:
  - 9 process flows, max. 355 process steps;
  - Avg. line yield: 93%.
- Tool group and operator:
  - 104 tool groups, 228 tools;
  - 46 single processing tool groups, 58 batch processing tool groups;
  - 9 operator groups.
- Availability:
  - Failures: clock time-based exponentially distributed random failures;
  - Avg. downtime per tool: 13.6%;
  - All downtimes are modeled as non-preemptive.
- Process time:
  - Constant per wafer, per lot or per batch process times;
  - Load and unload times;
  - Transport times not modeled.
- Setup and batch:
  - Setup avoidance to minimize setup time;
  - Different batch IDs to form batches, minimum and maximum batch size.

4 Performance Analysis

We conducted the simulation with Factory Explorer from Wright Williams & Kelly (WWK), a commercial simulation package for factory models. The simulation of MIMAC6 was run for 18 months. The first 6 months were considered as warm-up periods, and not taken into account for statistics. MIMAC6 was simulated with six dispatching rules: the WB&DDC, MOD, ODD, CR, MIVS and FIFO under 75, 85 and 95% fab capacity loading and with target due date flow factor ranging from 1.5 to 2.9 in steps of 0.2 respectively. Seventy five percent fab loading is considered as a low loading. In this situation, most of the work centers have a low WIP and the lots go through the fab smoothly even though FIFO is applied as dispatching rule, which means WIP may not need to be balanced at all. We are curious that whether the WB&DDC can take effect to smooth the manufacturing process further under this low fab loading case. In contrast, when the fab runs under a high loading like 95% loading, the manufacturing environment is extremely complex, e.g. some work centers like bottleneck have extraordinary long queue and the WIP fluctuates oftentimes. Can the WB&DDC overcome the WIP imbalance and avoid high WIP taking place? How much improvement can the WB&DDC achieve? These are what we concern. Here the simulation results of 95% fab loading are analyzed in detail, the simulation results of 75 and 85% fab loading are listed in the appendix with less detail. We also present the best average cycle times, the best cycle time variances and the best cycle time upper 95% percentiles of all rules under 75, 85, and 95% fab loading, respectively.

In this study, the minimal workload level for all work centers and the maximal workload level for the bottleneck with 95% fab loading are obtained as following: (1) one year simulation runs of the factory model with 95% fab loading and FIFO dispatching was carried out and the 'minimum' and 'maximum' workload levels for each work center were acquired; (2) the 'minimum' workload was self-increased and self-decreased from 2 to 20% in steps of 2% like 'mimimum' $\pm 2\% \times$ 'mimimum′, $\pm 4\% \times$ 'mimimum′, $\pm 6\% \times$ 'mimimum′ ... $\pm 20\% \times$ 'mimimum', and used as the minimum workload level for all work centers for the WB&DDC approach. The maximum workload level for the bottleneck was calculated the same as minimum workload level. By simulation experiment, 'mimimum'-12%$\times$ 'mimimum' as minimum workload for each work centers and 'maximum'-16%$\times$ 'maximum' as maximum workload for the bottleneck can achieve the best performance for the WB&DDC approach in the following. The same procedure was implemented to gain the target minimum and maximum workload levels for the 75 and 85% fab loading.

Average cycle time. Table 1 shows the average cycle time and the respective half width of the 95% confidence interval for each product. Figure 1 shows the average cycle time evolution of the fab for different target due date flow factors with six dispatching rules (WB&DDC, MOD, ODD, CR, MIVS and FIFO). We observe that the average cycle time of ODD and CR is considerably large when the target due date is set too tight. Especially for the CR rule, there is a cycle time degradation when the due date flow factor is changed from 2.1 to 1.9. It also tells us that it is not a trivial task to assign an appropriate target due dates for each product to facilitate applying due-date oriented dispatching. The MOD succeeds in avoiding large cycle times with tight target due dates. Because there is no due date control mechanism for MIVS and FIFO, their curves do not change for different target due date values. It is evident that MIVS succeeds in reducing cycle time compared to FIFO. Our proposed WB&DDC has a similar trend as MOD. Furthermore, it outperforms MOD, MIVS and FIFO for loose target due dates. At a due date flow factor of 2.1, it reaches its minimum average cycle time which is better than all other rules. Therefore, the WB&DDC approach is more robust than ODD, CR and MOD with respect to different target due date flow factor settings and still superior to MIVS and FIFO with loose target due date.

Table 1 Average cycle time comparison for different products (95% capacity loading)

		Due Date Flow Factor (DDFF)							
	Product	1.5	1.7	1.9	2.1	2.3	2.5	2.7	2.9
	B5C	36.9 ± 1.6	37.0 ± 1.5	36.3 ± 1.0	35.9 ± 0.7	35.6 ± 0.5	35.8 ± 1.1	35.8 ± 0.5	35.8 ± 0.5
	B6HF	35.8 ± 1.9	35.9 ± 1.7	34.9 ± 1.3	34.5 ± 1.2	34.7 ± 1.0	34.9 ± 1.4	35.9 ± 1.3	35.8 ± 1.4
WB&	C4PH	26.2 ± 1.4	25.4 ± 1.4	23.9 ± 1.0	22.9 ± 0.8	22.6 ± 0.8	22.5 ± 0.8	22.5 ± 0.8	22.3 ± 0.7
DDC	C5F	31.9 ± 1.5	31.7 ± 1.5	30.8 ± 1.1	30.7 ± 0.9	30.1 ± 0.9	31.9 ± 1.1	32.7 ± 1.1	33.3 ± 1.3
	C5P	26.6 ± 1.4	25.9 ± 1.4	24.5 ± 1.0	24.9 ± 0.8	24.7 ± 0.8	24.8 ± 0.8	24.9 ± 0.8	24.9 ± 0.7
	C5PA	30.4 ± 1.5	29.9 ± 1.5	27.6 ± 1.2	27.9 ± 1.0	27.7 ± 0.9	27.9 ± 1.0	28.0 ± 1.0	28.2 ± 1.0
	C6N3	33.3 ± 1.7	33.0 ± 1.7	30.3 ± 1.4	29.7 ± 1.3	29.7 ± 1.1	29.4 ± 1.1	30.1 ± 1.1	30.2 ± 1.2
	C6N2	30.2 ± 1.6	29.5 ± 1.5	26.9 ± 1.2	25.5 ± 1.0	25.3 ± 0.9	26.3 ± 0.9	25.5 ± 0.9	25.6 ± 1.0
	OX2	29.3 ± 1.5	28.8 ± 1.5	27.4 ± 1.1	26.5 ± 1.0	26.2 ± 0.8	26.3 ± 0.9	26.4 ± 0.8	26.4 ± 0.7
	Summary	30.6	30.2	28.6	28.1	27.9	28.1	28.3	28.5
	B5C	38.4 ± 1.8	38.1 ± 1.3	37.8 ± 1.3	37.7 ± 1.2	36.9 ± 0.5	37.2 ± 0.5	38.9 ± 0.5	39.1 ± 0.4
	B6HF	34.8 ± 1.5	34.6 ± 1.3	34.8 ± 1.2	33.4 ± 1.3	33.5 ± 1.0	35.9 ± 0.6	36.1 ± 0.7	36.9 ± 1.8
	C4PH	26.2 ± 1.6	25.1 ± 1.3	24.5 ± 1.1	24.1 ± 0.9	23.5 ± 0.9	23.3 ± 0.8	23.6 ± 0.8	23.7 ± 0.7
	C5F	33.8 ± 1.6	33.6 ± 1.4	33.6 ± 1.2	33.3 ± 1.1	33.2 ± 0.8	33.6 ± 0.5	33.8 ± 0.5	34.7 ± 0.5
MOD	C5P	28.3 ± 1.4	27.7 ± 1.3	27.3 ± 1.1	26.7 ± 1.1	25.6 ± 0.8	25.4 ± 0.7	26.0 ± 0.8	26.3 ± 0.7
	C5PA	30.1 ± 1.5	29.3 ± 1.3	29.1 ± 1.2	28.7 ± 1.1	27.4 ± 1.1	28.5 ± 1.1	29.0 ± 1.2	29.5 ± 1.2
	C6N3	32.9 ± 1.8	31.4 ± 1.5	31.0 ± 1.6	30.8 ± 1.3	30.5 ± 1.2	30.7 ± 1.3	30.6 ± 1.3	31.6 ± 1.4
	C6N2	29.8 ± 1.6	28.8 ± 1.4	27.6 ± 1.3	27.5 ± 1.2	27.1 ± 1.0	27.0 ± 1.0	27.3 ± 1.0	27.4 ± 1.2
	OX2	28.1 ± 1.4	27.2 ± 1.2	26.8 ± 1.0	27.0 ± 0.8	26.5 ± 0.8	27.0 ± 0.8	27.5 ± 0.7	27.7 ± 0.9
	Summary	30.9	30.0	29.7	29.2	28.7	28.8	29.4	29.8
	B5C	55.7 ± 6.5	43.9 ± 3.4	38.1 ± 1.6	36.2 ± 0.9	35.1 ± 0.6	35.3 ± 0.4	36.6 ± 0.5	37.9 ± 0.4
	B6HF	54.7 ± 6.6	42.8 ± 4.5	36.9 ± 1.9	34.9 ± 1.5	34.2 ± 1.2	34.5 ± 1.3	35.8 ± 1.6	36.9 ± 1.5
	C4PH	46.1 ± 6.5	33.5 ± 3.4	27.6 ± 1.9	24.8 ± 1.2	23.3 ± 0.8	23.3 ± 0.6	23.4 ± 0.7	23.9 ± 0.8

(Continued)

Table 1 (Continued)

		Due Date Flow Factor (DDFF)							
	C5F	52.6 ± 6.6	40.3 ± 3.3	35.6 ± 1.8	33.6 ± 1.2	32.9 ± 0.8	32.9 ± 0.4	33.5 ± 0.5	34.6 ± 0.5
ODD	C5P	48.5 ± 6.6	35.1 ± 3.3	29.3 ± 1.9	26.7 ± 1.2	25.4 ± 0.8	25.5 ± 0.6	25.8 ± 0.6	26.2 ± 0.7
	C5PA	49.8 ± 6.6	37.5 ± 3.4	31.9 ± 1.9	29.4 ± 1.3	28.2 ± 1.1	28.3 ± 1.1	28.7 ± 1.2	28.6 ± 1.3
	C6N3	52.6 ± 6.6	39.4 ± 3.5	33.0 ± 2.0	31.4 ± 1.5	30.4 ± 1.2	30.5 ± 1.2	30.9 ± 1.3	30.6 ± 1.4
	C6N2	50.1 ± 6.6	37.6 ± 3.5	30.9 ± 2.0	27.2 ± 1.3	26.9 ± 1.1	26.7 ± 0.9	26.9 ± 1.0	27.3 ± 1.2
	OX2	49.8 ± 6.6	37.3 ± 3.4	29.7 ± 1.8	27.1 ± 1.1	26.2 ± 0.7	26.8 ± 0.6	27.2 ± 0.8	27.9 ± 0.9
	Summary	50.8	38.1	31.9	29.5	28.5	28.6	29.1	29.7
	B5C	74.4 ± 10.5	71.4 ± 10.3	65.0 ± 9.4	35.7 ± 1.6	35.1 ± 0.4	35.8 ± 0.5	36.9 ± 0.5	37.9 ± 0.6
	B6HF	72.0 ± 9.9	69.2 ± 9.9	61.8 ± 9.4	34.3 ± 1.9	33.9 ± 0.9	34.5 ± 1.1	34.8 ± 1.4	35.5 ± 1.7
	C4PH	48.1 ± 6.1	48.3 ± 6.2	43.2 ± 5.6	23.2 ± 1.0	23.6 ± 0.6	24.1 ± 0.7	25.2 ± 0.8	28.4 ± 0.9
CR	C5F	68.0 ± 10.0	56.1 ± 9.9	58.7 ± 8.8	32.3 ± 1.5	32.5 ± 0.9	34.9 ± 1.2	36.1 ± 1.3	28.7 ± 1.6
	C5P	62.1 ± 9.6	58.6 ± 9.6	51.8 ± 8.5	25.5 ± 1.1	25.0 ± 0.5	26.3 ± 0.7	28.2 ± 0.8	30.4 ± 0.9
	C5PA	65.1 ± 9.8	63.5 ± 9.9	55.7 ± 8.9	28.2 ± 1.5	27.8 ± 0.9	28.5 ± 1.0	28.6 ± 1.1	30.3 ± 1.3
	C6N3	71.0 ± 10.7	69.4 ± 10.7	60.9 ± 9.7	29.2 ± 2.0	28.7 ± 1.1	28.4 ± 1.0	28.9 ± 1.1	29.3 ± 1.1
	C6N2	66.6 ± 10.1	63.9 ± 10.1	54.5 ± 9.1	26.1 ± 1.6	26.0 ± 0.9	25.9 ± 0.8	25.9 ± 0.9	28.3 ± 1.0
	OX2	62.9 ± 9.7	61.6 ± 9.8	53.5 ± 8.6	25.7 ± 1.3	25.9 ± 0.9	26.9 ± 0.8	27.5 ± 0.9	30.3 ± 0.5
	Summary	65.5	63.2	55.6	28.4	28.2	28.9	29.6	30.5
	B5C	34.9 ± 1.6	34.9 ± 1.6	34.9 ± 1.6	34.9 ± 1.6	34.9 ± 1.6	34.9 ± 1.6	34.9 ± 1.6	34.9 ± 1.6
	B6HF	32.3 ± 1.6	32.3 ± 1.6	32.3 ± 1.6	32.3 ± 1.6	32.3 ± 1.6	32.3 ± 1.6	32.3 ± 1.6	32.3 ± 1.6
	C4PH	24.7 ± 1.2	24.7 ± 1.2	24.7 ± 1.2	24.7 ± 1.2	24.7 ± 1.2	24.7 ± 1.2	24.7 ± 1.2	24.7 ± 1.2
	C5F	31.6 ± 1.5	31.6 ± 1.5	31.6 ± 1.5	31.6 ± 1.5	31.6 ± 1.5	31.6 ± 1.5	31.6 ± 1.5	31.6 ± 1.5
MIVS	C5P	25.7 ± 1.0	25.7 ± 1.0	25.7 ± 1.0	25.7 ± 1.0	25.7 ± 1.0	25.7 ± 1.0	25.7 ± 1.0	25.7 ± 1.0

(Continued)

Table 1 (Continued)

		Due Date Flow Factor (DDFF)							
	C5PA	27.6 ± 1.1	27.6 ± 1.1	27.6 ± 1.1	27.6 ± 1.1	27.6 ± 1.1	27.6 ± 1.1	27.6 ± 1.1	27.6 ± 1.1
	C6N3	30.2 ± 1.3	30.2 ± 1.3	30.2 ± 1.3	30.2 ± 1.3	30.2 ± 1.3	30.2 ± 1.3	30.2 ± 1.3	30.2 ± 1.3
	C6N2	26.8 ± 1.1	26.8 ± 1.1	26.8 ± 1.1	26.8 ± 1.1	26.8 ± 1.1	26.8 ± 1.1	26.8 ± 1.1	26.8 ± 1.1
	OX2	26.4 ± 1.2	26.4 ± 1.2	26.4 ± 1.2	26.4 ± 1.2	26.4 ± 1.2	26.4 ± 1.2	26.4 ± 1.2	26.4 ± 1.2
	Summary	28.5	28.5	28.5	28.5	28.5	28.5	28.5	28.5
	B5C	37.8 ± 1.8	37.8 ± 1.8	37.8 ± 1.8	37.8 ± 1.8	37.8 ± 1.8	37.8 ± 1.8	37.8 ± 1.8	37.8 ± 1.8
	B6HF	32.5 ± 1.7	32.5 ± 1.7	32.5 ± 1.7	32.5 ± 1.7	32.5 ± 1.7	32.5 ± 1.7	32.5 ± 1.7	32.5 ± 1.7
	C4PH	24.2 ± 1.1	24.2 ± 1.1	24.2 ± 1.1	24.2 ± 1.1	24.2 ± 1.1	24.2 ± 1.1	24.2 ± 1.1	24.2 ± 1.1
	C5F	33.3 ± 1.7	33.3 ± 1.7	33.3 ± 1.7	33.3 ± 1.7	33.3 ± 1.7	33.3 ± 1.7	33.3 ± 1.7	33.3 ± 1.7
FIFO	C5P	27.0 ± 1.3	27.0 ± 1.3	27.0 ± 1.3	27.0 ± 1.3	27.0 ± 1.3	27.0 ± 1.3	27.0 ± 1.3	27.0 ± 1.3
	C5PA	28.0 ± 1.2	28.0 ± 1.2	28.0 ± 1.2	28.0 ± 1.2	28.0 ± 1.2	28.0 ± 1.2	28.0 ± 1.2	28.0 ± 1.2
	C6N3	31.5 ± 1.4	31.5 ± 1.4	31.5 ± 1.4	31.5 ± 1.4	31.5 ± 1.4	31.5 ± 1.4	31.5 ± 1.4	31.5 ± 1.4
	C6N2	28.1 ± 1.1	28.1 ± 1.1	28.1 ± 1.1	28.1 ± 1.1	28.1 ± 1.1	28.1 ± 1.1	28.1 ± 1.1	28.1 ± 1.1
	OX2	25.4 ± 1.1	25.4 ± 1.1	25.4 ± 1.1	25.4 ± 1.1	25.4 ± 1.1	25.4 ± 1.1	25.4 ± 1.1	25.4 ± 1.1
	Summary	29.6	29.6	29.6	29.6	29.6	29.6	29.6	29.6

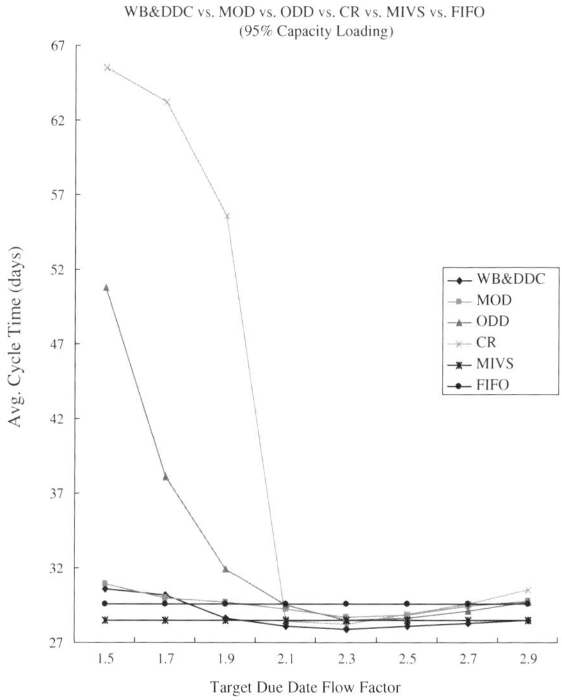

Fig. 1 Average cycle time comparison (95% capacity loading)

Cycle time variance. Table 2 shows the cycle time variance and the corresponding half widths of 95% confidence intervals for each product. Figure 2 depicts the cycle time variance performance. As we can see, the ODD rule has a smooth and excellent cycle time variance curve which is consistent with previous studies. MOD provides worse results than ODD for tight due dates and similar results as ODD for loose due dates, because MOD performs like SPT for tight due dates and like ODD for loose due dates. The CR rule seems to lose control because its cycle time variance curve changes badly and ranges from 0.6 to 3.2. Because there is no special mechanism for MIVS and FIFO to minimize cycle time variance both rules are worse than ODD. The WD&DDC approach has the same control performance as ODD with respect to cycle time variance because the ODD rule is integrated as an internal rule for lots belonging to the same priority class. The WD&DDC approach achieves shorter cycle times and, in addition, a precise prediction of the production completion time as well. This is a promising result for customer-oriented companies because they will be able to provide an accurate lead time commitment to their customers.

Cycle time upper percentile 95%. This performance measure provides a cycle time value below which 95% of the lots' cycle times fall. It is another important indicator for cycle time distributions. Figure 3 looks quite similar to average cycle time curve

Table 2 Cycle time variance comparison for different products (95% capacity loading).

		Due Date Flow Factor (DDFF)							
	Product	1.5	1.7	1.9	2.1	2.3	2.5	2.7	2.9
	B5C	0.6 ± 0.2	0.5 ± 0.2	0.5 ± 0.1	0.3 ± 0.1	0.3 ± 0.2	0.3 ± 0.2	0.6 ± 0.5	0.7 ± 0.4
	B6HF	0.6 ± 0.2	0.5 ± 0.2	0.5 ± 0.1	0.3 ± 0.1	0.6 ± 0.4	0.7 ± 0.2	0.9 ± 0.5	0.9 ± 0.4
WB&	C4PH	0.7 ± 0.2	0.6 ± 0.2	0.4 ± 0.2	0.3 ± 0.2	0.3 ± 0.2	0.3 ± 0.1	0.4 ± 0.1	0.6 ± 0.3
DDC	C5F	0.9 ± 0.2	0.8 ± 0.2	0.5 ± 0.2	0.4 ± 0.1	0.5 ± 0.2	0.7 ± 0.5	1.0 ± 0.4	1.0 ± 0.3
	C5P	0.4 ± 0.1	0.4 ± 0.1	0.3 ± 0.1	0.2 ± 0.1	0.3 ± 0.1	0.4 ± 0.2	0.4 ± 0.2	0.6 ± 0.3
	C5PA	0.5 ± 0.2	0.5 ± 0.1	0.5 ± 0.3	0.3 ± 0.2	0.4 ± 0.2	0.7 ± 0.7	0.5 ± 0.3	0.6 ± 0.4
	C6N3	0.9 ± 0.3	0.8 ± 0.3	0.7 ± 0.3	0.6 ± 0.3	0.8 ± 0.5	0.9 ± 0.2	0.9 ± 0.4	1.2 ± 0.4
	C6N2	0.7 ± 0.2	0.7 ± 0.2	0.5 ± 0.2	0.4 ± 0.1	0.5 ± 0.1	0.6 ± 0.2	0.6 ± 0.2	0.7 ± 0.2
	OX2	0.6 ± 0.2	0.6 ± 0.2	0.4 ± 0.2	0.3 ± 0.2	0.3 ± 0.1	0.5 ± 0.3	0.6 ± 0.3	0.6 ± 0.3
	Summary	0.65	0.59	0.47	0.33	0.43	0.56	0.66	0.76
	B5C	1.7 ± 0.5	1.3 ± 0.4	1.0 ± −.3	0.7 ± 0.2	0.3 ± 0.1	0.3 ± 0.2	0.3 ± 0.3	0.3 ± 0.2
	B6HF	2.1 ± 0.9	1.2 ± 0.4	1.1 ± 0.4	0.5 ± 0.2	0.8 ± 0.7	0.6 ± 0.2	1.1 ± 1.1	1.2 ± 1.2
	C4PH	1.8 ± 0.4	1.5 ± 0.3	1.1 ± 0.2	0.7 ± 0.4	0.4 ± 0.1	0.3 ± 0.1	0.3 ± 0.2	0.5 ± 0.4
	C5F	1.6 ± 0.4	1.0 ± 0.2	1.0 ± 0.2	0.5 ± 0.2	0.3 ± 0.1	0.4 ± 0.2	0.4 ± 0.2	0.5 ± 0.1
MOD	C5P	1.0 ± 0.2	1.0 ± 0.2	0.6 ± 0.1	0.4 ± 0.2	0.2 ± 0.1	0.3 ± 0.1	0.3 ± 0.1	0.3 ± 0.2
	C5PA	1.2 ± 0.4	0.9 ± 0.2	0.8 ± 0.3	0.8 ± 0.6	0.5 ± 0.4	0.4 ± 0.3	0.6 ± 0.6	0.9 ± 1.1
	C6N3	1.7 ± 0.5	1.3 ± 0.3	1.3 ± 0.4	0.9 ± 0.3	0.7 ± 0.4	0.8 ± 0.3	1.0 ± 0.7	1.3 ± 1.0
	C6N2	1.5 ± 0.3	0.9 ± 0.3	1.1 ± 0.3	0.7 ± 0.3	0.6 ± 0.3	0.6 ± 0.3	0.9 ± 1.6	0.8 ± 0.7
	OX2	1.6 ± 0.3	1.5 ± 0.5	1.2 ± 0.3	0.6 ± 0.1	0.3 ± 0.2	0.6 ± 0.8	0.6 ± 0.6	0.9 ± 1.0
	Summary	1.55	1.17	1.0	0.65	0.43	0.48	0.62	0.75
	B5C	0.6 ± 0.2	0.3 ± 0.2	0.2 ± 0.1	0.2 ± 0.1	0.2 ± 0.1	0.2 ± 0.1	0.4 ± 0.3	0.4 ± 0.1
	B6HF	0.7 ± 0.1	0.3 ± 0.1	0.3 ± 0.1	0.4 ± 0.2	0.7 ± 0.8	0.5 ± 0.1	0.9 ± 0.6	1.4 ± 1.5
	C4PH	0.7 ± 0.2	0.3 ± 0.2	0.3 ± 0.2	0.3 ± 0.1	0.2 ± 0.2	0.3 ± 0.2	0.4 ± 0.1	0.6 ± 0.3
	C5F	0.6 ± 0.2	0.2 ± 0.1	0.2 ± 0.1	0.3 ± 0.1	0.2 ± 0.1	0.3 ± 0.2	0.5 ± 0.2	0.6 ± 0.1

(Continued)

Table 2 (Continued)

		Due Date Flow Factor (DDFF)							
ODD	C5P	0.7 ± 0.1	0.3 ± 0.1	0.2 ± 0.1	0.2 ± 0	0.3 ± 0.1	0.3 ± 0.1	0.3 ± 0.1	0.4 ± 0.1
	C5PA	0.7 ± 0.1	0.3 ± 0.2	0.4 ± 0.3	0.3 ± 0.3	0.4 ± 0.4	0.5 ± 0.5	1.0 ± 1.0	1.0 ± 0.9
	C6N3	0.9 ± 0.2	0.5 ± 0.1	0.5 ± 0.4	0.6 ± 0.3	0.5 ± 0.1	0.9 ± 0.6	1.2 ± 0.8	1.5 ± 1.1
	C6N2	0.8 ± 0.1	0.5 ± 0.2	0.4 ± 0.3	0.5 ± 0.3	0.4 ± 0.1	0.6 ± 0.3	0.9 ± 0.7	0.9 ± 0.5
	OX2	0.7 ± 0.1	0.4 ± 0.2	0.3 ± 0.2	0.2 ± 0.1	0.3 ± 0.3	0.3 ± 0.1	0.6 ± 0.5	1.1 ± 1.0
	Summary	0.7	0.34	0.31	0.32	0.35	0.44	0.67	0.87
	B5C	2.4 ± 0.8	2.0 ± 0.6	2.8 ± 1.2	0.5 ± 0.4	0.8 ± 0.7	1.5 ± 0.7	1.8 ± 0.6	2.5 ± 0.9
	B6HF	3.2 ± 0.9	3.1 ± 1.3	2.9 ± 0.9	0.8 ± 0.9	1.5 ± 1.2	2.4 ± 0.7	2.9 ± 0.8	4.9 ± 1.5
	C4PH	2.7 ± 0.7	3.2 ± 1.0	2.1 ± 0.6	0.4 ± 0.4	0.4 ± 0.2	0.8 ± 0.4	1.2 ± 0.5	1.9 ± 0.4
CR	C5F	2.5 ± 0.5	2.5 ± 0.7	2.2 ± 0.7	0.5 ± 0.5	1.0 ± 0.8	2.6 ± 1.2	3.8 ± 1.4	7.7 ± 2.6
	C5P	1.4 ± 0.3	1.6 ± 0.3	1.4 ± 0.4	0.4 ± 0.2	0.5 ± 0.2	0.9 ± 0.5	1.3 ± 0.4	2.0 ± 0.4
	C5PA	1.8 ± 0.3	1.8 ± 0.4	1.5 ± 0.3	0.6 ± 0.6	0.9 ± 0.7	1.3 ± 0.4	1.9 ± 0.6	2.8 ± 0.6
	C6N3	3.5 ± 0.8	3.1 ± 0.9	2.9 ± 1.1	0.9 ± 0.4	1.1 ± 0.4	2.1 ± 0.6	2.7 ± 0.6	2.9 ± 0.6
	C6N2	2.6 ± 0.7	2.4 ± 0.6	2.2 ± 0.7	0.7 ± 0.3	0.8 ± 0.3	1.2 ± 0.3	1.8 ± 0.5	2.2 ± 0.5
	OX2	2.8 ± 0.5	2.6 ± 0.8	2.0 ± 0.6	0.7 ± 0.6	0.5 ± 0.4	1.2 ± 0.5	1.2 ± 0.3	1.5 ± 0.3
	Summary	2.53	2.45	2.21	0.61	0.83	1.53	2.05	3.15
	B5C	1.4 ± 0.3	1.4 ± 0.3	1.4 ± 0.3	1.4 ± 0.3	1.4 ± 0.3	1.4 ± 0.3	1.4 ± 0.3	1.4 ± 0.3
	B6HF	2.9 ± 0.8	2.9 ± 0.8	2.9 ± 0.8	2.9 ± 0.8	2.9 ± 0.8	2.9 ± 0.8	2.9 ± 0.8	2.9 ± 0.8
	C4PH	2.2 ± 0.5	2.2 ± 0.5	2.2 ± 0.5	2.2 ± 0.5	2.2 ± 0.5	2.2 ± 0.5	2.2 ± 0.5	2.2 ± 0.5
	C5F	1.7 ± 0.4	1.7 ± 0.4	1.7 ± 0.4	1.7 ± 0.4	1.7 ± 0.4	1.7 ± 0.4	1.7 ± 0.4	1.7 ± 0.4
MIVS	C5P	0.9 ± 0.3	0.9 ± 0.3	0.9 ± 0.3	0.9 ± 0.3	0.9 ± 0.3	0.9 ± 0.3	0.9 ± 0.3	0.9 ± 0.3
	C5PA	1.0 ± 0.2	1.0 ± 0.2	1.0 ± 0.2	1.0 ± 0.2	1.0 ± 0.2	1.0 ± 0.2	1.0 ± 0.2	1.0 ± 0.2
	C6N3	2.0 ± 0.4	2.0 ± 0.4	2.0 ± 0.4	2.0 ± 0.4	2.0 ± 0.4	2.0 ± 0.4	2.0 ± 0.4	2.0 ± 0.4
	C6N2	1.5 ± 0.2	1.5 ± 0.2	1.5 ± 0.2	1.5 ± 0.2	1.5 ± 0.2	1.5 ± 0.2	1.5 ± 0.2	1.5 ± 0.2
	OX2	2.0 ± 0.3	2.0 ± 0.3	2.0 ± 0.3	2.0 ± 0.3	2.0 ± 0.3	2.0 ± 0.3	2.0 ± 0.3	2.0 ± 0.3

(Continued)

Table 2 (Continued)

		Due Date Flow Factor (DDFF)							
	Summary	1.73	1.73	1.73	1.73	1.73	1.73	1.73	1.73
	B5C	1.6 ± 0.5	1.6 ± 0.5	1.6 ± 0.5	1.6 ± 0.5	1.6 ± 0.5	1.6 ± 0.5	1.6 ± 0.5	1.6 ± 0.5
	B6HF	2.8 ± 0.9	2.8 ± 0.9	2.8 ± 0.9	2.8 ± 0.9	2.8 ± 0.9	2.8 ± 0.9	2.8 ± 0.9	2.8 ± 0.9
	C4PH	1.8 ± 0.4	1.8 ± 0.4	1.8 ± 0.4	1.8 ± 0.4	1.8 ± 0.4	1.8 ± 0.4	1.8 ± 0.4	1.8 ± 0.4
	C5F	2.0 ± 0.4	2.0 ± 0.4	2.0 ± 0.4	2.0 ± 0.4	2.0 ± 0.4	2.0 ± 0.4	2.0 ± 0.4	2.0 ± 0.4
FIFO	C5P	1.0 ± 0.3	1.0 ± 0.3	1.0 ± 0.3	1.0 ± 0.3	1.0 ± 0.3	1.0 ± 0.3	1.0 ± 0.3	1.0 ± 0.3
	C5PA	1.2 ± 0.5	1.2 ± 0.5	1.2 ± 0.5	1.2 ± 0.5	1.2 ± 0.5	1.2 ± 0.5	1.2 ± 0.5	1.2 ± 0.5
	C6N3	1.5 ± 0.3	1.5 ± 0.3	1.5 ± 0.3	1.5 ± 0.3	1.5 ± 0.3	1.5 ± 0.3	1.5 ± 0.3	1.5 ± 0.3
	C6N2	1.4 ± 0.5	1.4 ± 0.5	1.4 ± 0.5	1.4 ± 0.5	1.4 ± 0.5	1.4 ± 0.5	1.4 ± 0.5	1.4 ± 0.5
	OX2	1.6 ± 0.3	1.6 ± 0.3	1.6 ± 0.3	1.6 ± 0.3	1.6 ± 0.3	1.6 ± 0.3	1.6 ± 0.3	1.6 ± 0.3
	Summary	1.65	1.65	1.65	1.65	1.65	1.65	1.65	1.65

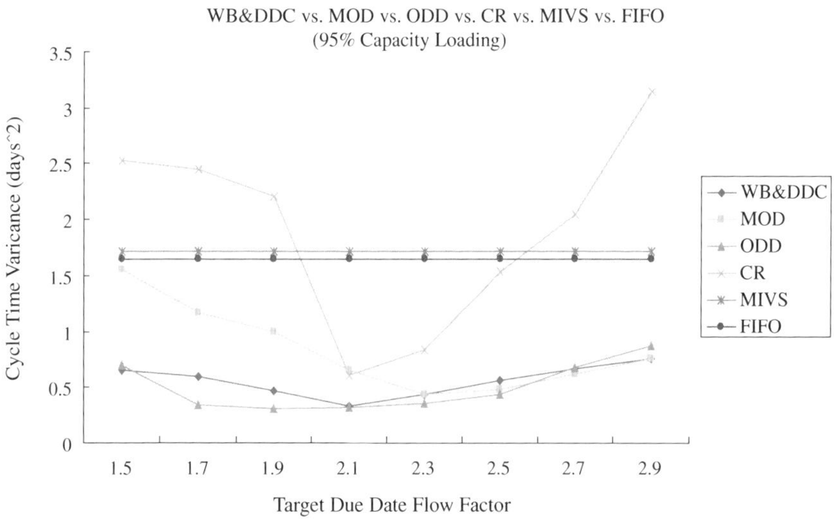

Fig. 2 Cycle time variance comparison (95% capacity loading)

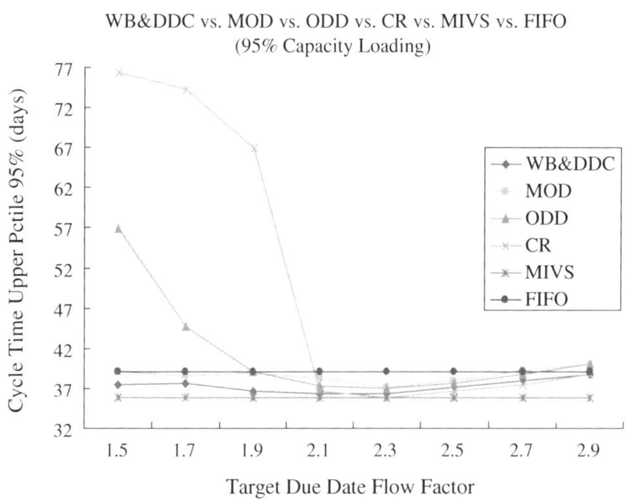

Fig. 3 Cycle time upper percentile 95% comparison (95% loading)

in Fig. 1. The WB&DDC approach shows superior results over MOD, ODD, CR, and FIFO. However, it is outperformed by MIVS.

Percent tardy lots. Figure 4 illustrates the on-time delivery percentage performance. If the target due date is defined too tight, 100% of lots are delayed. If the target due date is too loose, there are no tardy lots. Therefore, we only focus on difference of selected rule with due date flow factor 1.9, 2.1, and 2.3. For other flow factors,

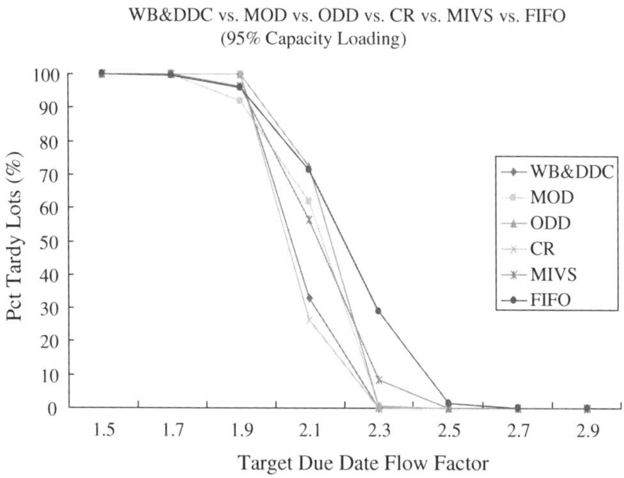

Fig. 4 Percent tardy lots comparison (95% capacity loading)

the on-time delivery percentage is either 100 or 0%. Starting from flow factor 1.9, most of the rules drop from 100% of tardy lots. For flow factor 2.1, the WB&DDC approach reduces the percentage of tardy lots remarkably compared to MOD, ODD, MIVS, and FIFO except CR. When the flow factor is changed to 2.3, the percentage of tardy lots of WB&DDC, MOD, ODD, and CR decreased to 0%, but MIVS and FIFO still show some delayed lots.

Average tardiness for tardy lots. In Fig. 5, due to their sensitivity to tight due dates, ODD and CR produce considerable tardiness for tardy lots with tight due date flow factors of 1.5, 1.7, and 1.9. We see that the WB&DDC approach has a relatively flat tardiness curve and that it has more robust behavior of the average tardiness for tardy lots than most other rules (except MIVS and FIFO with tight due date flow factor 1.5 and 1.7). For a flow factor of 2.1, the CR rule has a better percentage of tardy lots than WB&DDC, but WB&DDC has less tardiness than CR.

With regard to the 75 and 85% loading cases, in general, the results are to be similar with 95% loading case. Considering the average cycle time, WB&DDC has the same trend as MOD, ODD, and CR rules but outperforms them. If the target due date is too tight or too loose, the average cycle time is relatively large compared to the medium target due date case. Unlike 95% loading, WB&DDC only outperforms MIVS and FIFO rules when the target due date is appropriate, because it seems that WB&DDC takes care of the bottleneck more efficiently under 95% loading. Looking at the cycle time variance, WB&DDC is always superior to MOD, CR, MIVS, and FIFO. Although WB&DDC achieves a better average cycle time and cycle time variance than MOD, it is quite interesting to see that WB&DDC is outperformed by MOD with respect to the percentage of tardy lots and the average tardiness of tardy lots for tight target due dates. When the target due date is tight, most of the lots tend to

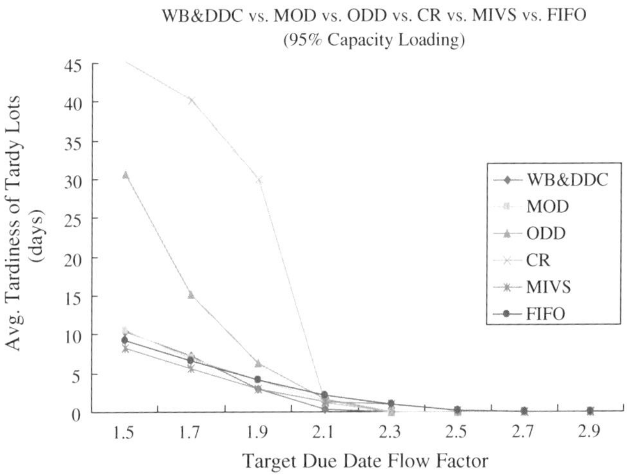

Fig. 5 Average tardiness of tardy lots comparison (95% capacity loading)

be late. In this situation, SPT plays a major role in MOD, which actually breaks up the ties of operation due date and speeds up those lots with shorter process times to achieve better cycle times and to yield a better tardiness performance. Especially under a low fab loading, the WIP does not need to be balanced at all because lots go through work centers quite smoothly anyways and due date control seems to be more important.

5 Summary

Dispatching rules are the most common and popular techniques for operational control of complex fabrication facilities. Different dispatching rules have different performance objectives. While due date-oriented rules such as CR, ODD, and MOD focus on on-time delivery and lateness performance, WIP-oriented rules such as MIVS focus on WIP balance for specified operations or work centers, or even balance the whole factory. There is no doubt that an appropriate target due date can achieve shorter cycle time as well as on-time delivery. However, setting target due date is not an easy task. The biggest issue for due date-oriented rules such as CR and ODD is that they are not robust with respect to different due date flow factor values, particularly when working with tight due dates. On the one hand, over emphasized due date control may lead to tremendously high inventory level that causes worse on-time delivery performance. On the other hand, WIP-oriented rules prefer WIP balance over due date control that sometimes is not acceptable for customer-oriented companies. Therefore, to achieve both low inventories and good due date performance simultaneously, a compromise between WIP balance and due date control is necessary.

We propose an approach called Workload Balance & Due Date Control (WB&DDC) which combines WIP balance and due date control with components for bottleneck workload control, feeding non bottleneck, acceleration of delayed lots, and lots close to their due dates. We compared WB&DDC with five classic dispatching rules (MOD, ODD, CR, MIVS and FIFO) considering average cycle time, cycle time variance, cycle time 95%percentile, percent tardy lots, and average tardiness of tardy lots as major performance measures under 75, 85, and 95% fab loading with target due date flow factors ranging from 1.5 to 2.9 in steps of 0.2, respectively. The simulation results indicate that the WB&DDC approach is superior and robust to CR, ODD, and MOD rules with regard to average cycle time, considering different target due date flow factor changes. In contrast to ODD and CR, it is successful in avoiding high inventory levels for tight due dates. WB&DDC also outperforms MIVS and FIFO for average cycle times with appropriate target due date. In addition WB&DDC achieves the same excellent cycle time variance as ODD, which is a major advantage and better than all other rules. Although WB&DDC is not the best approach regarding on-time delivery performance, it is still better than ODD. Furthermore, WB&DDC produces less tardiness than other rules except of the tight due date case. Our proposed WB&DDC approach achieves shorter cycle time, lower cycle time variance, better on-time delivery, and lateness performance simultaneously. The disadvantage is that the optimal target workload levels, for instance, the minimum workload level for the non-bottleneck work centers, the minimum and maximum workload level for the bottleneck work center have to be set experimentally based on the factory model with FIFO dispatching described above. In our future study, we intend to develop a back-propagation neural network to determine the target workload level for work centers. Neural networks have the advantage of being trained with real or simulation data instead of having to develop complex models and algorithms [6]. In this study, one-year simulations of MIMAC6 are carried out with different dispatching rules including FIFO, ODD, CR, EDD, and SPT to generate training data for the neural network. From each simulation the following results are considered: (1) minimum (maximum) wafer WIP, (2) coefficients of variation of work center inter-arrival times, (3) coefficients of variation of process times, (4) numbers of tools in a work center, (5) maximum process rates, and (6) percent online of work center. This data forms the input to the back-propagation neural network to determine the minimum and maximum workloads for all work centers.

Appendix

Figures 6, 7, 8, 9, 10, 11, 12, 13, 14, 15, 16, 17 and 18

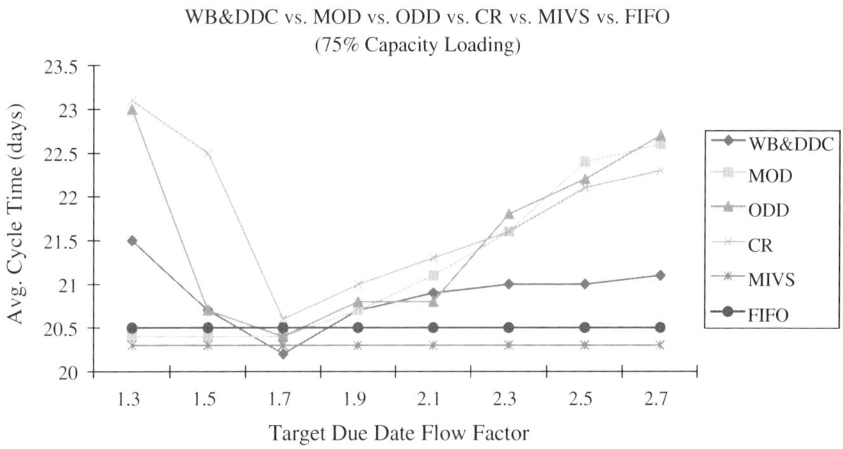

Fig. 6 Average cycle time comparison (75% capacity loading)

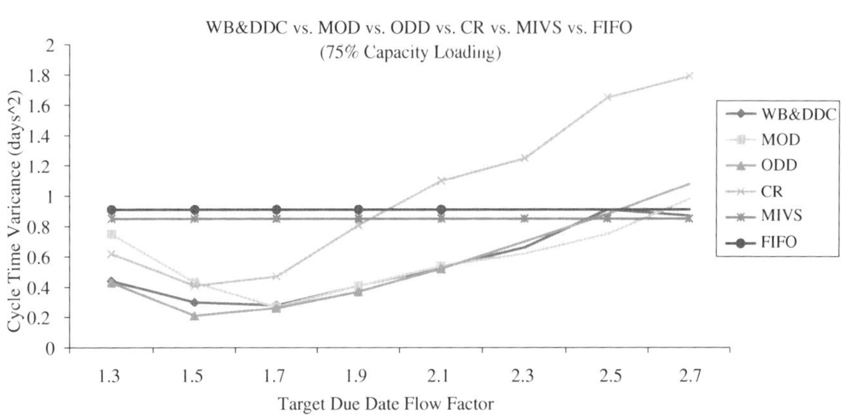

Fig. 7 Cycle time variance comparison (75% capacity loading)

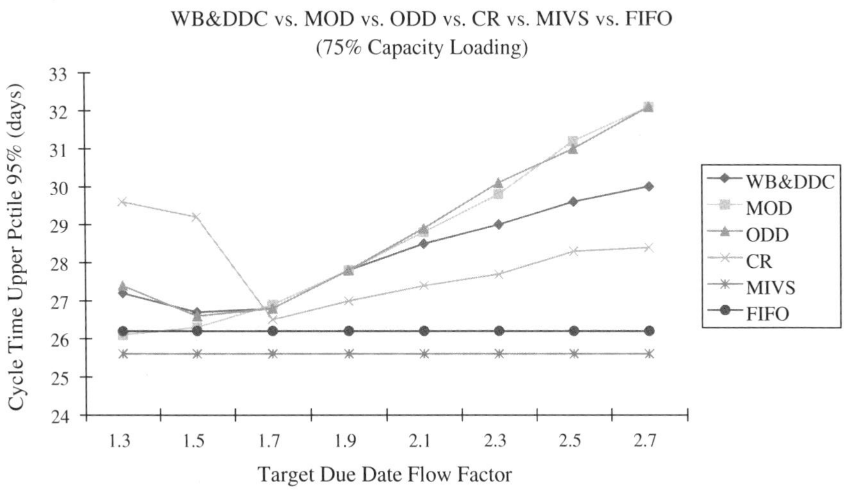

Fig. 8 Cycle time upper percentile 95% comparison (75% capacity loading)

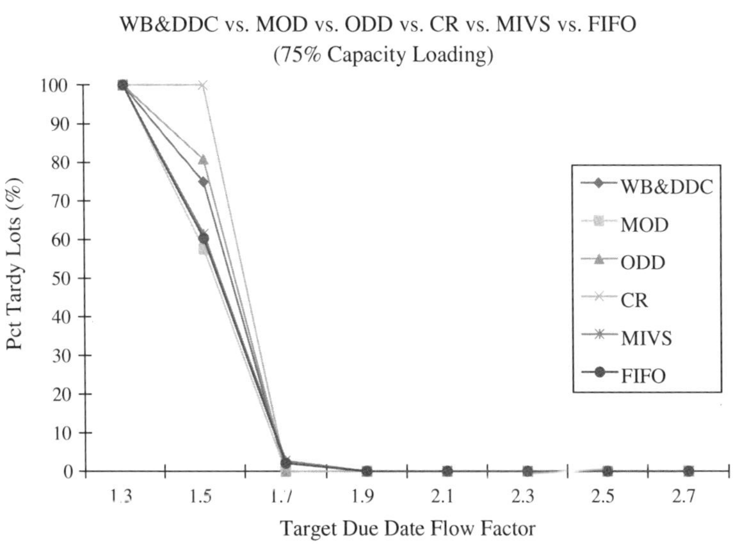

Fig. 9 Percent tardy lots comparison (75% capacity loading)

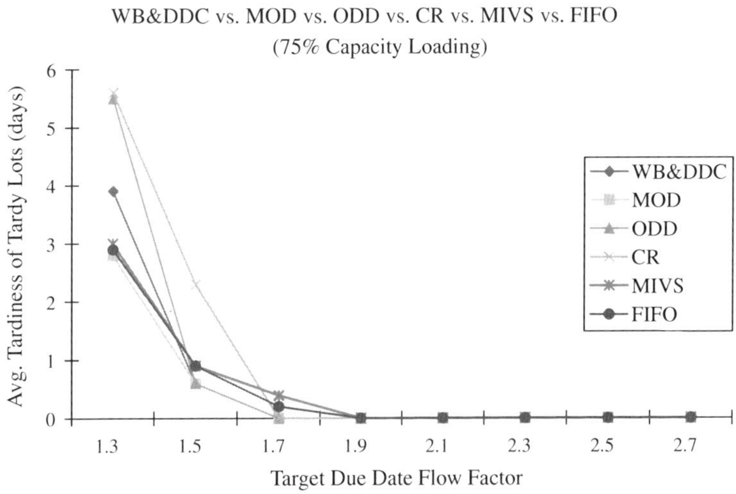

Fig. 10 Average tardiness of tardy lots comparison (75% capacity loading)

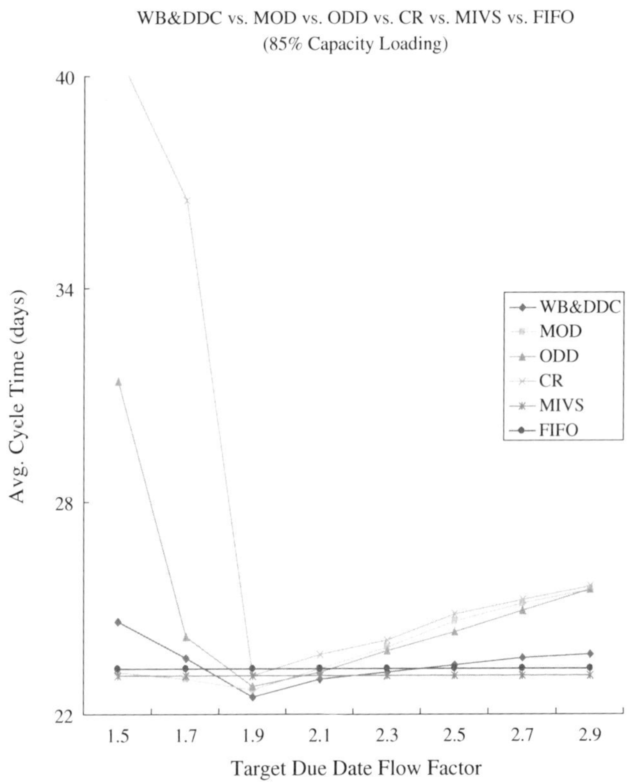

Fig. 11 Average cycle time comparison (85% capacity loading)

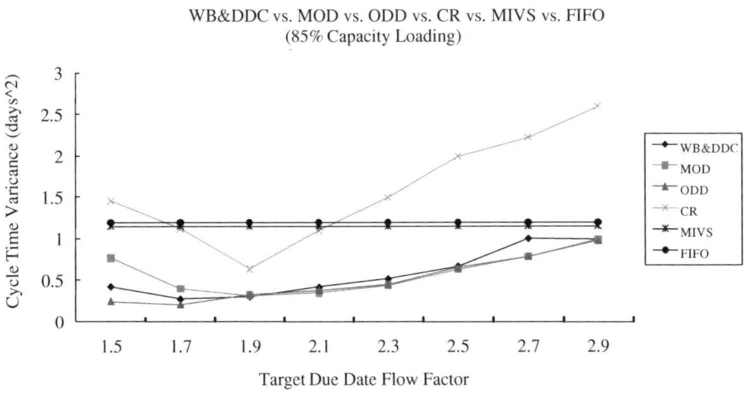

Fig. 12 Cycle time variance comparison (85% capacity loading)

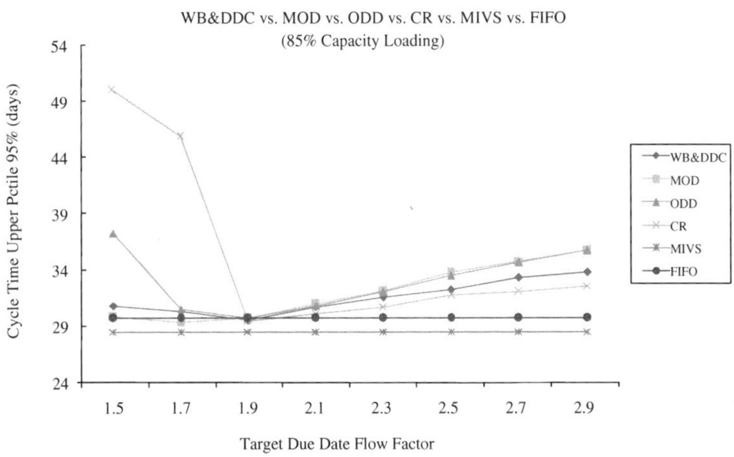

Fig. 13 Cycle time upper percentile 95% comparison (85% capacity loading)

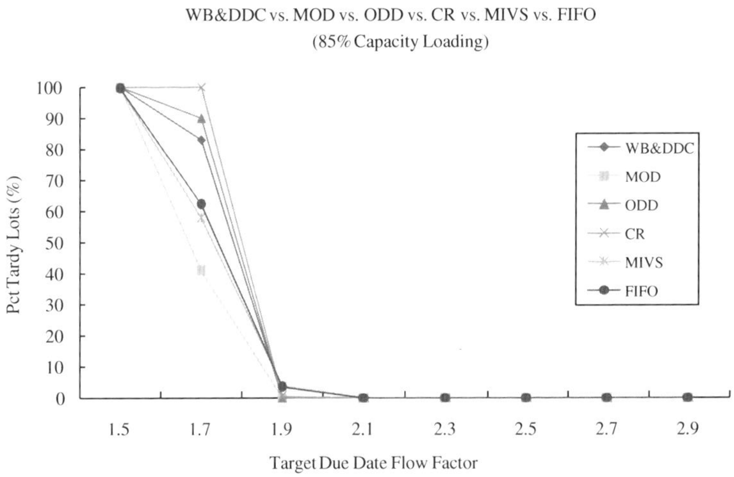

Fig. 14 Percent tardy lots comparison (85% capacity loading)

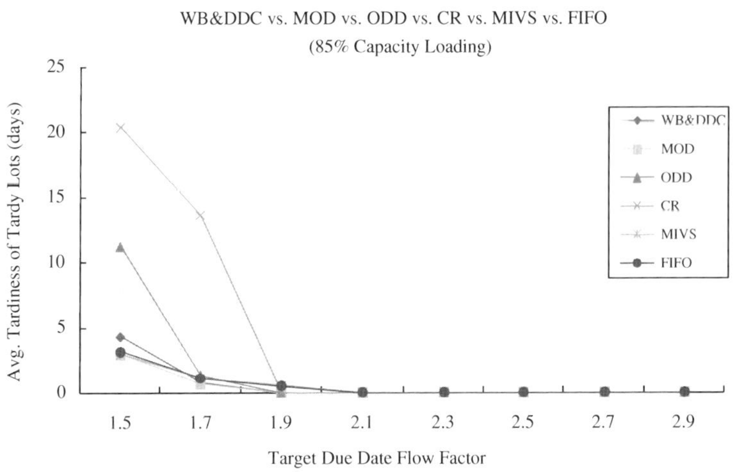

Fig. 15 Average tardiness of tardy lots comparison (85% capacity loading)

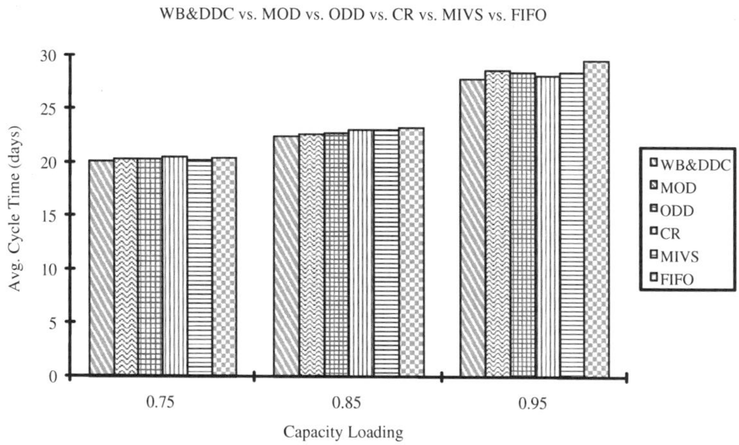

Fig. 16 Best average cycle time comparison of different fab capacity loading

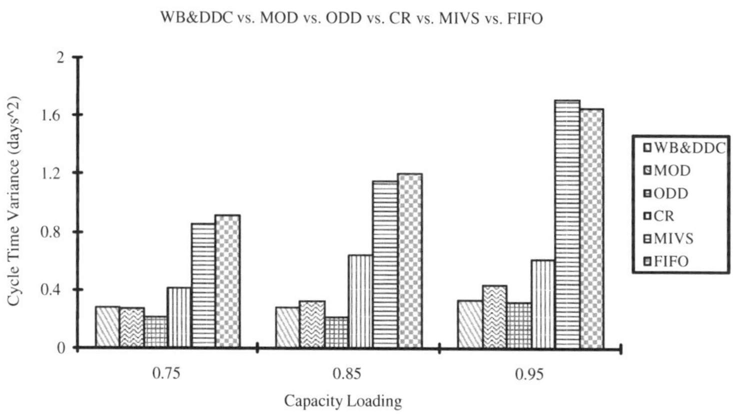

Fig. 17 Best cycle time variance comparison of different fab capacity loading

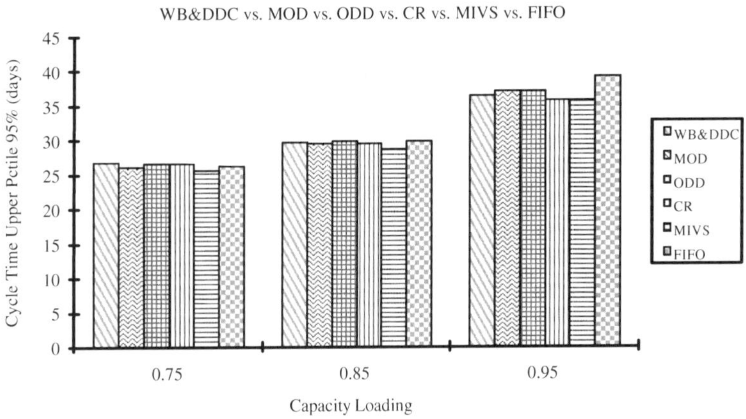

Fig. 18 Best cycle time upper percentile 95% comparison of different fab capacity loading

References

1. Atherton LF, Atherton RW (1995) Wafer fabrication: factory performance and analysis. Kluwer, Boston
2. Bertrand JWM (1983) The use of work load information to control job lateness in controlled and uncontrolled release production systems. J Oper Manag 3:79–92
3. Baker KR, Bertrand JWM (1981) A comparison of due-date selection rules. AIIE Trans 13: 123–13
4. Baker KR, Bertrand JWM (1981) An investigation of due date assignment rules with constrained tightness. J Oper Manag 1:109–120
5. Burman DY, Gurrola-Gal FJ, Nozari A, Sathaye S, Sitarik JP (1986) Performance analysis techniques for ic manufacturing lines. AT&T Tech J 65:46–57
6. Chambers M, Mount-Campbell CA (2002) Process optimization via neural network metamodeling. Int J Prod Econ 79:93–100
7. Collins DW, Palmeri V (1997) An analysis of the "k-step ahead minimum inventory variability policy using sematech semiconductor manufacturing data in a discrete-event simulation model. In: 6th International Conference on Emerging Technologies and Factory Automation Proceedings, pp 520–527.
8. Dabbas RM, Fowler JW (2003) A new scheduling approach using combined dispatching criteria in wafer fabs. IEEE Trans Semicond Manuf 16:501–510
9. Elvers DA (1973) Job shop dispatching rules using various delivery date setting criteria. Prod Invent Manag 4:62–70
10. Hopp WJ, Spearman ML (2001) Factory physics: foundations of manufacturing management, 2nd ed. Irwin McGraw-Hill, London
11. Kuo CJ, Liu CM, Chi CY (2008) Standard wip determination and wip balance control with time constraints in semiconductor wafer fabrication. J Qual 15:409–423
12. Li S, Tang T, Collins DW (1996) Minimum inventory variability scheduler with applications in semiconductor manufacturing. IEEE Trans Semicond Manuf 9:1–5
13. Little JDC (1992) Are there 'Laws' of manufacturing. manufacturing systems: foundations of world-class practice, pp 180–188
14. Fowler JW, Hogg GL, Mason SJ (2002) Workload control in the semiconductor industry. Prod Plan Control 13:568–578

15. Fowler J, Robinson J (1995) Measurement and improvement of manufacturing capacities (MIMAC): final report. Technical Report 95062861A-TR, Austin, TX: SEMATECH
16. Glassey CR, Resende MGC (1988) Closed-loop job release control for VLSI circuit manufacturing. IEEE Trans Semicond Manuf 1:36–46
17. Goldratt EM (1984) The goal. Great Barrington, MA
18. Goldratt EM, Cox J (1986) The goal: a process of ongoing improvement. North River Press, New York
19. Ham M, Fowler JW (2007) Balanced machine workload dispatching scheme for wafer fab. Advanced semiconductor manufacturing conference, pp 390–395
20. Kalisch S, Ringel R, Weigang J (2008) Managing wip and cycle time with the help of loop control. In: Proceedings of the 2008 winter simulation conference, pp 2298–2304
21. Marek RP, Elkins DA, Smith DR (2001) Understanding the fundamentals of kanban and conwip pull systems using simulation. In: Proceedings of the 2001 winter simulation conference, pp 921–929
22. Monden Y (1981) What makes the toyota production system really tick. Ind Eng 13:36–46
23. Muhlemann AP, Lockett AG, Farn CI (1982) Job shop scheduling heuristics and frequency of scheduling. Int J Prod Res 20:227–241
24. Pai FY (2004) Wip management model for semiconductor back-end manufacturing. J Am Acad Bus 5:357–363
25. PanWalker SS, Iskandar WW (1977) A survey of scheduling rules. Oper Res 1:45–61
26. Perdaen D, Armbruster D, Kempf K, Lefeber E (2008) Controlling a reentrant manufacturing line via the push-pull point. Int J Prod Res 46(16):4521–4536
27. Rose O (2002) Some issues of the critical ratio dispatch rule. In: Proceedings of the 2002 winter simulation conference, pp 1401–1405
28. Rose O (2003) Comparison of due-date oriented dispatch rules in semiconductor manufacturing. In: Proceedings of the 2003 industrial engineering research conference, pp 18–20
29. Spearman ML, Zazanis MA (1992) Push and pull production systems: issues and comparisons. Oper Res 40:521–532
30. Spearman ML, Woodruff DL, Hopp WJ (1990) CONWIP: a pull alternative to kanban. Int J Prod Res 28:879–894
31. Vepsalainen APJ, Morton TE (1987) Priority rules for job shops with weighted tardiness costs. Manag Sci 33:1035–1047
32. Wein LM (1988) Scheduling semiconductor wafer fabrication. IEEE Trans Semicond Manuf 1:115–129
33. Wight OW (1970) Input/output control: a real handle on lead time. Prod Invent Manag J 11:9–31

Controlling a Re-entrant Manufacturing Line via the Push–Pull Point

Dominique Perdaen, Dieter Armbruster, Karl G. Kempf and Erjen Lefeber

Abstract A reduced model of a re-entrant semiconductor factory exhibiting all the important features is simulated, applying a push dispatch policy at the beginning of the line and a pull dispatch policy at the end of the line. A commonly used dispatching policy that deals with short-term fluctuations in demand involves moving the transition point between both policies, the push–pull point (PPP) around. It is shown that with a mean demand starts policy, moving the PPP by itself does not improve the performance of the production line significantly over policies that use a pure push or a pure pull dispatch policy, or a CONWIP starts policy with pure pull dispatch policy. However, when the PPP control is coupled with a CONWIP starts policy, then for high demand with high variance, the improvement becomes approximately

---

Previously published in Perdaen, Dominique, Armbruster, Dieter, Kempf, Karl and Lefeber, Erjen (2008) 'Controlling a re-entrant manufacturing line via the push-pull point', International Journal of Production Research, 46 16:4521–4536. Reproduced here with permission.

D. Perdaen · E. Lefeber
Department of Mechanical Engineering,
Eindhoven University of Technology,
P.O. Box 513, 5600 MB, Eindhoven,
The Netherlands

E. Lefeber
e-mail: A.A.J.Lefeber@tue.nl

D. Armbruster (✉)
School of Mathematical and Statistical Sciences,
Arizona State University,
Tempe, AZ 85287-1804, USA
e-mail: armbruster@asu.edu

K. G. Kempf
Decision Engineering Group,
Intel Corporation, 5000 W. Chandler Boulevard,
Chandler, AZ 85226-3699, USA
e-mail: karl.g.kempf@intel.com

D. Armbruster and K. G. Kempf (eds.), *Decision Policies for Production Networks*,
DOI: 10.1007/978-0-85729-644-3_5, © Springer-Verlag London 2012

a factor of 4. The unexpected success of a PPP policy with CONWIP is explained using concepts from fluid dynamics that predict that this policy will not work for perishable demand. The prediction is verified through additional simulations.

Keywords Re-entrant production · CONWIP · Dispatch policy

1 Introduction

A very important feature of the production of semiconductor wafers is the re-entrant line: Wafers are produced in layers and hence after one layer is finished a wafer returns to the same set of machines for processing of the next layer. Modern semiconductors may have on the order of 20–30 such layers. It is typical for wafers to spend several weeks in such a re-entrant production line, much of the time waiting for available machines. Process control in such long production lines with thousands of wafer and hundreds of processing steps making tens of different products is a special challenge. Most of the time the demand fluctuates on a much faster timescale than the factory cycle time, making it very difficult to use starts policies to react to the demand fluctuations. Typically, for a product with a constant mean demand, the mean demand is started. Due to stochasticity in the production and due to variation in the demand there is nevertheless a large mismatch in daily outputs and demand. In practice, to reduce the mismatch, production targets over a certain time horizon are given and wafers at the end of the production process are sped up or slowed down using dispatch policies. We are not concerned here with longer and larger fluctuations that might require an adjustment of the starting rate to cover changes of the desired WIP level as discussed in [14].

The combination of lot release and dispatching strategies is called Workload (or Flow) Control. An overview of state-of-the-art published research on workload control as applied to semiconductor industry is provided in [7]. A thorough overview of the literature on order release as a flow control is provided in [4], whereas [12] and [5] are two thorough surveys of the dispatching literature. Commonly used dispatching policies include: First-In, First-Out (FIFO), Earliest Due Date (EDD), Weighted Shortest Processing Time (WSPT), Least Slack (LS) and Least Setup Cost (LSC). In the seminal paper [16] many of these lot sequencing rules as well as a variety of input controls have been evaluated using simulation models of representative but fictitious semiconductor fabs. The main conclusion was that order release is more important than dispatching (30–40% change versus less than 10%), though there is an important connection between these decisions. Dynamic scheduling studies were done by [3] who implemented learning of dispatch rules in their simulation environment. Pure push and pull dispatch policies were studied by [2].

Most of the time demand fluctuates on a much faster timescale than the factory cycle time. Unfortunately, almost no literature exists on how to deal with the impact of a production surge or short-term increase in wafer starts that occurs when unexpected orders are received by a fab that is operating close to its designed capacity. In [6, 9, 11] some preliminary investigations into the surge problem have been done.

In order to deal with these short-term variations in demand we consider a dispatching policy which to the authors' knowledge has not been considered in the literature before, but which is used in practice. We simulate a reduced model of a re-entrant semiconductor factory exhibiting all the important features, applying a push (dispatch) policy at the beginning of the line and a pull (dispatch) policy at the end of the line. Here a push (pull) policy refers to the fact that a machine that is able to process more than one step gives priority to the earlier (later) step. Push policies are also known as first-buffer-first-served and pull policies are known as shortest-expected-remaining-process-time policies. We use a push policy upstream and a pull policy downstream. The step at which we switch from a push to a pull policy is called the push–pull point (PPP). Its dynamics is the control variable. Our objective (metric) is to reduce the mismatch between daily outputs and demand over a long time interval. We assume that over that time interval the demand has a constant mean demand and varies stochastically around the mean. By focussing on the output, this study complements the important work by [10] who were not concerned with output but with the behavior of the mean and variance of the cycle times as a function of different scheduling policies.

We show that with a policy that starts the mean demand, moving the PPP by itself does not improve the performance of the production line significantly over a pure push, a pure pull policy or a pure CONWIP starts policy [13] with pure pull dispatch. However, when the PPP dispatch control is coupled with a CONWIP starts policy, then for high demand with high variance, the improvement becomes approximately a factor of 4. We explain the unexpected success of a PPP policy with CONWIP using concepts from fluid dynamics that predict that this policy will not work for perishable demand. We verify this prediction.

2 The Factory Model

Our basic factory model consists of 26 production steps executed on nine machine sets. Table 1 contains all the specifications of this model. The first six machines are called diff1, diff2, litho1, etch clean, etch1 and ion impl, corresponding to production steps associated with diffusion, photolithography, etching and ion implantation respectively. They are associated with the transistor section of the production line and a wafer performs four loops through these machines in a specific order as indicated in Table 1. The last three machine sets are called metal dep, litho2 and etch2, corresponding to production steps that generate metal layers for interconnection of the transistors. The wafer loops through the metalization section of the production line twice. The transistor and metal loops are completely disjoint and do not share equipment. Rows 1–26 in Table 1 correspond to the 26 production steps. The entries in each row indicates the machine set that performs the step and the processing time spent in a machine in the set. For instance, step 3, 6, 10 and 14 are all performed on the photolithography machine litho1 with cycle times of 1, 1.25, 1 and 1.25 h, respectively.

Table 1 Factory model

	Diff 1	Diff 2	Litho 1	Etch clean	Etch 1	Ion impl	Metal dep	Litho 2	Etch 2	
Step	0	1	2	3	4	5	6	7	8	Station #
1				0.25						Clean wafer
2	8.00									Grow a layer
3			1.00							Pattern it
4					1.00					Etch away some
5		6.00								Grow a layer
6			1.25							Pattern it
7						2.50				Implant ions
8				0.50						Remove mask
9	7.00									Grow a layer
10			1.00							Pattern it
11					1.00					Etch some away
12				0.25						Clean wafer
13		5.00								Grow a layer
14			1.25							Pattern it
15						3.50				Implant ions
16				0.50						Remove mask
17								1.50		Pattern contact
18									1.75	Etch contact
19							2.25			Layer metal
20								1.00		Pattern metal
21									2.25	Etch metal
22								1.50		Pattern contact
23									2.00	Etch contact
24							2.25			Layer metal
25								1.00		Pattern metal
26									2.50	Etch metal
	15.00	11.00	4.50	1.50	2.00	6.00	4.50	5.00	8.50	Total hours required per lot
	750	550	900	300	400	1200	900	1000	1700	Total hours needed per week
	0.80	0.75	0.90	0.60	0.75	0.85	0.85	0.90	0.55	Average availability
	134.40	126.00	151.20	100.80	126.00	142.80	142.80	151.20	92.40	Total hours avail per machine
	5.58	4.37	5.95	2.98	3.17	8.40	6.30	6.61	18.40	Minimum num. tools needed
	1.25	1.25	1.00	1.25	1.50	1.10	1.25	1.05	1.10	Constraint degree desired
	6.98	5.46	5.95	3.72	4.76	9.24	7.88	6.94	20.24	Number of tools needed
	7	6	6	4	5	10	8	7	21	Number of tools installed

The second part of Table 1 is a spreadsheet calculation to determine the required number of machines (tools) to have a production target of 200 lots per week, given availability rates of the machines and desired levels of contraints for a given machine set. Consider for instance the last 8 rows in the column litho1: A wafer spends a total of 4.5 h in litho1. Hence to produce 200 wafers per week we need 900 h per week of machine time. Assuming that a litho1 machine is 90% available and a work week of 168 h this machine works for 151.2 h per week and hence we need 5.95 machines of that type. Since this is a very expensive machine, it is planned to be the bottleneck and hence has a constraint factor of 1.0. As a result six machines will be installed. Taking into account that the diffusion machines batch four wafers per machine cycle we reach the installation targets in the last row in a similar way for all columns.

This model is implemented as a discrete event simulation in χ [15, 8] a specification language developed at the Eindhoven University of Technology. Stochasticity enters the simulation at various levels: The time that a machine is in service, and the time that it is not, is distributed by a Weibull-distribution [8] with a mean "in service" time of 10 process times and a variance of 50%. The demand is randomly generated and is fixed for a simulation.

The actual processing times are pulled out of an Exponential-distribution [8] with the mean equal to the process times in Table 1. Note that, while the raw processing times of semiconductor processing machines are narrowly distributed, the unloading of machines depends on the availability of human operators and is highly variable. Nevertheless using an exponential distribution probably constitutes a worst case scenario for a practical model. Overall the stochastic parameters are fixed in a way, such that simulations of the model generate an outflux variance of 20% around the nominal influx of 200 per week, i.e. the throughput varies between 160 and 240 wafers per week.

3 The Push–Pull Point Algorithm

The goal of the PPP policy is to reduce the mismatch between fluctuating demands and the stochastically varying outflux of the factory. This policy divides the production line in two parts. Upstream of the PPP, priorities are assigned using a push strategy, downstream they are assigned according to a pull strategy. In conflicts across the PPP we always give priority to the steps in the pull-part. Figure 1 shows a typical priority assignment.

The PPP is moved depending on the demand: Given a demand period and a distribution of the work in progress (WIP) over the queues of all production steps (the WIP-profile), we place the PPP at such a point that the WIP downstream from the PPP is equal to the demand in the chosen demand period. When the demand increases, more products have to be pulled out of the line moving the PPP upstream. When the demand decreases, the PPP will shift downstream.

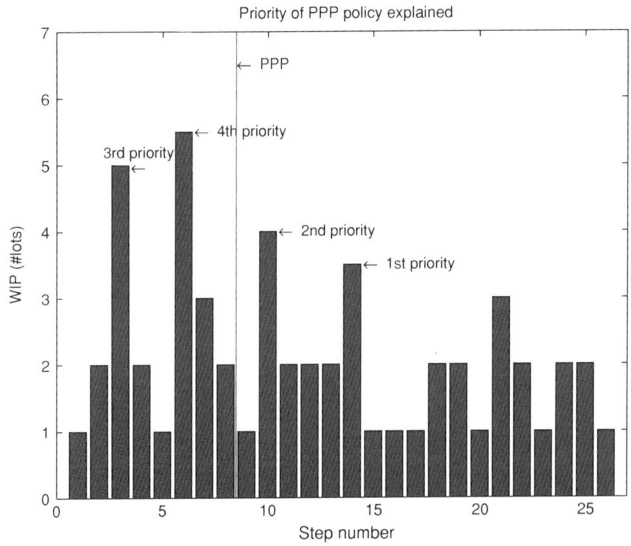

Fig. 1 The priority distribution when the PPP-policy is used

The possible success of such a strategy is based on three important facts:

- The clearing function [1], i.e. the throughput as a function of the load in the factory in steady state is significantly higher for a production line run completely with a push dispatch policy than for one run completely with a pull dispatch policy. Hence by increasing or decreasing the part of the production line that is run in pull policy we temporarily should increase or decrease the outflux. We show below the details of this effect for our model production line.
- The location of the push–pull point determines the average shape of the WIP profile in steady state. In particular, on average WIP decreases in the queues downstream of the PPP and increases upstream from the PPP. Figure 1 shows this schematically for the queues in front of the photolithography machines for a fixed PPP point. Figure 2 shows that this is true to a large extent for simulations on average, even when the PPP point is dynamically moved.
- The cycle time through the factory and the time between readjustments of the PPP have to be related. In particular, if adjusting the PPP according to demand on average places the PPP approximately in the middle of the production line adjusting to higher and lower demand by changing the PPP should be feasible.

4 Results

To determine the effectiveness of the PPP strategy we compare it to simulations with a starts policy of the mean demand and dispatch policies of pure push, pure pull as well as a CONWIP starts strategy using a dispatch policy of pure pull. We have

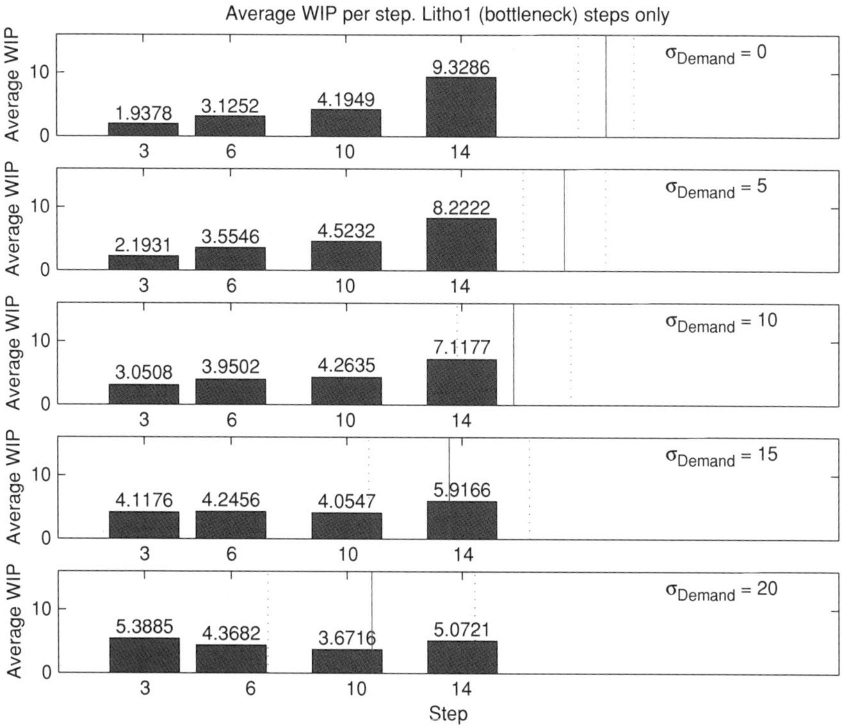

Fig. 2 Average queue length at the litho1 steps. The vertical line and its two dashed sidebars are the average position of the PPP plus/minus 2σ. As the PPP point moves upstream the WIP in the last two photolithography steps decreases and the WIP in the first two photolithography steps increases

also combined the PPP strategy with CONWIP as a starts policy. In all simulations we employ a FIFO policy within a given queue for a given production step. We run 500 simulations per data point. The demand $d(t)$ for each simulation is generated independently by choosing a demand for a two day period out of a normal distribution (throwing away the rare events that gave negative demands) with an average of 180 lots per week. The demand is not perishable, which means that the backlog or the inventory of the previous demand period is taken into account for the present demand period. The PPP is adjusted every 2 days (one demand period). Since the cycle time for our simulation factory is in the order of 5 days, the two day readjustment time places the PPP well inside the production line. The simulation-time for every single run is 144 weeks. The different control strategies are compared using the absolute value of the mismatch between output and demand over each demand period. Mismatch $m(t)$ and costs are given as

$$m(0) = 0 \tag{1}$$

$$m(i) = m(i-1) + d(i) - o(i) \tag{2}$$

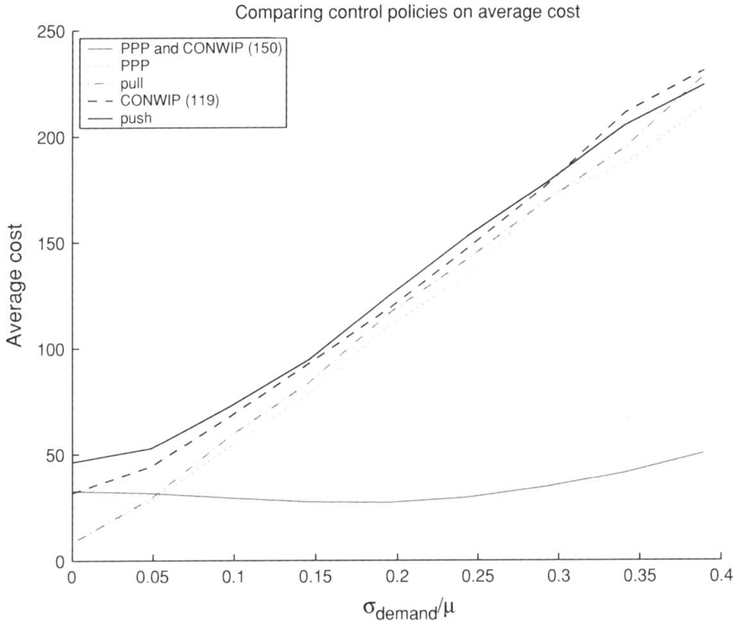

Fig. 3 Average costs per simulation for different control strategies as a function of the coefficient of variation of the demand

Table 2 Variance of cost as a function of the variation of the demand

$\sigma_{\text{demand}}/\mu$	0	0.05	0.1	0.15	0.2	0.25	0.3	0.35	0.4
σ^2_{cost}	3.0	3.7	6.6	7.9	9.7	27.8	75.1	216.6	578.1

$$\text{cost}(t) = \sum_{i}^{t} |m(i)|. \tag{3}$$

Here $o(t)$ is the output of the factory plus backlog and storage, i.e. over and under-production cost the same 1$ per lot per demand interval (2 days).

Figure 3 shows the average costs over 500 simulations as a function of the variance in the demand for all the different strategies. Table 2 shows the variances for the nine simulation points in Fig. 3.

The results are surprising: Pure push, pure pull, regular PPP (all with mean demand starts policy) and a CONWIP starts policy (pure pull dispatch policy) with a WIP level of 119 lots all increase monotonically with the demand variation and have very similar average cost. In contrast to that, a policy that combines the starts policy of a CONWIP rule and a WIP of 150 lots with the PPP control policy has almost constant costs over a wide range of demand variations. In addition the costs for high

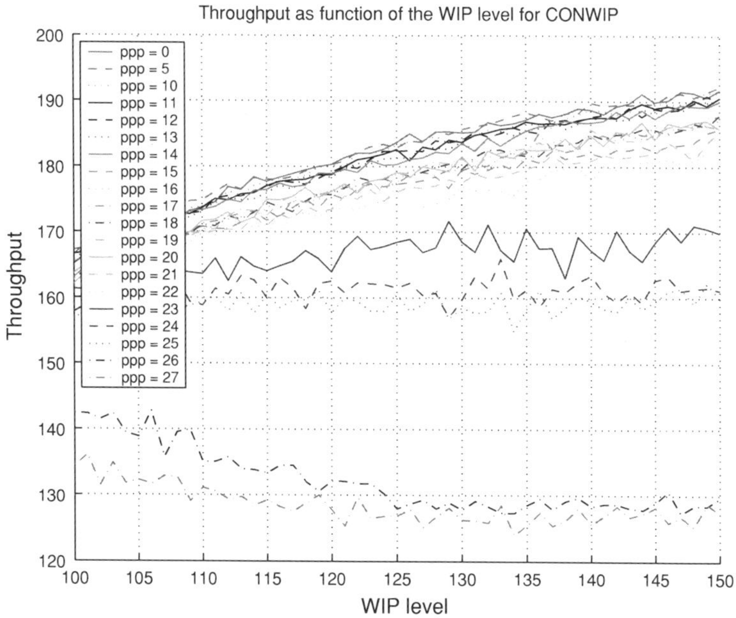

Fig. 4 Throughput as a function of total WIP for CONWIP policies with fixed push–pull points

demand variations are significantly lower for the PPP with CONWIP than for the other policies—50$ versus more than 200$.

5 Analysis of the PPP-CONWIP Policy

Figure 4 begins to explain the success of the PPP-CONWIP policies. It shows the clearing functions for CONWIP policies with different fixed push–pull points. The curve indicated with $ppp = 0$, corresponding to a pure pull dispatch policy, gives the highest throughput of all possible policies. The curve labeled $ppp = 27$ is a pure push dispatch policy that gives the lowest throughput of all. The intermediate curves indicated by $ppp = x$ denote a dispatch policy where the push–pull point has been fixed at step x. Note that for a complete push policy the throughput actually decreases with an increase in WIP. This is the result of an interplay between the back loaded WIP distribution of the push policy and the batching in the diffusion steps. Figure 4 also explains the choice of a CONWIP starts policy with a WIP level of 119 lots for a pure pull dispatch policy used in Fig. 3: The top curve in Fig. 4 represents a pure pull dispatch policy. The associated WIP level in steady state for a throughput of 180 lots/week is 119 lots which we use as the desired WIP level [14].

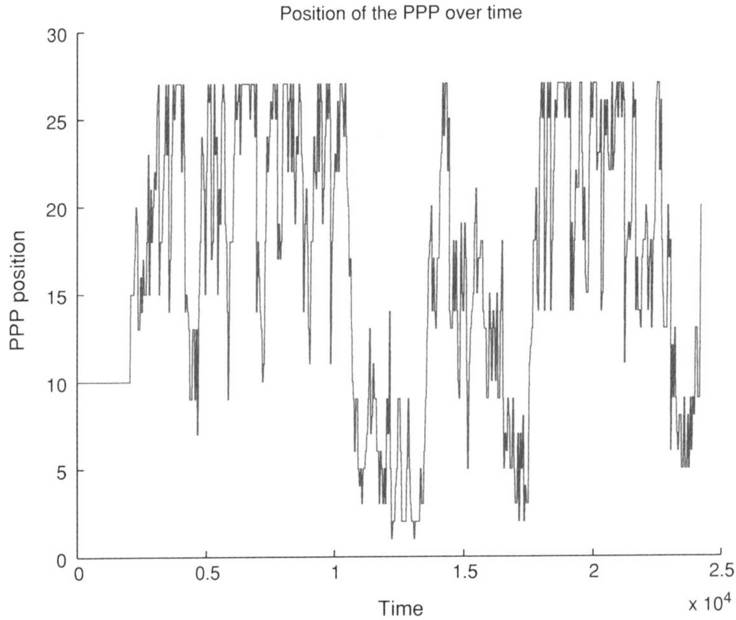

Fig. 5 Time evolution of the push–pull point as a function of time for a PPP-CONWIP policy with WIP level 130

These clearing functions suggests one reason for the success of the PPP-CONWIP policy: By using a CONWIP starts policy with a high WIP level and switching the PPP, we can change the outflux in the factory by a significant amount. For instance, for the WIP level of 150 lots we can get throughputs between approximately 130 and 190 per week. Note also that there is no good push–pull point for a WIP level of 150 that creates the throughput of 180 per week that we are using for our simulations. A PPP at stage 1–15 creates a throughput much higher and a PPP at stage 15–26 creates a throughput much lower than 180 per week. As a result, a completely deterministic demand cannot use a fixed PPP even though the demand is constant and hence has to jump back and forth, creating extra backlog or overproduction cost. This is the reason for the slight increase in cost for the PPP-CONWIP policy with WIP level 150 in Fig. 3 for low demand variation.

A different issue explains the failure of the pure PPP dispatch policy to be much better than a regular pull dispatch policy. Assume a push–pull point in the middle of the production line and an increase in demand. In response we will move the PPP upstream and clear out more of the WIP than we usually do over the demand period. However, we will only *start the average* amount. Consequently, WIP goes down and a second increase in demand will move the PPP rapidly further upstream. As a result we easily reach the point where the PPP is at the beginning of the line and the policy becomes a pure push dispatch policy. We cannot further increase the outflux than that. Similarly, a demand signal that has several periods below average will eventually move the PPP to the end of the factory and hence constitute a pull policy. We cannot reduce the outflow further than that. A CONWIP starts policy reduces the instances that the push–pull point is at one of the extremes of the production line by instantaneously starting more when more was pulled out of the factory and starting less if more was left in the factory. Figures 5 and 6 show the position of the PPP as a function of time for a PPP-CONWIP and a free PPP policy, respectively. Clearly the free PPP policy gets locked into pure push or pure pull policies much more often than the PPP-CONWIP.

We can illustrate the difference between free PPP and PPP-CONWIP policies with the following illustration based on fluid flows. For the purpose of this illustration let us consider the average behavior of a large number of lots as they move through the factory. We assume that the average speed $v(t)$ of a lot for a factory that is in steady state is constant over all production steps and depends on the dispatch policy. In particular, the average cycle time for a lot under a pull (dispatch) policy is shorter than for a lot produced under a push (dispatch) policy. Hence the associated average velocity for a pull policy is higher than that for a push policy. Let us consider a continuum of production steps and a continuum of lots such that we can define a WIP density $\rho(x, t)$ that describes the density of lots at stage x at time t. Then the throughput of the factory becomes $\lambda(x, t) = \rho(x, t)v$. In steady state, the throughput is constant and hence we get a constant WIP profile $\rho(x) = \frac{\lambda}{v}$ that does not depend on t because we are looking at steady state and does not depend on x, because we assume v to be constant. This is certainly not exactly true but a good approximation for the purpose of this illustration. Now, for a PPP policy we can consider the upstream part of the production line as a homogeneous push line and the downstream part as a homogeneous pull line, each with its own constant velocity with $v_{\text{push}} < v_{\text{pull}}$. Since the throughput is the same everywhere and since $\rho v = \lambda$ has to hold, we get a jump in the WIP profile at the push–pull point by the amount

$$\frac{\rho_{\text{push}}}{\rho_{\text{pull}}} = \frac{v_{\text{pull}}}{v_{\text{push}}}. \tag{4}$$

Figure 7a shows the constant throughput and the discontinuous WIP profile.

Assume we now move the PPP upstream by an amount Δx instantaneously. The queues that were just upstream of the PPP and hence had the lowest priority on the line now move up in priority and therefore speed up. In other words, part of the WIP profile that used to be in the push region and had a high WIP level now is in the pull region. As the velocity in the pull region is higher, the product of $\rho_{\text{push}} v_{\text{pull}} > \lambda$, i.e. we create a flux bump. Similarly we create a flux dip by moving the PPP downstream. The flux changes are

$$q \cdot \Delta x = \lambda \frac{v_{\text{pull}}}{v_{\text{push}}}, \tag{5}$$

$$q \cdot \Delta x = \lambda \frac{v_{\text{push}}}{v_{\text{pull}}}, \tag{6}$$

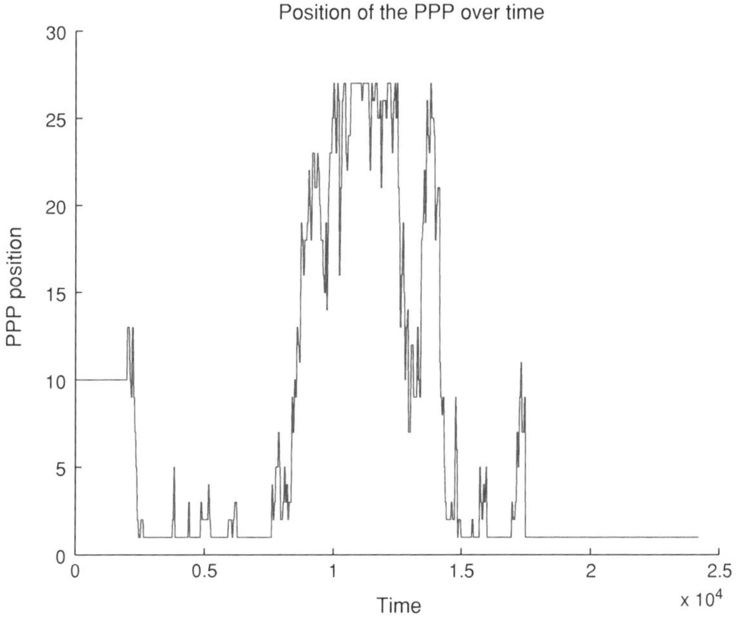

Fig. 6 Time evolution of the push–pull point as a function of time for a free PPP policy

for the flux bump and flux dip, respectively. Keeping the PPP at its new location the flux bump is downstream from the PPP and hence moves downstream with the constant speed v_{pull} pulling a WIP bump with it until they both exit the factory. During the time they exit they will increase the outflux. Depending on the remaining processing time from the push–pull point to the end of the production line, the increase in outflux may or may not happen within the demand time interval. Figure 7b and c show this time evolution. After the WIP/flux bump has exited, the total WIP in the factory is lower and hence in order to satisfy the same demand, the push pull point will have to move yet further upstream driving it toward the beginning of the factory.

In contrast, the time evolution of the flux bump for the PPP-CONWIP policy is illustrated in Fig. 8.

As the CONWIP starts policy is implemented by matching the starts to the outflux, once the WIP bump moves out of the factory, the starts will be increased to create a new WIP bump. In that way, the total throughput will stay high until the PPP point is moved downstream again. That will happen when the backlog has moved to zero and the sum of actual backlog and actual demand has decreased. In that way we have a policy that reverts all the time to a match between demand and outflux. This explanation can be checked by running the simulation with a perishable demand protocol: We only register whether there is a mismatch of the current outflux and the current demand but do not try to make up for that mismatch on the next time interval. For such a model the PPP-CONWIP policy should not be better than the

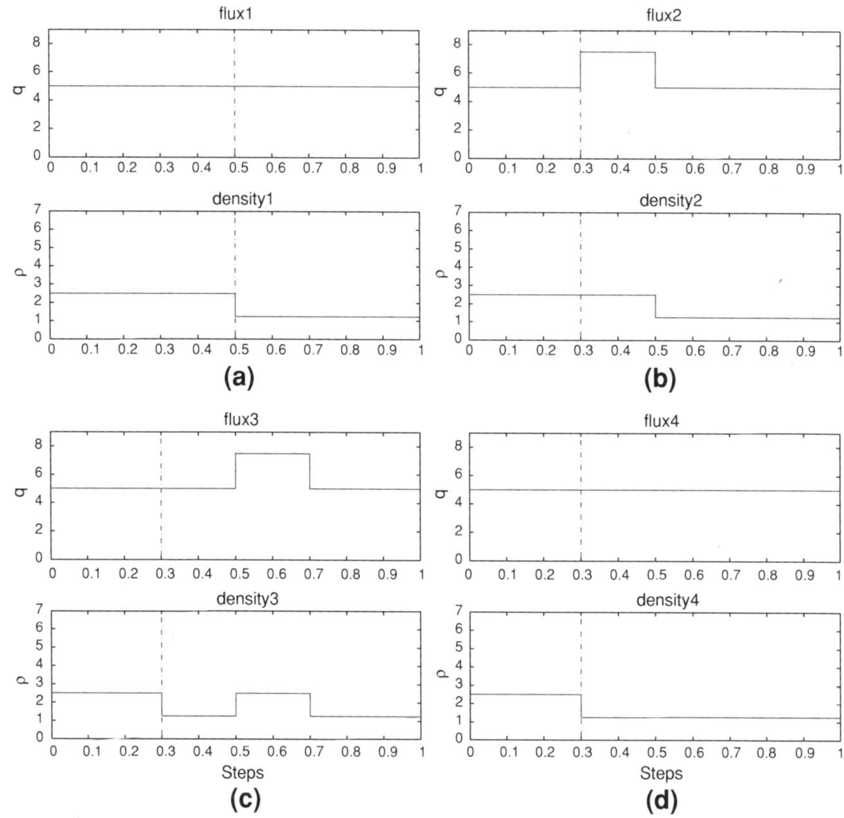

Fig. 7 Stages of creating a flux-bump

free PPP policy. The only thing that matters is whether the flux bump or flux dip that is created arrives at the end of the factory within the demand time window. Our simulations confirm this: PPP and PPP-CONWIP policies behave very similarly and do not improve the performance of the production line appreciably with perishable demand.

6 Conclusion

We have studied process control in a reduced model of a re-entrant semiconductor factory using discrete event simulations. We showed that when running a factory with a push dispatch policy at the beginning of the factory and a pull dispatch policy at the end of the factory while using an average demand starts policy, the transition

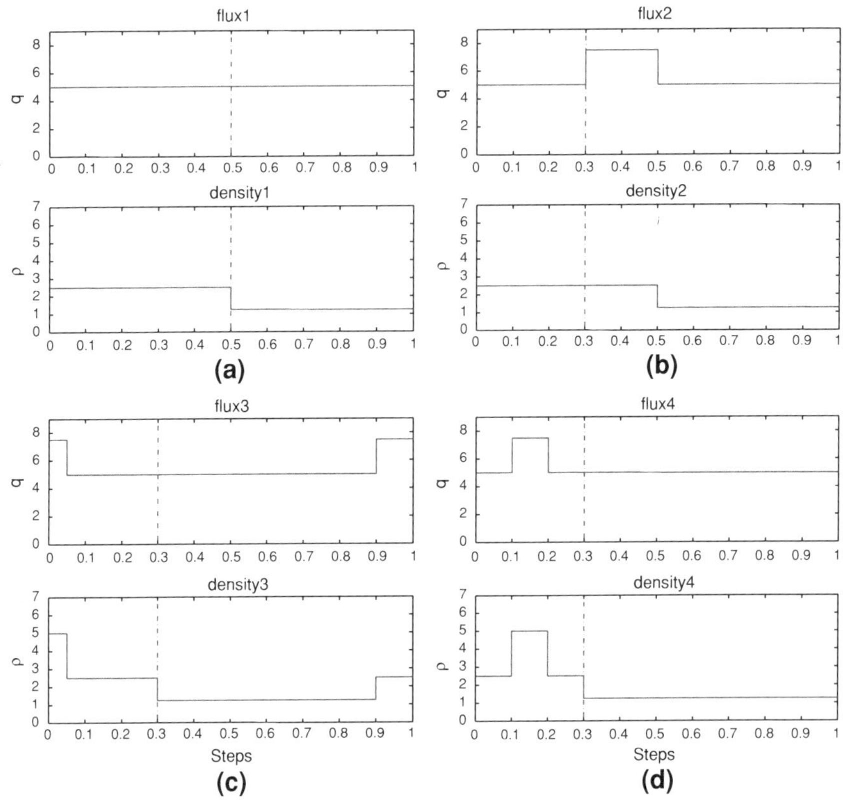

Fig. 8 Stages of creating a flux-bump for a PPP-CONWIP policy

point (the PPP) can be used to reduce the mismatch between stochastic outfluxes of the factory and stochastic demands.

We have two results that are of immediate practical interest:

1. A pure PPP dispatch policy that reaches into the factory from the end and pulls out the desired demand will not significantly reduce the mismatch between outflux and demand for a demand signal that has a constant average and varies stochastically around that average.
2. A PPP dispatch policy coupled with a CONWIP starts policy adjusted for a WIP level that allows maximal flux changes through moving the PPP will significantly reduce the mismatch for a production with non-perishable demand.

Process control in these re-entrant production lines is very difficult since only starts policies and dispatch rules are the obvious control actuators that influence the outflux of the factory. However, as a byproduct of this study we have identified another control parameter: The actual WIP profile will be very important for the success of a PPP policy. It seems likely that very homogeneous WIP profiles are better for the control action of the PPP policy than the WIP profile that we have currently examined. Those WIP profiles are determined by the level of constraint we are choosing for a particular machine set. It will be an interesting further study to determine the interplay of the constraint levels and the PPP policy.

Acknowledgments The research of D.A. was supported by NSF grants DMS-0604986 and DMS-0204543. We thank Ton Geubbels for help in developing the χ—model.

References

1. Asmundsson J, Rardin RL, Uzsoy R (2003) Tractable non-linear capacity models for aggregate production planning. Technical Report IN 47907–1287, Purdue University, School of Industrial Engineering, West Lafayette
2. Atherton RW, Dayhoff JE (1986) Signature analysis: Simulation of invetory, cycle time and throughput tradeoffs in wafer fabrication. IEEE Trans Compon Hybrids Manuf Technol 9(4):498–507
3. Aytug H, Koehler GJ, Snowdon JL (1994) Genetic learning of dynamic scheduling within a simulation environment. Comput Oper Res 21(8):909–925
4. Bergamaschi D, Cigolini R, Perona M, Portioli A (1997) Order review and release strategies in a job shop environment: A review and classification. Int J Prod Res 35(2):399–420
5. Blackstone JH, Phillips DT, Hogg GL (1982) A state-of-the-art survey of dispatching rules for job shop operations. Int J Prod Res 20(1):27–45
6. Dummler MA (2000) Analysis of the instationary behavior of a wafer fab during product mix changes. In: Proceedings of the winter simulation conference, pp 1436–1442
7. Fowler JW, Hogg GL, Mason SJ (2002) Workload control in the semiconductor industry. Prod Plan Control 13(7):568–578
8. Hofkamp AT, Rooda JE (2002) χ Reference manual. Eindhoven University of Technology, Systems Engineering Group, Department of Mechanical Engineering. http://se.wtb.tue.nl/documentation
9. Kato K (1996) Lot start evaluation system using simulator. In Proceedings of international symposium on semiconductor manufacturing, Tokyo, Japan, pp 293–296
10. Lu S, Ramaswamy D, Kumar PR (1994) Efficient scheduling policies to reduce mean and variance of cycle time in semiconductor plants. IEEE Trans Semicond Manuf 7:374–388
11. McKiddie R (1995) Some 'no-panic' help for wafer-start surges. Semicond Int, 115–120
12. Panwalkar SS, Iskander W (1977) Capacity planning and control. Oper Res 25(1):45–61
13. Spearman ML, Woodruff DL, Hopp WJ (1994) CONWIP: A pull alternative to kanban. Int J Prod Res 28(5):879–894
14. Sterman JD (2000) Business dynamics, systems thinking and modeling for a complex world. McGraw-Hill, New York
15. Vervoort J, Rooda JE (2003) Learning χ. Technical report, Eindhoven University of Technology, Systems Engineering Group, Department of Mechanical Engineering. http://sc.wtb.tue.nl/documentation
16. Wein LM (1988) Scheduling semiconductor wafer fabrication. IEEE Trans Semicond Manuf 1(3):115–129

JEDI: Just-in-Time Execution and Distribution Information Support System for Automotive Stamping Operations

Oleg Gusikhin and Erica Klampfl

Abstract Stamping is one of the most complex operations in the automotive supply chain, providing over 400 end items to dozens of assembly plants and service facilities. This operation consists of a complex network of blankers, presses, and subassemblies. Stamping is affected by much variability, such as unexpected machine and tool down time, quality concerns, and customer requirement fluctuations. These facilities typically run a tight schedule, and supply chain visibility is a critical factor in efficient operations. The data pertaining to operations is distributed across several systems including material requirements planning (MRP), plant floor automation, and logistics management. As a result, decision makers are faced with too much data and not enough information. This leads to time loss and effort spent in consolidating and comprehending the data. This chapter describes the Just-in-time Execution and Distribution Information (JEDI) system that collects and integrates relevant data from a set of disparate systems and generates a set of spreadsheet models that represent the stamping production and supply chain status. JEDI not only presents the information in an intuitive way, but also provides what-if analysis capability and decision support for scheduling and distribution.

1 Introduction

This chapter addresses scheduling in a complex manufacturing environment within the automotive supply chain. Specifically, we concentrate on the scheduling of automotive stamping operations. The main goal of stamping operations is to satisfy customer requirements posted using the electronic data interchange (EDI). Demand for individual plant operations is propagated using material requirements

O. Gusikhin (✉) · E. Klampfl
Ford Research & Advanced Engineering,
MD3137, 2101 Village Road,
Dearborn, MI 48121, USA
e-mail: ogusikhi@ford.com

D. Armbruster and K. G. Kempf (eds.), *Decision Policies for Production Networks*,
DOI: 10.1007/978-0-85729-644-3_6,

planning (MRP). The plant strives to follow an optimized cycle plan using safety stock to compensate for demand and production fluctuations. In the last decade, the competitive pressure to become a just in time (JIT) manufacturer has resulted in a substantial decrease in inventory at the plants that used to compensate for problems common to the manufacturing environment, such as machine failure and quality problems. In the absence of inventory cushions, plants need to rectify effects of such events through changes in the schedule.

The data for scheduling and manufacturing execution control is scattered across multiple corporate business and plant floor systems. These data may have inconsistencies, errors, and may not reflect the latest changes in the inventory status. Plant personnel typically manage the schedule with pencil and paper and utilize local Excel files. This leads to time lost and effort spent in consolidating and comprehending the data. There were a number of attempts to implement automatic stamping scheduling systems that were not successful because they overlooked the challenges related to ensuring the quality of the input data and complexity and breadth of operational decisions available to plant schedulers. Besides the data accuracy itself, the given input data might not warrant a feasible solution and might require deviations from the normal business practices, such as overtime, premium freight, non-optimal shipment batches, delaying shipment of service parts, outsourcing jobs, etc. Capturing and formalizing all these decisions is either impossible or may lead to an intractable model. In most cases, traditional scheduling approaches focus on optimization or heuristic methods for finding a scheduling solution with a given set of input data; however, in practice, establishing quality input data usually requires substantial user involvement. As a result, there is often a gap between the advancements of optimization capabilities and existing plant floor scheduling practices.

A system for effective and efficient support of scheduling and manufacturing execution must accomplish the following to close this gap:

- consolidate relevant data and organize it into meaningful information;
- support data validation and verification by making each input data element easily traceable to the original source;
- provide an intuitive and clear representation of the actual decision-making environment with visibility into the demand, supply chain, scheduling, and production constraints;
- allow for what-if and sensitivity analysis;
- provide a highly interactive interface with immediate feedback on the effect of decisions.

Then such a system can be an efficient front-end to powerful optimization algorithms.

This chapter introduces the just-in-time execution and distribution information (JEDI) system, which is a decision support system that allows plant floor personnel to customize, visualize, and manipulate the scheduling data for supply chain visibility and what-if scenario analysis capability. JEDI provides visibility to the schedulers so that they can interactively change the input data (e.g. part demand, or inventory counts) when appropriate to enable feasible scheduling. JEDI leverages the scheduler's expertise and enhances the scheduler's capabilities, by allowing simultaneous analysis of the schedule and distribution options, such as using premium freight, and immediate visualization of the impacts of the decisions on both the upstream and downstream supply chain operations. JEDI also provides an interface to a number of optimization algorithms that can be called on demand. The focus of this chapter, however, is on the models to integrate and manipulate the data. For optimization methods related to JEDI, refer to [1–4].

The chapter is organized as follows. We first present an overview of both the automotive supply chain data flow and automotive stamping. Second, we present the automotive supply chain spreadsheet model and its Excel implementation. Then, we discuss the decision support interface illustrated with some usage scenarios. Finally, we review how JEDI integrates with other stamping and enterprise-wide systems. We conclude with a short summary and system benefits.

2 Automotive Supply Chain Data Flow

Successful relationships between Original Equipment Manufacturers (OEM)s and suppliers are dependent on effectively communicating data between all levels of the supply chain. Most suppliers are not dedicated to one OEM, and similarly OEMs interact with multiple suppliers. Therefore, having a means for standard communication between all parties is required. The relationship between the automakers and supply base is governed by a long-term contract, while individual transactions are handled through an EDI. The key of EDI is that it follows a standard and can be thought of as a language for communicating structured documents [15]. There are two main standards: American National Standards Institute (ANSI) X12 and Electronic Data Interchange For Administration, Commerce and Transport (EDIFACT). ANSI X12 is the EDI standard used in the United States, and EDIFACT is the international EDI standard developed under the United Nations and used by most of the rest of the world. For a comparison of the two see MEMA [14]. The North American automotive EDI has been developed by the Automotive Industry Action Group (AIAG), using the ANSI X12 format.

The following North American EDI transactions are related to scheduling, manufacturing execution, and logistics: 1. Material Release — 830 [6, 10]; 2. Shipping Schedule — 862 [8, 9]; and 3. Production Sequence — 866 [7] that supports In-line vehicle sequencing (ILVS) [11]. These EDI transactions are critical in that they describe how the demand information is posted into the supply chain.

The 830 provides the "weekly" or planning release that is calculated and issued to suppliers weekly. It authorizes labor, materials, or other resources within a specified timeframe and provides the requirement forecast beyond that. The 862 provides the "daily" or ship release schedule. It is calculated and issued to suppliers daily, covering around two weeks of consecutive calendar days of requirements. This shipping schedule transaction set enables customers to convey precise shipping schedule requirements to a supplier and supplements the planning schedule transaction set (i.e., 830).

For suppliers who provide in-sequence parts, the 866 is calculated and issued to suppliers daily, covering short-term requirements in the vehicle rotation sequence. The use of 866 EDI transactions facilitate the JIT manufacturing practice by providing OEMs with a mechanism to issue precise shipping sequence requirements.

For the first tier assembly plant suppliers, such as stamping, the customer releases are generated from the assembly line schedule. An assembly plant has its own schedule, which depends on customer orders, plant and supply chain constraints, etc. The customer releases are generated based on this schedule and other inputs, such as balance on hand (BOH), parts in transit, and logistics constraints. For scheduling of stamping operations, the 862 release is a primary source for the customer requirements data, and the 830 release is required for planning beyond the 862 release timeframe.

3 Automotive Stamping

Stamping is one of the most complex operations in the automotive supply chain. Individual stamping plant daily requirements may include thousands of parts making over 400 different end items (i.e., part type that represents a finished product that is shipped to a customer) to dozens of assembly plants and service facilities. In general, automotive stamping plants are comprised of three main areas: blankers, presslines, and subassemblies. The *blanking press* uses a large sheet roll of metal (e.g. steel, aluminum) to cut *blanks*, which are pieces of sheet metal slightly larger than the desired part (see Fig. 1). These are then sent to the presslines (see Fig. 2), which consists of several *dies* that form the three-dimensional part. Example parts are inner and outer door panels and hoods. Once the parts are made, they are sent to welding subassemblies (see Fig. 3) or directly as end items to be shipped to the assembly plants or service facilities.

Figure 4 shows the complexity involving only one stamping part, the front floor panel assembly, that must be shipped to six customers. Note that this assembly consists of five subassemblies, three of which must also be shipped to three customers. One can extrapolate from this figure for only one part the complexity in a stamping environment with hundreds of parts.

The pressline area shown in Fig. 2 is the bottleneck operation since it has the most binding constraints [1]. Each pressline is capable of making roughly 5–15 different parts, with some parts having the ability to be made on multiple presslines. Small stamping facilities have around four presslines, where large stamping facilities have over 50 presslines. There are usually long changeovers involving the need of indirect labor for die changeover preparation. Typically, the pressline schedule is implemented first, and blanker and assembly are subsequently scheduled. Ideally, stamping would operate to a repeatable cycle plan that is optimized for the minimum cost of inventory, direct labor, and indirect labor services die changeovers [3].

Fig. 1 Blanker

Fig. 2 Pressline

However, the execution of this plan is often affected by problems typical for any manufacturing operation, such as machine breakdowns, quality problems, etc. In the absence of large inventory cushions, these problems must be compensated for by the plant schedulers: they must modify the existing cycle plan, for example, by reducing the batch sizes and working overtime. This type of change creates a ripple effect through the complex supply chain network, such as the one in Fig. 4, and may lead to the inability to satisfy assembly plant shipping requirements.

Fig. 3 Welding assembly

The job of the scheduler is very difficult due to the complexity of stamping operations and the multitude of data that needs to be analyzed. The scheduler must first get and consolidate data from many different systems, including the corporate MRP 3270 terminal emulator (e.g. Fig. 5), numerous plant floor systems, paper reports from the plant floor, and radio and phone communication. The MRP screen in Fig. 5 shows how the data is available, but not in an integrated or easy to use and manipulate interface. Hence, systems such as the MRP are not designed for decision support. As a result, decision makers are faced with too much data and not enough information. There is a need for a decision support system that will help to analyze and modify the data and support the scheduling for all operations and specifically the presslines. Additional complications arise because the bottleneck area, the press lines, is not the last area in the process. Consequently, the build requirements for the press line or blanker have to be exploded from customer releases through the bill of material (BOM). The net demand generated by the MRP for individual parts does not allow for distinguishing between actual assembly plant consumption demand from the demand raised by the need for safety stock or transportation optimization. Thus, the decision support system needs to combine scheduling support and MRP logic of BOM explosion.

The JEDI system, discussed in this chapter, integrates and consolidates supply chain and production data and generates decision support models as spreadsheet models, which provide a natural representation for the multi-period, multi-product scheduling problem at hand. JEDI implements MRP logic as a spreadsheet model of basic functions; thus, it allows on-the-fly analysis of how changes in the input data affect the scheduling demand in upstream operations. It is implemented using Microsoft Excel, which is the most commonly used system at the plant, and hence reduces the need for training and facilitates system acceptance.

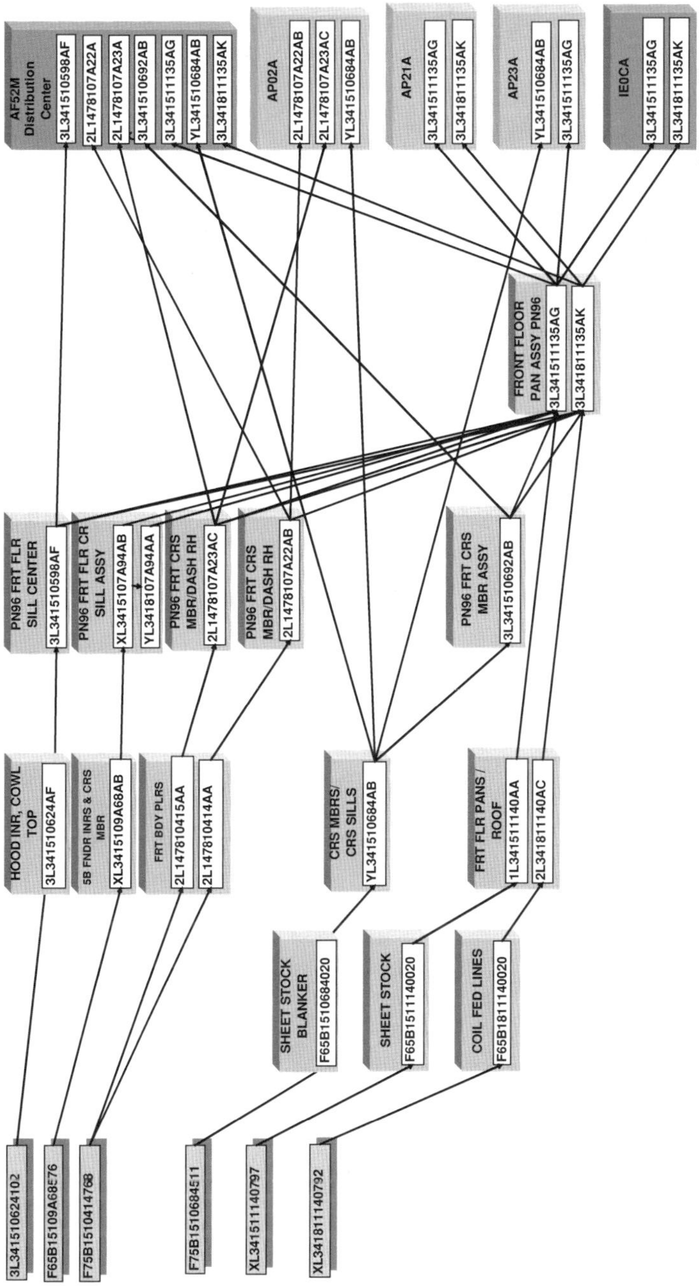

Fig. 4 Stamping complexity

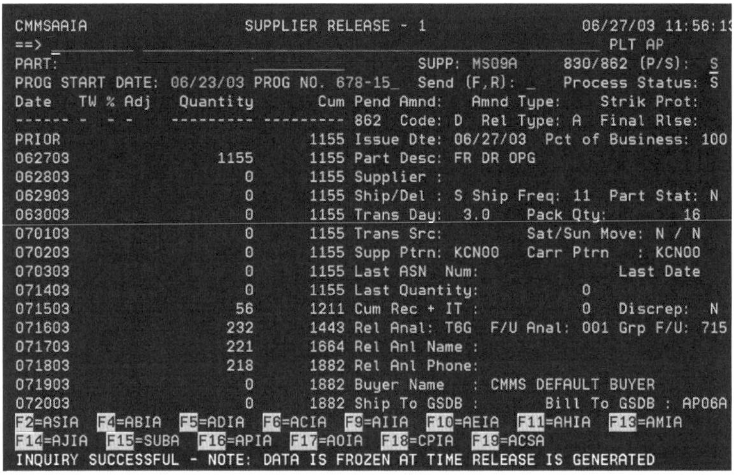

Fig. 5 Corporate MRP screen with 862 data

4 Spreadsheet Model of Automotive Stamping

As we stated previously, stamping is driven by the schedule of bottleneck operations (e.g. presses), while the requirements for the press operations are driven by the customer releases on the end items. To obtain press line parts requirements, the customer releases on end items are propagated through the BOM explosion into the net requirements for the make parts (i.e. components produced at a facility that are used in a higher level items), running at the presses. Thus, our decision support model needs to integrate BOM explosion calculus with the machine finite capacity scheduling. In this section, we first describe the mathematical model that is based on recursive calculations of the net requirements of the component parts at any level in the BOM from the gross requirements of the parts in which they are used. Second, we illustrate the implementation of this model in Excel and describe the algorithm to automatically generate such a model.

4.1 BOM Explosion Calculus and Scheduling

To simplify the overview of the model, we make following assumptions:

- we assume zero lead time for all of the orders between shipping and presses, since in most cases assembly can expedite the parts through the lines;
- part demand needs to be met by the end of each time bucket;
- the parts are assigned to a specific machine (i.e. the same part does not run on different machines);

- a part must run in a single batch in a given time bucket (i.e. there can not be more than one changeover for a part in a given time bucket).

Note that the JEDI implementation addresses the cases where these assumptions are not valid.

In the model we will use the following notation for the part, customer, machine, and time bucket sets.

ρ = number of parts.
τ = number of time buckets.
ξ = number of customers.
μ = number of machines.
P = $\{1, \ldots, \rho\}$ set of parts.
T = $\{1, \ldots, \tau\}$ set of time buckets.
C = $\{1, \ldots, \xi\}$ set of customers.
M = $\{1, \ldots, \mu\}$ set of machines.
$A_p \subset P \bigcup C$ = set of items for which the make part p is an immediate successor in the BOM. For an end-item p it is the set of customers for part p.
$P_m \in P$ = the set of parts assigned to machine m.

Then, we define the input values and introduce the calculated parameters that keep track of the inventory position, the balance on hand, the machine capacity, and the net demand.

r_p = the hourly production rate to make part p.
h_{pt} = the hours scheduled to make part $p \in P$ in time bucket $t \in T$
l_{pt} = the hours of changeover for part $p \in P$ scheduled in time bucket $t \in T$
Q_{mt} = the number of hours available for machine $m \in M$ in time bucket $t \in T$
D_t^p = the net demand for part $p \in P$ in time bucket $t \in T$
G_t^p = the gross demand for part $p \in P$ in time bucket $t \in T$
B_t^p = the balance for part $p \in P$ in time bucket $t \in T$, which represents either the projected inventory or demand in time bucket t.
S_t^p = the scheduled quantity of part $p \in P$ in time bucket $t \in T$
I_t^p = the projected inventory on hand for part $p \in P$ in time bucket $t \in T$
$\bar{I}_t^p$ = the inventory position for part $p \in P$ in time bucket $t \in T$, which represents either the projected inventory or cumulative demand in time bucket t.
$\bar{I}_0^p = I_0^p = B_0^p$ = initial balance on hand for part $p \in P$.

The gross demand for part $p \in P$ in time bucket $t \in T$ equals the sum of the net demands coming from all successors to part p: $G_t^p = \sum_{\tilde{p} \in A_p} D_t^{\tilde{p}}$, where $D_t^{\tilde{p}} \leq 0$. If part p is an end-item, then the demand D_t^p is the customer release. However, if p is a make part, then the gross demand will be the sum of net demands from the successor part in the BOM. In this case, the net demand for the part is be calculated using BOM explosion calculus from the customer releases as follows.

We introduce B_t^p to be the balance for part $p \in P$ in bucket $t \in T$. We define D_t^p to be the net demand of part $p \in P$ in bucket $t \in T$. Note that if there is any net demand for a part in a given bucket, then the value is always less than or equal to zero. That is,

$$D_t^p = \begin{cases} B_t^p & \text{if } B_t^p < 0 \\ 0 & \text{otherwise} \end{cases}$$

This logic can be represented, for example, by the following formula:

$$D_t^p = \min(B_t^p, 0). \tag{1}$$

We define I_t^p to be the projected inventory on hand for part $p \in P$ in time bucket $i \in T$: if there is any projected inventory on hand for a part in a given bucket, then the value is always greater than or equal to zero. Hence,

$$I_t^p = \begin{cases} B_t^p & \text{if } B_t^p > 0 \\ 0 & \text{otherwise} \end{cases}$$

Similar to (1), this logic can be represented, for example, by the following formula:

$$I_t^p = \max(B_t^p, 0). \tag{2}$$

B_t^p can be thought of as the non-zero part balance, which will either be the net demand or the projected inventory on hand and is calculated as

$$B_t^p = I_{t-1}^p + \sum_{\tilde{p} \in A_p} D_t^{\tilde{p}}, \tag{3}$$

where $I_{t-1}^p \geq 0$ and $\sum_{\tilde{p} \in A_p} D_t^p \leq 0$.

Using the formulas in (1) and (2), the material balance equation in (3) can be reformulated as follows:

$$B_t^p = \max(B_{t-1}^p, 0) + \sum_{\tilde{p} \in A_p} \min(B_t^{\tilde{p}}, 0) \tag{4}$$

Note that B_0^p is the existing balance on hand for every $p \in P$ and is always non-negative. Also, in the case that $p \in P$ is an end-item, the net demand is the customer release (i.e. 862) and is represented as a negative number.

We can recursively apply formula (4) to go from the customer demand to the net requirements of the parts assigned to the machine that we will schedule. Then, for each bucket $t \in T$, machine $m \in M$, and part $p \in P_m$ (i.e. parts running on machine m), we calculate the inventory position as

$$\bar{I}_t^p = \bar{I}_{t-1}^p + G_t^p + S_t^p, \tag{5}$$

where $\bar{I}_0^p$ is the initial balance on hand for part p, and $G_t^p = \sum_{\tilde{p} \in A_p} \min(B_t^{\tilde{p}}, 0) \leq 0$ is the gross demand for part p in time bucket t. S_t^p is the quantity of parts $p \in P_m$ scheduled on machine $m \in M$ in time bucket $t \in T$. The inventory position $\bar{I}_t^p$ gives the cumulative demand up to the current time bucket and takes into account parts scheduled for the given time bucket, while B_t^p gives the demand only for the given time bucket.

The desired schedule should ensure that for any time buckets within the period for which a schedule exists, $\bar{I}_t^p$ is at least non-negative or ideally close to a preset safety stock number. In other words, for any period in which we have a schedule, we should not have any unsatisfied demand. If for a certain time bucket $\bar{I}_t^p$ is negative and $\bar{I}_{t+1}^p$ is non-negative, then this indicates that certain orders are potentially late against the shipping demand and requires the scheduler's attention.

The quantity S_t^p of parts $p \in P_m$ that can be scheduled in any time bucket $t \in T$ is bound by the finite capacity of machine $m \in M$. We let r_p be the rate at which part $p \in P_m$ can be made per hour on machine $m \in M$ and h_{pt} be the hours scheduled to make part $p \in P_m$ on machine $m \in M$ in time bucket t. Then, the quantity of parts $p \in P_m$ scheduled in time bucket $t \in T$ on machine $m \in M$ is

$$S_t^p = h_{pt} r_p. \tag{6}$$

We consider that each part is assigned to a dedicated machine, but we must guarantee that in each time bucket every machine is not over its maximum capacity, Q_{mt}, defined by the number of hours available. If we let l_{pt} be the hours of change-over required for part $p \in P_m$ in time bucket $t \in T$ on machine $m \in M$, then we can satisfy the condition that the machine's maximum capacity is not exceeded by the following constraint:

$$\sum_{p \in P_m} (h_{pt} + l_{pt}) \leq Q_{mt} \quad \forall\ m \in M, \quad t \in T. \tag{7}$$

In addition to the above constraint, a valid schedule would need to satisfy other constraints such as no overlapping jobs on the same machine in the same time bucket and that the run hours are always preceded by changeover hours. These constraints can be enforced through customized data input or highlighted through Excel conditional formatting. Note that the goal of this model is not to serve as a basis for scheduling optimization but to provide a visual representation of the relations between the data and constraint violations in the decision support system, which we demonstrate in the next sections.

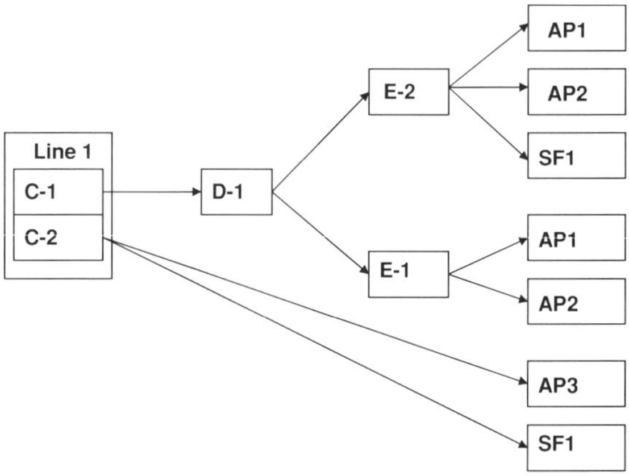

Fig. 6 Pegging tree for work center Line 1

4.2 Excel Implementation

For illustration purposes, consider the example presented in Fig. 6. Assume we have two parts, C-1 and C-2, which run on work center Line 1. The first part, C-1, is a component, which is required by another component, D-1. This part, in turn, is used in two different end items, E-1 and E-2, which are shipped to several customers, AP1, AP2, and SF1. The second part, C-2, is an end-item that is directly shipped to two different customers, AP3 and SF1. The customers whose names begin with "AP" are assembly plants, and those beginning with "SF" are service facilities.

Figure 7 shows the excel implementation of the model presented in Fig. 6. Note that the parentheses are used to represent negative numbers. The rows associated with the BOM structure for each of the parts assigned to the work center (e.g. C-1 and C-2) are grouped together. Rows 3–11 represent the demand chain rooted in the part C-1, and rows 14–16 represent the demand chain rooted in the part C-2. The customer demand is organized into daily buckets. The customer requirements, the 862 shipping release, are populated in the cells corresponding to different time buckets (see rows 3, 4, 6, 7, 8, 14, and 15 with a gray background).

The two end items E-1 and E-2 are associated with part C-1. Rows 5 and 9 contain the demand net on-hand inventory for these end items, which is calculated from the customer release and the existing on-hand inventory derived from equation (3) in Sect. 4.1 . The existing on-hand inventory for these end items are in cells F5 and F9, respectively. For example, the daily net demand on 7/31 in cell G5 is calculated using equation (4) as "= MAX(F5,0) + MIN(G4,0) + MIN(G3,0)," where F5 corresponds to the inventory in the previous period, G3 is the 862 shipping release associated with the assembly plant AP1, and G4 is the shipping release to assembly plant AP2. As a result, each cell associated with the demand of part E-1 will contain either the projected inventory for the given day in the case of a positive number or the net demand for this day in the case of a negative number.

Microsoft Excel - Line_Schedule.xls

File Edit View Insert Format Tools Data Window Help

J12 =I12 + (J10-ABS(J10))/2+J13

	A	B	C	D	E	F	G	H	I	J	K	L	M	N	O
1	Line 1	Rate	Chng	Balance On Hand			7/31	8/1	8/2	8/3	8/4	8/5	8/6	8/7	8/8
2							Thu	Fri	Sat	Sun	Mon	Tue	Wed	Thu	Fri
3	AP1						(380)	(2600)			(956)	(956)	(956)	(956)	(956)
4	AP2										(961)	(961)	(961)	(961)	(959)
5	Part E-1					29	(351)	(2600)	-	-	(1917)	(1917)	(1917)	(1917)	(1915)
6	SF1							(100)							
7	AP1										(764)	(764)	(764)	(764)	(764)
8	AP2										(764)	(764)	(764)	(764)	(766)
9	Part E-2					190	190	90	90	90	(1438)	(1528)	(1528)	(1528)	(1530)
10	Part D-1				861		510	(2090)	-	-	(3355)	(3445)	(3445)	(3445)	(3445)
11	Part C-1			3312			3312	1222	1222	1222	(2133)	(3445)	(3445)	(3445)	(3445)
12	IP						4512	2422	2422	2422	(933)	(4378)	(7823)	(11268)	(14713)
13	Schedule						1200	-	-	-	-	-	-	-	-
14	SF1							(70)							(100)
15	AP3										(900)	(600)	(600)	(1200)	(3300)
16	Part C-2			5442			5442	5372	5372	5372	4472	3872	3272	2072	(1328)
17	IP						5442	5372	5372	5372	4472	3872	3272	2072	(1328)
18	Schedule						-	-	-	-	-	-	-	-	-
19	Start						15:00								
20	Chng Hrs		4				4.00								
21	Run Hrs	300					4.00								
22	Start														
23	Chng Hrs		4												
24	Run Hrs	300													
25	Chng Total						4.0		-	-	-	-	-	-	-
26	Run Total								-	-	-	-	-	-	-
27	Total Hours						4.0	-	-	-	-	-	-	-	-

Line1 / Line2 / Line3 / Line4

Fig. 7 Excel implementation of work-center Line 01

The combined demand from parts E-1 and E-2 constitute the gross demand for part D-1. The net demand for D-1 is calculated in row 10 and is then used to calculate the net demand for part C-1 in row 11. The net demand for part C-2 is calculated directly from the shipping releases to service facility SF1 and assembly plant AP3 in rows 14 and 15, respectively.

Rows 12 and 17 contain the inventory positions for parts C-1 and C-2, respectively, which are calculated using Eq. (5): this is the cumulative demand minus the inventory on hand plus the cumulative parts scheduled up to this period that are assigned to the given work center. The schedule for parts C-1 and C-2 are entered as the quantity of run hours in rows 21 and 24 starting from column G. Run hours are converted into the quantity of parts using the hourly rate in cells B21 and B24: for example, the formula to calculate the number of parts of type Part C-1 scheduled in cell G13 is "=B21* G21" that results in the value of 1,200.

We include in the schedule for any part the number of changeover and run hours. For example, for parts C-1 and C-2, the number of changeover hours are in cells C20 and C23, respectively, and the number of run hours are in cells B21 and B24, respectively. The item associated with "Start" in column A is an informational field containing the start time of the changeover if different from beginning of the day. For example, cell G19 contains the start time of the changeover of part C-1 to be at 15:00. Rows 25 and 26 provide a summary for the total changeover and total run hours for the day, while row 27 summarizes the total work center hours scheduled for the day to make sure that the hour limits are not exceeded, such as 24 hours for a three shift operation. To visualize the constraint violation for hours available, we can implement conditional formatting that will change the cell background in row 27 to red when the value in the cells exceeds the number of hours available.

4.3 Automatic Model Generation

Generation by hand of such models described in the previous section would be prohibitively time consuming and error prone. The way to address these issues from manual generation is to automate the generation of such models from the MRP data. In doing so, the models can be formatted and protected so that the users can only modify the cells for which they have permission based on their job function. Furthermore, this ensures the models would match predefined templates that would allow storing all modified data back into the database.

As we can see in Fig. 6 , the data behind the model has a tree structure with a root at the given make part and leaves associated with customers. The model is generated using data from the MRP system, including the BOM, parts and their associated work center, and end items and their associated customers shown in the tables in Fig. 8. Based on these tables, we create a new table that defines a pegging tree for each part by listing pairs of consecutive nodes in the tree structure with a root in the given part. Figure 9 provides an example of such a table for parts C-1 and C-2. The column Root has a reference to the part ID that defines the root of the tree. Other columns are "Node" (i.e. a part or customer ID that is downstream from the root), "Node Prev" (i.e. node that immediately precedes the specified Node), "lineage" (i.e. concatenation of the unique node ids from the root to the given node, and "depth" (i.e. how many levels are between the root and the given node). Then for the given work center, we can create a query that includes all the rows from this table associated with the parts at the given work center sorted in descending order by lineage. Sorting this way guarantees that the order of the rows in the resulting set satisfies that the calculations for the given row are derived from the values in the rows preceding the given row in the result set.

At first we determine the maximum number of levels for the given set of parts and determine the starting column in Excel to start generating the requirements. Figure 10 provides a schematic of the algorithm used to generate the model. The algorithm reads one row at a time starting with the first customer, c4. The algorithm generates an appropriate set of rows in Excel. For a row that is associated with customer requirements, the cells are merely inputs that will be subsequently populated with customer release data. For assembly plant customers, the algorithm will generate additional rows for the assembly plant status.

After processing the initial row, the algorithm puts the references to the Excel row in the last-in first-out stack and proceeds to the next row. The next two rows are other customers, c3 and c2, for the same end item, p5. In this case, the algorithm generates appropriate rows in Excel and puts the appropriate references to rows associated with c3 and c2 in the stack. The next entry is part p5: the algorithm creates Excel

Part

ID	Part
p1	C-2
p2	C-1
p3	D-1
p4	E-1
p5	E-2

BOM

PartID	NextPartID
p2	p3
p3	p4
p3	p5

WorkCenter - Part

WorkCntr	PartID
Line 1	p1
Line 1	p2

Customer

ID	Customer	Type
c1	AP3	AP
c2	AP2	AP
c3	AP1	AP
c4	SF1	SF

Part - Customer

PartID	CustomerID
p1	c1
p1	c4
p4	c2
p4	c3
p5	c2
p5	c3
p5	c4

Fig. 8 MRP tables

Root	Node_Prev	Node	Lineage	Depth
p2	p5	c4	p2->p3->p5->c4	3
P2	p5	c3	p2->p3->p5->c3	3
p2	p5	c2	p2->p3->p5->c2	3
p2	p5	p5	p2->p3->p5	2
p2	p5	c3	p2->p3->p4->c3	3
p2	p5	c2	p2->p3->p4->c2	3
p2	p3	p4	p2->p3->p4	2
p2	p2	p3	p2->p3	1
p2	*null*	p2	p2	0
p1	p1	c4	p1->c4	1
p1	p1	c1	p1->c1	1
p1	*null*	p1	p1	0

Fig. 9 Pegging table containing the data set for Line 01

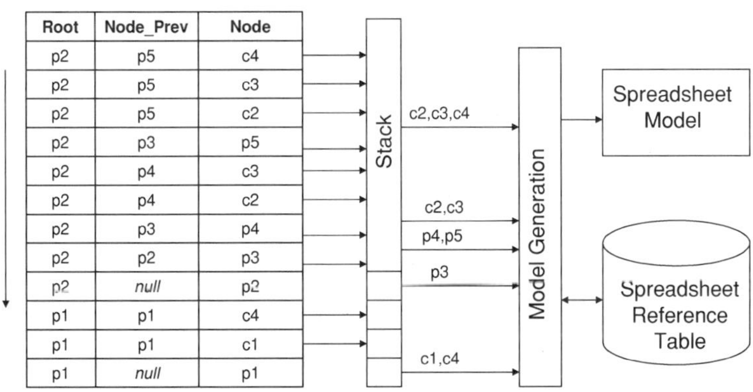

Fig. 10 Model generation algorithm schematic

rows associated with make part p5 and creates appropriate formulas by extracting the reference to children nodes of p5 from the stack and generating a formula that includes the sum of the net requirements from c4, c3, and c2. The algorithm puts a reference to the p5 row in the stack. The two next entries in the data set are the reference to the customers c3 and c2 for the end item p4. The algorithm follows the above logic until it reaches the entry with part p3. Here, the algorithm extracts the references to the children of p3, p4, and p5 and generates appropriate formulas with gross demand of p3 as the sum of the net demands from p4 and p5. The algorithm also builds a cross reference table that references the part numbers and customers with the position in Excel. This table will be used to create a scheduling portion of the spreadsheet model and to populate actual data from the systems. When the algorithm finishes performing this logic for all the parts assigned to the workcenter, it then generates the rows for schedule input using a cross-reference table. After generating the rows and formulas, the algorithm performs post-processing that formats and structures different elements of the spreadsheet to improve clarity and visibility of the information as described in the next section.

5 Spreadsheet-Based Decision Support System

The type of model described in Sect. 4.2 could be too cumbersome for the scheduler, especially when the number of parts assigned to a workcenter is relatively large (e.g. 10 or more parts). To address this, we exploit the Excel rich formatting capabilities to modify the representation described in Sect. 4.2 to provide a clearer view of the model together with supplemental information for decision support. First, we can hide all rows for intermediate parts for which the user will not provide input (e.g. rows 5, 9 and 10 in Fig. 7). Also, we use the Excel group function to group all rows associated with the supply chain representation of the individual parts.

As a result, we can get a clear view of the work center load and schedule with the capability to drill down on individual parts. For example, the Capacity view in Fig. 11 shows Line 01 with 10 parts representing inventory positions for every part and the schedule for parts. Different work centers are represented by individual excel worksheets: for example, you can see in Fig. 11 worksheet tabs associated with fifteen workcenters. The Capacity view shows all parts assigned to the given work center with associated projected inventory positions (the first row for each part) and scheduled quantity (second row for each part) grouped in daily buckets. The positive numbers in the inventory position row represent the projected balance on hand for the given part, while negative numbers denoted with parentheses represent the cumulative demand exploded from customer requirements through the BOM structure and associated inventory levels. The capacity view provides clear visualization of the capacity load and potential problems that could impact satisfying customer demand for every part in the work center.

The user can explore the individual parts in detail by clicking the '+' next to the part. The part information will expand to provide a supply chain and pegging view for

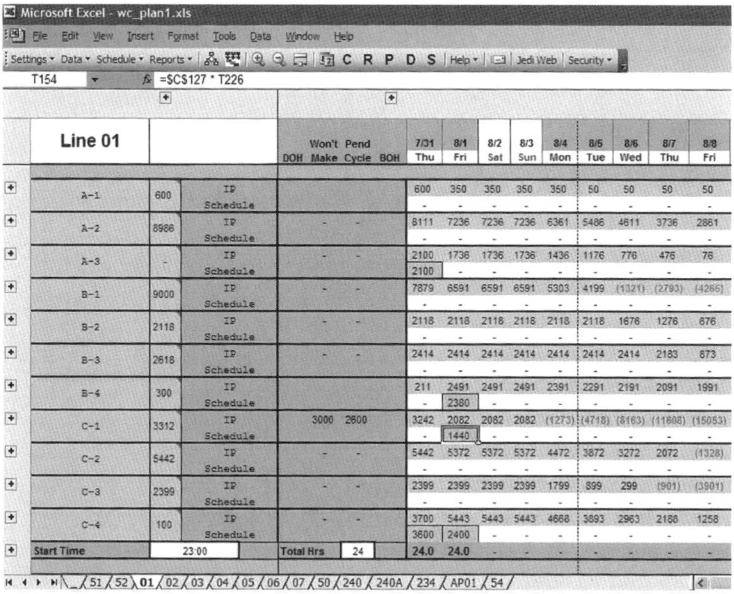

Fig. 11 Capacity view

individual parts. Figure 12 presents a supply chain view together with the BOH for every intermediate part: for example, we can see the expanded view for the part C-1 from Fig. 6. This view allows the scheduler to check and modify the BOH for every part in the chain. For critical parts, the user can inquire after the latest information on available parts from the floor (through radio) and modify the BOH in the model, accordingly. In this representation, all hidden part names from the BOM structure are defined in the comment field associated with appropriate cells and can be viewed by mousing over: see the comment highlighted for part E-1 in Fig. 12 .

For assembly plant customers, the system provides more detailed information pertaining to the given assembly plant status and consumption schedule. Specifically, this information includes the preferred transportation mode (e.g. rail or truck) together with the transit time. It also provides the assembly plant status information, such as

- days on hand: the number of days that can be covered by the existing assembly plant BOH
- will not make: parts in transit that are deemed to be late
- pending cycle: the variance between a part's physical count and the plant's record
- BOH: parts at the assembly plant warehouse and parts in transit less parts that will not make it on time.

If we look at the expanded view for part C-1, we first see the balance on hand for part C-1 = 3312. Next we see the balance on hand for part D-1 = 861 and subsequently

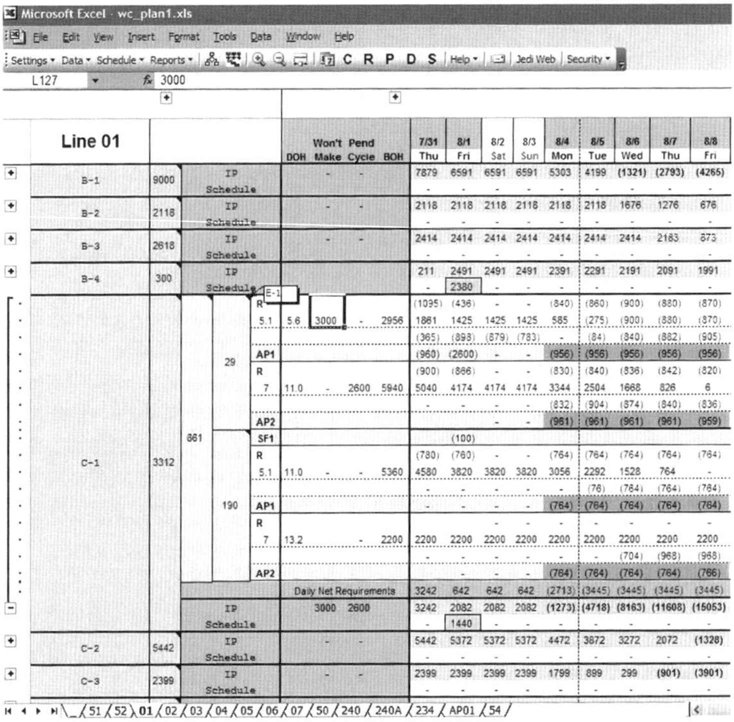

Fig. 12 Pegging supply chain view

the balance on hand for parts E-1 = 29 and E-2 = 190. Associated with the parts E-1 and E-2 is the information related to assembly plant demand and the 862 release schedule. In this example, we are shipping part E-1 to assembly plants AP1 and AP2. The rows with AP1 and AP2 in them contain the 862 release schedule. The information related to the assembly plant consumption schedule is organized in three rows above the 862 release schedule row. The first row shows the assembly plant consumption schedule. The second row shows the consumption schedule net the assembly plant BOH. Finally, the third row shifts the demand net inventory based on the transportation time associated with the given part-customer combination. This gives a base demand number (i.e., the minimum number of parts that a supplier must provide to satisfy the assembly plant's consumption) that can be compared to the customer release on the next row. The transportation time shift is implemented as a custom formula in Excel. We also implemented conditional formatting for the assembly plant demand to highlight the cases when the cumulative base demand of the assembly plant exceeds the cumulative customer release. This allows schedulers to compare the shipping release against the actual consumption requirement to validate the accuracy of the data and in case of shortages in capacity, modify the shipping release to non-optimal shipping alternatives that satisfy the assembly consumption schedule.

Furthermore, looking at the first customer in Fig. 12 AP1 uses rail transportation denoted by a "R" with a transportation time of 5.1 days. The inventory available to the assembly plant covers 5.6 calendar days. Also, part Part C-1 has 3,000 parts in transit to AP1 (in the "won't make" column) that according to the transportation records will not make it to the customer on time, which causes an inflated customer release on the first two days. The user can analyze the details of this situation, which could reveal that the delay will be only for few minutes and that all of the parts can be considered "on time." Based on this insight, the user can manually modify the customer requirements and gain a completely different perspective for the demand even before looking at the work center scheduling. The user can also analyze other what-if scenarios, such as how would requirements change if we can use a truck with 1.2 days instead of rail with 5.1 days. This way, the scheduler can see the tradeoff in shipping using a truck with less transportation time versus the existing preferred rail mode of transportation requiring more transportation time and can modify the original 862 release if the decision is for a transportation deviation.

In addition, Fig. 12 shows that customer AP2 for E-1 has 2,600 parts in pending cycle. When booked, it will affect the assembly plant's BOH with an immediate jump in the customer's shipping releases. This pending cycle column gives early warning of potential part shortage and allows the scheduler to take preventive actions to avoid overtime and premium freight for shipping, when possible. Our internal studies showed that 70% of premium freight transportation is due to pending cycle booking process.

This additional assembly plant information allows comparison between the actual requirements by the assembly plant and the current MRP generated customer releases: the scheduler can then correct potential errors in the BOH at the assembly plant or for intermediate parts and conduct different what-if analysis. These scenario analysis can address what happens if we use alternative faster modes of transportation, such as truck instead of rail or the effect from booking of pending cycle parts. As a result of this detailed analysis, the scheduler can modify the original customer release.

Finally, to improve editing and visibility into the daily schedule, we developed an Excel add-in that can provide a block diagram for the schedule representation (see Fig. 13). This add-in reads and interprets the data pertaining to the setup start time, changeover duration, and run time. This chart shows the detailed time required for changeover and to make each batch of parts. Each batch concatenates the changeover hours followed by the production run hours. The changeover time is the first time shown associated with a part number, and the second time associated with the same part number is the time required to make that part. We refer to this detailed portion of the scheduling chart as a "snake diagram" because as the length or position (i.e., start time) of the bars in the chart are modified, the bars instantly wrap around to the next day. For example, part C-1 cannot be finished being made on Friday, and so the needed time to make C-1 is reflected by the bar for that part on Monday, skipping the weekend. The snake diagram automatically skips days when the plant is shut

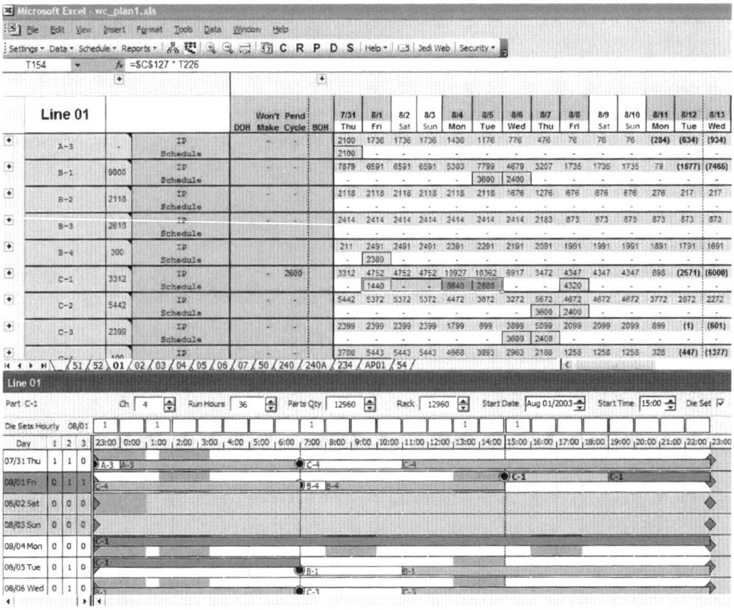

Fig. 13 Scheduling view

down as defined in the work center calendar, such as Sunday. If the plant wanted to conduct a what-if scenario to see if they should run the plant on the weekend with overtime, they could change Saturday or Sunday to be a working day, and the new batch parts for Line 1 would be reflected in the snake diagram. Any changes in the snake diagram are immediately connected to the appropriate Excel cells that show on the fly how changes to the schedule affect the part demand and inventory.

6 System Integration

JEDI is an integral part of a suite of plant floor and enterprise business systems that collectively assist in managing the stamping production. Figure 14 provides an overview of the interactions of JEDI with other key systems. These seamless interactions to and from JEDI provide the foundation for its success: all of the necessary data is available to the scheduler in one location in an easy to use interface, and the schedule information is automatically shared back to other systems reliant upon this data.

As previously mentioned, JEDI relies on data from the corporate MRP, and in turn, sends the schedule back to the MRP to drive the upstream supply chain. JEDI uploads all of the data that is used to define the structure of the problem (e.g. the BOM, customer, part: see Fig. 8) to build the model. This structural data is updated

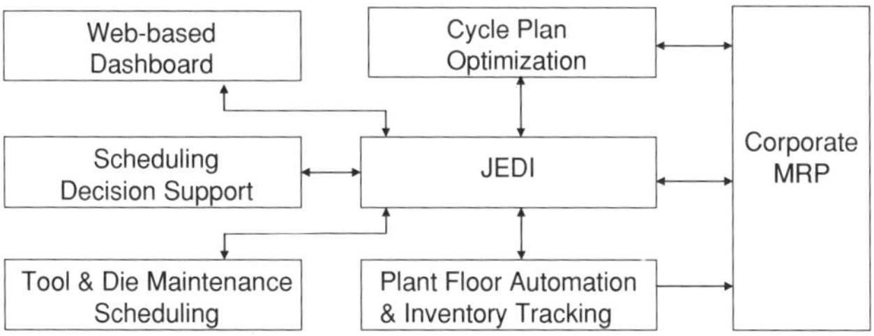

Fig. 14 System architecture

daily, comparing the the new data to the existing workcenter structure. If there are no changes in the problem structure, JEDI keeps the model and will update only the dynamic data (e.g. the BOH, demand, and schedule). If there are any changes, regeneration of a new model is triggered for the workcenter. Changes to a workcenter could include new part or customer introduction, part engineering level changes, changes in the BOM structure, and the removal of obsolete parts. When the model is completed, JEDI collects all relevant dynamic data from the MRP and plant floor systems and updates the model. This data includes customer releases, assembly plant consumption schedule and status, and parts in transit status.

Plant floor automation and inventory tracking systems provide up-to-date information on the BOH and status of critical manufacturing resourccs. Keeping track of the available inventory in a dynamically changing environment is a very challenging task. The methodological approach to proper integration is outlined in [13]. Data inaccuracies, incompleteness, and inconsistencies have to be rectified through intelligent integration of the information [12]. In recent years, the maturity of the RFID technology has tremendously helped improve plant floor data collection capabilities. Fodor et al. [5] describes the approach to track stamping rack location utilizing a forklift mobile RFID reader combined with forklift deadreckoning techniques. Directly tracking the location of the racks tremendously improves the accuracy and timeliness of the balance on hand data versus indirectly estimating it from production counts at the given workcenters.

Furthermore, JEDI supports collaborative scheduling and decision support. It allows schedulers for different areas of the plant (e.g. blankers, press, and assembly) to verify the feasibility of the interdependent schedules and collectively address any potential issues. For instance, we can substitute net demand exploded from customer releases with the actual assembly schedule, and this will allow the scheduler to see how the given press schedule supports the assembly schedule. Similarly, the blanker schedule can load the schedule from presses.

JEDI also facilitates collaborative scheduling between material planning and logistics and tool and die maintenance. When scheduling is done for preventive maintenance of tools and dies, there must be enough inventory to satisfy customer requirements while the die/tool is undergoing maintenance work. JEDI helps to coordinate die/tool scheduling to ensure inventory requirements are met.

The stamping complexity requires tight coordination between different shared resources, such as direct labor that can be reallocated between different workcenters based on the specific jobs and indirect labor services for die changeover, cranes, etc. Thus, it is important that based on the expected customer demand for different job types, we properly determine the needs for direct and indirect labor for upcoming time buckets, identify the needs for overtime, allocate resources among different shifts, and determine a feasible plan for each job and corresponding changeovers required. In addition, we would like to find the most cost-efficient plan that optimizes the tradeoff between labor cost and inventory: this is the goal of the cycle plan optimization module [4]. JEDI takes this optimized plan as a roadmap for the upcoming time period and develops corrective actions to compensate for events happening on the floor, such as parts shortages, machine breakdowns, customer requirement fluctuations, etc.

JEDI also provides a web-based plant and business unit management dashboard. The dashboards increase visibility of the stamping supply chain status, help quickly identify and collectively address critical issues, and facilitate information sharing between Material Planning and logistics, Manufacturing and Maintenance.

7 Summary

This chapter describes the JEDI support system designed and implemented for complex automotive suppliers, such as automotive stamping. JEDI serves as an interactive decision support system that allows schedulers to be an active part of the decision-making process, providing them the information that they need, consolidated in one location, available at the right time, and that can be manipulated within an intuitive system. It has filled a critical need as a front-end decision support system for seamless integration with scheduling optimization and other corporate systems.

The core element of JEDI is a spreadsheet model of the typical automotive supply chain with inputs mapped to MRP and automotive EDI standards. As such, the system and underlying model can be adapted to wide a range of automotive suppliers beyond stamping, such as powertrain, plastics or climate control. The system implementation leverages Microsoft Excel features of rich formatting and automation capabilities and takes advantage of familiarity of Excel to the plant user community.

JEDI has enabled early identification of problems by around two to three hours, allowing schedulers to address production problems in advance. JEDI is integrated with other enterprise and plant floor systems and supports collaborative scheduling between different interdependent manufacturing, distribution and maintenance departments: this has facilitated an improvement in data accuracy in various corporate systems and has enhanced collaboration between multiple stakeholders. Another important benefit is that it provides an efficient and effective interface to the mathematical scheduling models, leveraging optimization technologies while keeping the user in control of the solution.

The implementation of the system in a production environment has demonstrated significant benefits resulting in substantial financial savings associated with reduction in premium freight, overtime, inventory, and excessive material handling. The savings that are typically observed after the introduction of JEDI over the previous year are an average 30% reduction in overtime and a premium transportation reduction of 40% (which for some plants is over a million dollars a year). Some additional benefits that result from the reduction of excess inventory include a reduction in obsolete parts and excessive material handling and improved plant floor utilization.

Acknowledgments We gratefully acknowledge the invaluable contributions from many people at Ford Stamping, Material Planning & Logistics, Information Technology, and Research & Advanced Engineering. Specifically, we would like to thank John Batey, Edward Zvoch, James Higgins, Craig Morford, Mark Catri, and Robert Quick from Stamping, Rich Davidson, Kasey Kasemodel, and Serguei Vassiliev from IT, and Giuseppe Rossi from Research & Advanced Engineering. We would also like to thank Gloria Chou from Research & Advanced Engineering for providing study results on how pending cycle parts effect premium freight.

References

1. Barlatt A, Cohn A, Gusikhin O (2007) A hybridization of mathematical programming and search techniques for integrated operation and workforce planning. In: proceedings of the 2007 IEEE international conference on systems, man and cybernetics, pp 632–637
2. Barlatt A, Cohn A, Gusikhin O (2008) A hybrid approach for solving shift-selection and task-sequencing problems. Lect Notes Comput Sci 5015:288–292
3. Barlatt A, Cohn A, Gusikhin O (2010) A hybridization of mathematical programming and dominance-driven enumeration for solving shift-selection and task- sequencing problems. Comput Oper Res 37(7):1298–1307
4. Barlatt A, Cohn A, Gusikhin O, Fradkin Y, Morford C (2009) Using composite variable modeling to achieve realism and tractability in production planning: an example from automotive stamping. IIE Trans 41:421–436
5. Fodor M, Gusikhin O, Tseng E, and Wang, W (2009) Integration of mobile RFID and inertial measurement for indoor tracking of forklifts moving containers. In Proceedings of the Third Workshop on Intelligent Vehicle Control and Intelligent Transportation Systems, pages 120–129, Milan, Italy.
6. Ford (2002) 830 planning schedule with release capability. https://web.gsec.ford.com/GEC/edispecs/edispecs.asp. Ford Motor Company.
7. Ford (2003) 866 production sequence. https://web.gsec.ford.com/GEC/edispecs/edispecs.asp. Ford Motor Company.
8. Ford (2008) 862 shipping schedule. https://web.gsec.ford.com/GEC/edispecs/edispecs.asp. Ford Motor Company.
9. Grubar K (2002) Eaton corporation electronic data interchange (EDI) standards: 862 shipping schedule, Version 4010. http://www.eaton.com/ecm/groups/public/@pub/@eaton/@corp/documents/content/98065458.pdf.EDI
10. Grubar K (2006) Eaton corporation electronic data interchange (EDI) standards: 830 material release / forecast, Version 4010. http://www.eaton.com/ecm/groups/public@pub@eaton@corp/documents/content/98065456.pdf.EDI
11. Gusikhin O, Caprihan R, Stecke K (2007) Least in-sequence proba- bility heuristic for mixed-volume production lines. Int J Prod Res 46(3):647–673

12. Gusikhin O, Rossi G (2005a) The Knowledge Gap in Enter- prise Information Flow, chapter Improving Automotive Supplier Operations through Information Logistics, pp. 81–90. Jonkoping University
13. Gusikhin O, Rossi G (2005) Well-connected. APICS Perform Advant 15(2):32–35
14. MEMA (1997) EDI transaction comparison, ANSI vs EDIFACT. http://www.mematechnology.com/committees/standards/factedi.pdf. MEMA Technology Council
15. Nicol M (1996) Reaching your customers and suppliers: electronic data interchange- EDI. Manufact-Line: NIST/Michigan Manufacturing Technology Center for Michigan Small and Medium-sized Manufacturers, 2

A Control Theoretic Evaluation of Schedule Nervousness Suppression Techniques for Master Production Scheduling

Martin W. Braun and Jay D. Schwartz

Abstract In manufacturing operations, a Master Production Schedule (MPS) can be used to make mid-range planning decisions that not only influence the production decisions for a manufacturing facility, but serve as input into other decision systems to determine materials ordering, staffing, and other business requirements. With the advance of computing and data acquisition technologies, an MPS can be recomputed on a more frequent basis to make the production schedule more agile in meeting customer needs. However, uncertainty in the demand forecast or production model may also increase the possibility and/or severity of "schedule nervousness". The mitigation techniques of frozen horizon, move suppression, and schedule change suppression are evaluated to determine the robust stability margins of each approach at their performance-optimal tunings. Since an MPS is typically computed using Linear Programming these techniques are formulated in this manner, and therefore an empirical Nyquist stability analysis using Empirical Transfer Function Estimates (ETFE) is employed. The technique of move suppression is shown to provide better robust stability margins in the small-scale problem. Further evaluation is needed on scheduling problems of industrial size.

1 Introduction

Master production scheduling is becoming a critical decision system in a wide range of manufacturing industries, including high-tech [1, 2]. The main goal of an MPS is to integrate information from expected sales forecasts (demand), manufacturing

M. W. Braun (✉) · J. D. Schwartz
Intel Corporation, 5000 W. Chandler Boulevard,
Chandler, AZ 85226, USA
e-mail: martin.w.braun@intel.com

J. D. Schwartz
e-mail: jay.schwartz@intel.com

D. Armbruster and K. G. Kempf (eds.), *Decision Policies for Production Networks*,
DOI: 10.1007/978-0-85729-644-3_7,

statistics (capacity, throughput times, utilization targets), and other business rules to produce a build plan over a rolling horizon. Traditionally this technique assumes a linear mass balance model to approximate inventory behavior. As the frequency of MPS execution increases and the granularity of time periods in the horizon decreases (smaller sampling intervals), it is possible that un-modeled nonlinear or time-varying effects impact the quality of the supply model and resulting schedule. Furthermore, new updates for demand forecasts and/or raw material supply forecasts may happen more often due to new technology (e.g. inventory tracking with radio frequency identification) and business emphasis on information sharing in the supply chain. As a result, the phenomena of "schedule nervousness" may be observed. In this chapter, three different approaches to mitigate schedule nervousness are discussed and evaluated from a performance and robustness perspective using tools from control theory. For the industrial practitioner this analysis illustrates the potential impact a particular technique might have on the manufacturing system and allows the user to make an informed choice given the level of modeling or forecast uncertainty.

Traditionally, in the MPS literature, a frozen horizon approach is used to lock down the schedule over some period of time out into the future. At a minimum, the length of the frozen horizon typically covers the cumulative throughput time of the production system so that Work-In-Progress (WIP) is not impacted by schedule changes. Beyond the frozen horizon, changes in the schedule are limited only by business rules and mass balance constraints. A number of references have evaluated the frozen horizon technique and the length of the frozen horizon in concert with other variables to determine the effect on the schedule nervousness problem [3–5].

The use of move suppression is a common technique to robustify the closed-loop response of an algorithm to uncertainty in the model or forecasted information. Recently, stability and performance analysis of Model Predictive Control algorithms based on a 1-norm linear program objective function has received attention. A particularly interesting aspect of Linear Programming Model Predictive Control (LPMPC) is that if it is formulated as a sum of 1-norm penalties on the move velocity and error to target, the resulting control law may exhibit idle behavior or deadbeat behavior, depending on the location in the state-space of the closed-loop system. By recasting the problem in terms of an ∞-norm criteria in both the temporal and spatial dimensions (∞/∞) these behaviors are eliminated and the desired robust behavior is achieved, yet without the need for a quadratic program (QP) solver [6, 7]. By incorporating this method into the MPS formulation, it is possible to explore move suppression as traditionally applied for real time process control in the domain of production planning.

Schedule nervousness may be defined as excessive change in the production schedule for a given period and item from one planning interval to the next. Schedule nervousness has a number of deleterious effects, including excess setup costs, unnecessary staffing changes, and excessive order changes for materials suppliers. In the MPS literature, a number of measures of schedule nervousness have been proposed [8]. For the purposes of this paper, schedule stability will be measured as follows

$$S = \sum_{i}^{N} |u(k + i|k) - u(k + i|k - 1)|, \quad i = 2, \ldots, N, \tag{1}$$

where S is the schedule nervousness metric of interest, $u(k)$ is a manipulated variable such as factory starts at control interval k, $u(k + i|k)$ is the plan for factory starts during control interval $k + i$ in the current MPS at control interval k, $u(k + i|k - 1)$ was the plan for factory starts at control interval $k + i$ as was resolved in the previous MPS in control interval $k - 1$. Therefore, $u(k + i|k) - u(k + i|k - 1)$ is the change in scheduled factory starts for interval $k + i$ between the current MPS and the previous MPS. N is the total number of time intervals in the time series.

The above metric also serves as motivation for a third method to explore, a move suppression term designed to penalize the excess change in schedule for a particular point in time, from solve epoch to solve epoch. The term *schedule change suppression* is coined to describe this approach.

The goal of this chapter is to address the handling of schedule nervousness in a Master Production Schedule through traditional means (the frozen horizon approach), a formulation borrowed from optimization-based control theory (the move suppression approach), and a novel formulation (the schedule change suppression approach). An evaluation of the stability and performance of these techniques is the focus of this work.

In Sect. 2, the three different methods of mitigating schedule nervousness are formally stated. In Sect. 3, the Empirical Transfer Function Estimate and how it can be used for estimating robust stability bounds in the Nyquist stability framework are discussed. Section 4 examines the effectiveness of this analysis in the context of a simple single inventory problem. A case study is used to compare and contrast the proposed methods in Sect. 5. Conclusions are drawn in Sect. 6. Last, a few items of future work are discussed in Sect. 6.1.

2 Three Approaches for Mitigating Schedule Nervousness

To formulate the MPS problem as an LP, the parameters following the convention of [7, 9] are now described with the aid of Fig. 1. The planning horizon refers to the overall horizon for which the production schedule will be determined. The forecast horizon is the horizon for which demand or other forecasted information is available. The re-planning interval is the length of time changes in the incoming data to the MPS process are allowed to accumulate before another MPS is solved. The cumulative lead time is the sum of all serial lead times. The planning horizon is typically chosen to be significantly longer than the cumulative lead time.

From a systems perspective, the incoming data to the MPS process can be classified in several ways. For the purposes of this analysis, any data forecasted or measured that impacts the inventory levels, WIPs, or other dependent variables but are not directly manipulated can be classified as disturbances or disturbance forecasts, d.

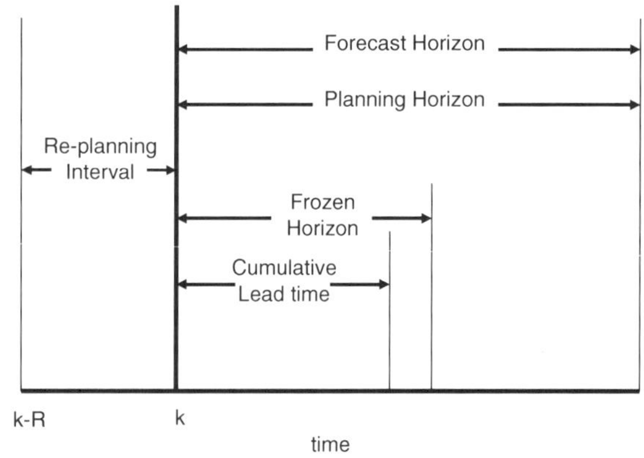

Fig. 1 Definitions for the MPS frozen horizon approach

Examples include product end demand (forecasts or actual consumption), materials or supply forecasts that the MPS does not control (forecasts or actual incoming material counts). Inventory targets, manufacturing utilization targets, and similar are treated as targets or setpoints T. Inventories, measures of WIP, in-transit shipments may be considered outputs or controlled variables of the physical system y. In the MPS problem, the distance of y from the corresponding target T is often minimized. This minimization, subject to a plethora of constraints is achieved by manipulating the input variables u to the physical system. These could be manufacturing starts, shipment quantities, and potentially other discrete decision variables such as the equipment type or even preventive maintenance cycles. The planning solve returns these decisions to the MPS systems, but this information may not be executed exactly and so the actual u used may be returned to the MPS at later intervals. The MPS system may hold additional or historical information that constitute an estimate of the state of the physical system x.

With the nomenclature of the production planning problem described, the presentation of the three methods of suppressing schedule nervousness can begin.

2.1 The Frozen Horizon Approach

One way to dampen schedule nervousness is to use a frozen horizon approach, in which the linear program (LP) is enhanced with constraints that force the manufacturing starts or inventory build targets to remain at the same values over a certain number of time periods known as the frozen horizon. In the case the LP is solving for a build plan, the frozen horizon is chosen at least as long as the manufacturing throughput time and potentially significantly longer. This eliminates the need for in-line product rate adjustments for current Work-In-Progress (WIP). Extending the frozen horizon beyond the throughput time may be necessary for further dampening of schedule instability brought about by demand uncertainty, materials supply uncertainty, or other concerns. Figure 1 depicts the moving horizon and definitions for the frozen horizon approach.

This is equivalent to the following formulation

$$\min_{\mathbf{u}} \sum_{n=1}^{p} |T_{k+n|k} - y_{k+n|k}| \tag{2}$$

$$s.t. \quad y_{k+1} = y_k + u_{k-TPT} - d_k \tag{3}$$

$$u_{k+i|k} - u_{k+i|k-1} = 0, \quad i = 1, \ldots, f \tag{4}$$

$$u_{k+i|k} \geq 0, \quad i = 1, \ldots, m \tag{5}$$

$$y_{k+n|k} \geq 0, \quad i = 1, \ldots, p \tag{6}$$

The production schedule **u** is a vector of manipulated variables that the optimization procedure adjusts to achieve the minimization objective. $T_{k+n|k}$ is the inventory target for control interval $k + n$, as understood by the current MPS in interval k. $y_{k+n|k}$ is the projected inventory level for interval $k + n$, as estimated during the current sampling interval k. The minimization problem seeks to minimize the sum of the absolute values of the inventory deviations over the planning horizon p. The minimization problem is subject to multiple constraints. Equation (3) represents the mass balance constraint where y_{k+1} is the projected inventory for the next control interval, y_k is the measured inventory in the current interval, u_{k-TPT} is the factory material that has finished processing and will be added to inventory in the current interval (TPT representing the cumulative lead time of the manufacturing system), and d_k is demand during the current interval. Equation (4) is the frozen horizon constraint, that no schedule changes are allowed within the frozen horizon. Equation (5) states that factory starts must be positive. Equation (6) enforces the constraint that inventory values must be positive. f is the frozen horizon parameter. m is the move horizon over which the solver will make factory starts decisions.

This MPS technique is typically employed in a receding horizon fashion. An advantage of the frozen horizon approach is that it is easy to understand and implement. The only real decision the user has to make is how long to make the frozen horizon window. One potential drawback is the impact freezing the horizon has on the ability of the decision variables to respond to near-term customer demand changes. Another potential drawback is that while the decision variables remain frozen, the magnitude of the change from one time period to the next is still unlimited by the problem formulation and a slew-rate constraint (a.k.a maximum move constraint or linearity constraint) may be needed. The variation in demand forecasts could still thrash the decision variables, yet the timing of the changes in the decision variables is stale when the solution is actually used by manufacturing.

2.2 The Move Suppression Approach

A classic approach to stabilize control laws is to employ a penalty on the change in value from one interval u_{k-1} to the next u_k in a receding horizon. As mentioned in the introduction, this can be employed using LP or Quadratic Programming (QP) problem formulations and has been successful in Model Predictive Control formulations in many industries. The user can adjust the speed of response of the closed-loop system by adjusting the weighting for this penalty. A larger weight slows the response of the closed-loop system and provides additional robustness to plant-model mismatch. Reducing the value of the weight enables the closed-loop system to respond in a more agile fashion, with the added risk of under damped or unstable behavior. Using the ∞/∞ approach, this formulation can be written as an LP

$$\min_{\mathbf{u}} \quad \theta \tag{7}$$

$$s.t. \quad y_{k+1} = y_k + u_{k-TPT} - d_k \tag{8}$$

$$|T_{k+n|k} - y_{k+n|k}| \leq \theta \tag{9}$$

$$|Q_{\Delta u} \Delta u_{k+i|k}| \leq \theta \tag{10}$$

where

$$\Delta u_{k+i|k} = u_{k+i|k} - u_{k+i-1|k} \tag{11}$$

The optimization seeks to minimize the parameter θ, which itself is the maximum of the absolute inventory deviation and absolute weighted manipulated variable change over the planning interval. Equation (8) represents the mass balance constraint where y_{k+1} is the projected inventory for the next control interval, y_k is the measured inventory in the current interval, u_{k-TPT} is the factory material that has finished processing and will be added to inventory in the current interval (TPT representing the cumulative lead time of the manufacturing system), and d_k is demand during the current interval. $T_{k+n|k}$ is the inventory target for control interval $k+n$, as understood by the current MPS in interval k. $y_{k+n|k}$ is the projected inventory level for interval $k+n$, as estimated during the current sampling interval k. Equation (9) enforces the constraint that the maximum absolute inventory deviation over the planning horizon is less than θ. Equation (10) enforces the constraint that the maximum absolute weighted starts change is less than θ. $Q_{\Delta u}$ is the weight used to penalize factory starts changes. Applying a value of 0 would lead to aggressive inventory control, but high levels of factory thrash. Increasing the penalty reduces thrash at the expense of greater inventory deviations from targets. Equation (11) explicitly defines the concept of a factory starts change (or thrash), the difference between u_{k+i} and u_{k+i-1} in the current MPS during interval k.

One advantage of this approach is that it directly reduces the "thrash" or dramatic changes in build plan that manufacturing personnel prefer to avoid. The effects on the closed-loop system of applying move suppression can be readily understood through simple simulation examples or case studies. For demand scenarios where ABC Analysis or similar are being used, this technique must be carefully applied to be sure that the schedule remains agile in serving A grade customers/products, while potentially smoothing the response to B/C grade customers/products.

A significant concern with using an ∞-norm instead of a 2-norm is that the ∞-norm may be particularly sensitive to outliers in forecast data. The optimization may choose to modify the decision variables to reduce as much as possible the effect of a few spurious data, whereas using a 2-norm would still enable the optimization routine to reduce the effects of other demand data or disturbances in the system in addition to the effort on the spurious data. In practice, one may use a number of outlier detection methods to reduce the impact of outlier data.

2.3 The Schedule Change Suppression Approach

This approach is a slight twist on the move suppression approach in that it penalizes changes in the schedule for a given period from one plan to the next. Conceptually, it is this quantity that manufacturing operations would like to minimize the most since it most closely matches the definition of schedule nervousness.

$$\min_{u_k} \quad \theta \tag{12}$$

$$s.t. \quad y_{k+1} = y_k + u_{k-TPT} - d_k \tag{13}$$

$$|T_{k+n|k} - y_{k+n|k}| \leq \theta \tag{14}$$

$$|Q_{\delta u} \delta u_{k+i|k}| \leq \theta \tag{15}$$

where

$$\delta u_{k+i|k} = u_{k+i|k} - u_{k+i|k-1} \tag{16}$$

While intuitively this method is appealing due to its similarity with the schedule nervousness metric, it may not achieve the schedule stability in the strict sense that a hard constraint (frozen horizon) might. In this way, it may not fully meet the needs of manufacturing for a stiff schedule, yet it may not provide the robustness of the move suppression approach. Similar to the move suppression approach, this formulation seeks to minimize the parameter θ. However, θ in this formulation is defined as the maximum of the absolute inventory deviation and absolute weighted schedule change. Schedule change is defined in Eq. (16), essentially the change in MPS between the previous production schedule and the one we are computing. $Q_{\delta u}$ is the weight for penalizing schedule changes. Setting the value to 0 would minimize inventory deviations at the expense of high levels of schedule nervousness. Increasing the penalty would reduce schedule nervousness at the expense of greater inventory deviations.

3 Empirical Robust Stability Analysis

From a control theoretic standpoint, there is always the fundamental tradeoff between robustness and performance. By moving more aggressively, a control policy is likely to provide better performance with the assumption it was built with an accurate model of the system it will control. However, there is always some level of uncertainty of how well the model describes the actual system. Additionally, external disturbances act on the system, forecasted information may be inaccurate, and lastly noise or delayed updates may impact the quality of the measurements of signals the policy is trying to control (in our case, inventories). An interesting aspect of working with inventory systems is that they also require a minimum level of agility in the planning system in order to stabilize what is an open-loop unstable system. Therefore, the MPS must be aggressive enough to respond to supply/demand rate imbalances, yet not too aggressive as to become closed-loop unstable due to model or supply/demand forecast uncertainty.

To help understand this tradeoff for our three different planning policies, it is proposed to empirically examine the closed-loop robust stability margins. In practice, one cannot perturb the planning policies actually being used in production to assess the impact on the stability of the inventory levels and customer satisfaction. Instead simulation studies can be employed to determine the robustness properties of the planning policies. While it is possible to increase the plant-model mismatch until the system is observed to go unstable, this is potentially quite time consuming for simulations of industrially relevant size. In this section, a method is proposed to empirically determine the robust stability margins for the closed-loop simulations under study.

3.1 The Empirical Transfer Function Estimate

For non-parametric identification of a dynamical model, consider the Empirical Transfer Function Estimate (ETFE) [10]. For a given input to a system $u(k)$, resulting in output $y(k)$, over time interval k from $1 \ldots N$, the ETFE can be written

$$\hat{H}(e^{i\omega}) = \frac{Y_N(\omega)}{U_N(\omega)} \tag{17}$$

where $Y_N(\omega)$ and $U_N(\omega)$ are simply the Discrete Fourier Transform of $y(k)$, and $u(k)$, for $k = 1 \ldots N$, respectively. While the ETFE provides an unbiased estimate of the frequency response of the system, the variance in the estimate for a given frequency asymptotically approaches the noise-to-input signal ratio as N increases.

In order to gain more resolution of the low frequencies of the ETFE, and provide some smoothing of the estimate albeit with the potential risk of biasing the estimate, it is also possible to apply the Blackman-Tukey method [11]. This procedure involves computing the cross spectrum of the output and input $\hat{\Phi}_{YU}(\omega)$, and the power spectrum of the input $\hat{\Phi}_U(\omega)$, and taking the ratio thereof

$$\hat{H}(e^{i\omega}) = \frac{\hat{\Phi}_{YU}(\omega)}{\hat{\Phi}_U(\omega)} \tag{18}$$

The power spectra can be computed with additional emphasis on particular frequencies, and windowed in a way to reduce error variance in tradeoff with potential bias error. Where necessary in this chapter, the Blackman-Tukey method is employed via the *spafdr* command in MATLAB®. Additional methods exist for smoothing the estimate. For the purposes of this analysis, an unbiased though noisy estimate is preferred to graphically assess the closed-loop stability bounds of the system. As a result, the additional process of order selection/model reduction is avoided. This is the main attractive point in using a non-parametric method for transfer function identification.

3.2 The Nyquist Stability Criterion

A number of methods exist for determining the robustness margins for discrete closed-loop system models (e.g. Lyupanov analysis, pole-zero analysis in the z-domain, etc.). Nyquist stability analysis is chosen since it does not require explicit computation of the poles of the closed-loop transfer function. Consider the closed-loop system shown in Fig. 2. In order to analyze the stability of the system the "broken" closed-loop transfer function $L(z^{-1})$ is computed,

$$L(z^{-1}) = H_c(z^{-1})H_p(z^{-1}). \tag{19}$$

The Nyquist diagram is plotted in the z-domain to examine the relationship with the critical point $(-1, 0)$. Inventory systems generally contain integrators in their transfer functions which results in open-loop system poles on but not outside of the unit circle in the z-domain. In the analysis in this work, the technique of differencing the input and output data is employed to provide a stationary data set of the ETFE analysis and provide a more clear identification of the stability boundaries. This also has the effect of removing the pure integrators in the system, leaving the unmodeled, yet stationary dynamics. Therefore, the special Nyquist stability criterion for open-loop stable systems may be used.

Theorem 1 *Special Nyquist Stability Criterion: The feedback system is asymptotically stable if the Nyquist curve does not encircle the critical point* $(-1, 0)$.

There are a number of metrics that can be used to measure the robust stability bounds in a Nyquist plot. The Gain Margin (GM) is the multiplicative factor by which $L(z^{-1})$ can be amplified in the Nyquist plot before it exceeds the critical point. The Phase Margin is the reduction in phase of $L(z^{-1})$ before $L(z^{-1})$ exceeds the critical point. For a simple production/inventory system with a cumulative lead

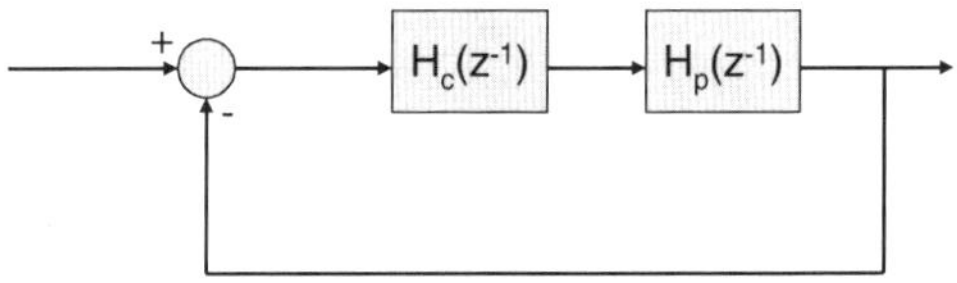

Fig. 2 Block diagram for development of the Nyquist stability criteria

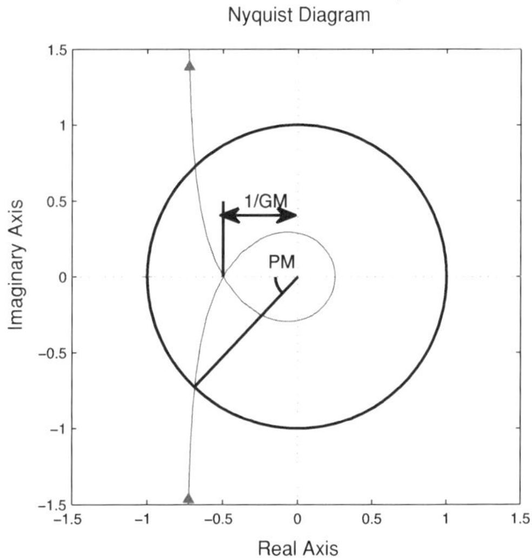

Fig. 3 Gain Margin (GM) and Phase Margin (PM) on a Nyquist diagram

time of 2 time units ($H_p = z^{-2}/(1 - z^{-1})$) and with a proportional controller ($K = 0.5$) making factory starts decisions, Fig. 3 illustrates the gain and phase margin. Additional discussion on Nyquist stability analysis can be found in [13]. The Nyquist plots in this section contain the Nyquist curve for both the positive and negative frequency range, hence the symmetry about the x-axis.

To provide a single measure of robust stability, it is also possible to measure the shortest distance from the critical point to $L(e^{j\omega})$. In the case of systems with constraints on the input variables (aka control signal saturation), there are also additional considerations for global asymptotic stability in the Nyquist plot. Specifically, as a sufficient condition for global asymptotic stability the curve $L(e^{j\omega})$ must not cross a boundary line of slope m in the Nyquist, whereas m may be arbitrarily defined although some authors have chosen values of m as discussed in [14]. When applying this criteria for systems with input signal constraints, only the Nyquist curve corresponding to the positive frequencies is plotted. As mentioned in the introduction, the LP nature of the objective function may in fact make the closed-loop system response nonlinear, and this analysis may not identify special cases of nonlinear instability. For the purposes of this analysis, the initial conditions are chosen such that the inventory levels are exceedingly high and the input signal saturations should be of minor impact to the performance of the systems under study, and the special Nyquist criteria above may be used.

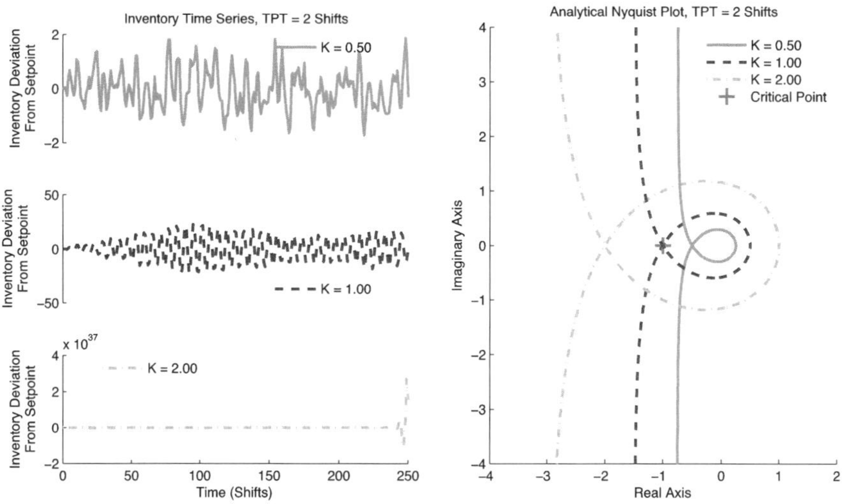

Fig. 4 Inventory time series results and analytical Nyquist plots for system ($H_p = z^{-2}/(1 - z^{-1})$) with a proportional controller ($K = 0.5$)

3.3 Combining the Empirical Transfer Function Estimate and the Nyquist Stability Criterion to Empirically Determine Robust Stability Measures

The ETFE can be analyzed with the Nyquist Stability Criterion in order to determine Gain Margin, Phase Margin, and the shortest distance to the critical point. This method of analysis is readily applicable to data collected from simulations of the combined planning policy, and inventory model.

Consider again, the system described in Fig. 3. When this system is controlled with a controller gain K of values 0.5, 1.0, and 2.0, the system is determined to be stable, marginally stable, and unstable, respectively. This is demonstrated with the simulation results shown in Fig. 4. The left plot shows the inventory levels over the simulation horizon; the right plot shows the analytically derived Nyquist plots for each of the three gain values. Notice that at a gain of 1.0, the system is on the border of instability, as predicted by the Nyquist plots in both Figs. 3 and 4. In Fig. 5, the Nyquist curve generated from the ETFE analysis proposed above demonstrates the effectiveness of the proposed approach. The Nyquist plots shown in Fig. 5, being generated from an empirical model, have more noise than their analytical counterparts but are still useful for analysis. This method for generating the Nyquist plots from ETFE data will be used throughout the remainder of the chapter.

In this particular example, the controller proportional gain was deliberately chosen to exceed the stability criteria and to demonstrate the potential for a closed-loop unstable system. In the next subsection, the three proposed approaches will be evaluated on the same example system, but with significant plant-model mismatch to

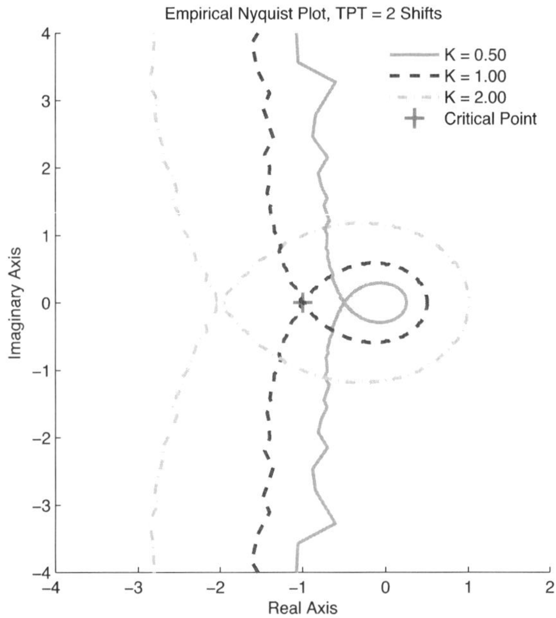

Fig. 5 ETFE Nyquist plot for system ($H_p = z^{-2}/(1 - z^{-1})$) with a proportional controller ($K = 0.5$)

assess the inherent robustness properties of the proposed approaches and verify the applicability of this stability analysis before it is applied to a more realistic system.

4 Analysis of the Three Approaches on a Single Input Single Output (SISO) System

Although real manufacturing systems exhibit nonlinear behavior, intuition can be formed by evaluating simple linear production-inventory systems. Consider again the production/inventory system with a cumulative lead of 2 time units $H_p = z^{-2}/(1 - z^{-1})$. In this subsection, the three schedule nervousness mitigation approaches will be examined for their abilities to attain performance-optimal metrics in spite of deterministic plant-model mismatch. Gain mismatch and delay mismatch will be introduced separately into a simulation. An exhaustive search was employed to determine tuning values which yielded the lower bound of stability, the performance-optimal tuning, and the upper bound of stability.

Because H_p is an integrating system, there is a minimum bandwidth requirement for the planning policy to meet in order for the system to be closed-loop stable. Due to the plant-model mismatch introduced, there will also be an upper bound on the bandwidth of the planning policy, beyond which the closed-loop system will exhibit instability. The planning policies subjected to noise in the inventory target signal drawn from a normal distribution with a mean of zero and a standard deviation of unity; the metrics for inventory tracking, factory thrash, and schedule thrash are computed across the entire simulation run. Note that varying the inventory target has effectively the same effect as varying the demand since these policies are not multi-degree-of-freedom policies (i.e. they will fundamentally respond to target changes and demand changes in the same manner).

The performance-optimal tunings are found by evaluating a range of tuning parameters for each of the policies and selecting the tuning that minimizes the ∞-norm of the target tracking error. All policies have a planning horizon of 10 intervals and a forecast horizon of 11 intervals. The frozen horizon is set to a value of 9 intervals in the frozen horizon policy. The schedule change suppression approach suppresses the changes over a horizon of 9 intervals as well.

The following subsections present simulation results with key control system performance metrics. $||T - y||_1$ is the 1-norm of the inventory deviation signal, or the sum of the absolute values of the inventory deviations over the course of the simulation. $||T - y||_2$ is the 2-norm of the inventory deviation signal, or the square root of the sum of the squared inventory deviation values. $||T - y||_\infty$ is the maximum absolute inventory deviation value that occurs during the simulation. Comparable metrics are also reported for factory thrash $||\Delta u||$ and schedule nervousness $||u_{k+1|k} - u_{k+1|k-1}||$. In all cases, lower values indicate improved performance. A value of zero indicates perfection: no inventory deviation, factory thrash, or schedule nervousness.

4.1 Delay Plant-Model Mismatch

In the first set of simulation results, the plant delay is set to four units (i.e. $H_p = z^{-4}/(1 - z^{-1})$), while the model in the planning policies is configured with a value of six. The simulation results for the frozen horizon approach show that while there are no schedule changes within the frozen horizon, the closed-loop system quickly becomes unstable as demonstrated in Fig. 6. Closer inspection of the results reveals that while the schedule remains frozen up to nine time periods into the horizon, the schedule change after the frozen horizon is essentially unconstrained and the policy will make what it thinks are optimal moves. Since there is no filtering action, the schedule that gets frozen is still aggressive, hence the unstable closed-loop response.

With move suppression, the planning policy is able to provide a stable closed-loop response, and achieve a somewhat sluggish tracking of the inventory targets. The performance-optimal tuning of $Q_{\Delta u} = 30$ produces the time series shown in Fig. 7. The plot shows how factory starts are adjusted over time to keep the inventory tracking its moving target. The move suppression effectively attenuates the effect of the error the algorithm observes between the expected response of the inventory and the actual response of the inventory. In this way the algorithm is acting robustly as expected from a control theoretic viewpoint.

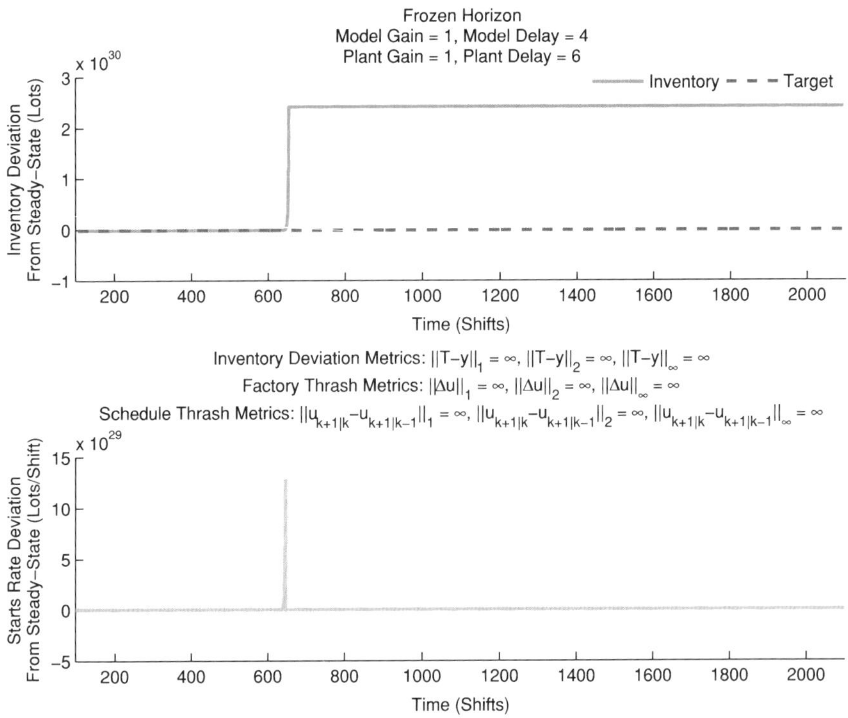

Fig. 6 Time series of the frozen horizon policy with delay plant-model mismatch of +50%

In Fig. 8, given the unstable result from the frozen horizon simulation, an unexpected result is found. The schedule change suppression approach is in fact able to produce a stable closed-loop response with a tuning $Q_{\delta u} = 3$. By enabling small but *penalized* changes, the schedule change suppression approach is able to provide some degree of filtering within the receding horizon, and mitigate some of the impact of the delay mismatch between the plant and the planning policy. The time series show how the schedule change suppression approach provides a more responsive inventory profile than the move suppression approach, however the inventory levels are still substantially out of phase with the inventory targets specified.

To compare the metrics of the move suppression approach vs. the schedule change suppression approach, consider Table 1. The metrics for the frozen horizon approach are not reported since the values are so large as to be of no comparative value. The inventory tracking metric is the maximum absolute inventory deviation value, the factory thrash metric is the maximum absolute change in factory starts from one interval to the next, and the schedule thrash metric is the maximum absolute change in value of scheduled starts for a particular time period from one solve epoch to the next. All metrics are computed in total over the simulation. The performance-optimal tuning (measured via the ∞-norm of the inventory deviation) and the range of stable tuning values were obtained via an exhaustive search. Tuning values were tested

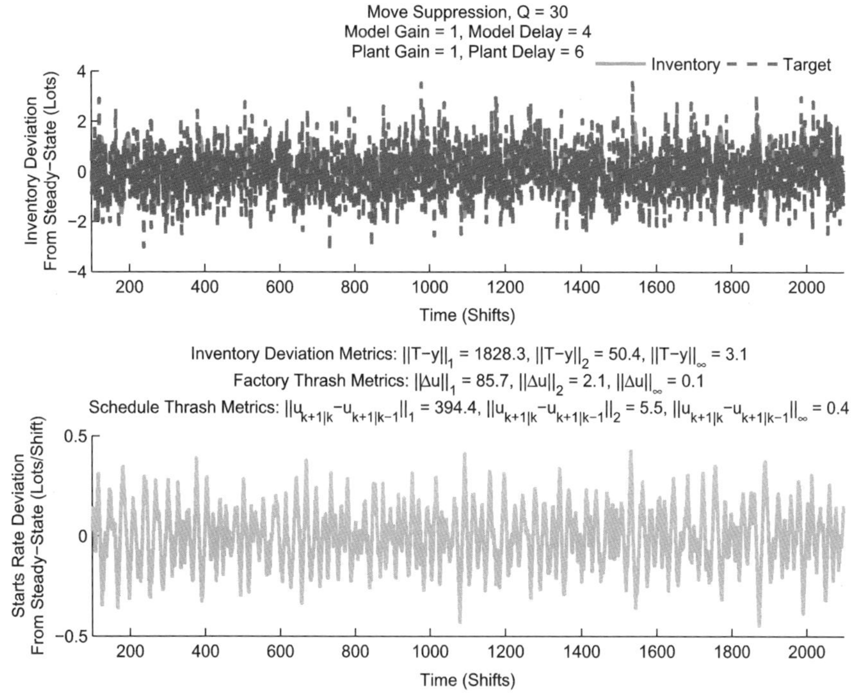

Fig. 7 Time series of the move suppression policy with delay plant-model mismatch of +50%

at increments of 0.1 for values less than 1, increments of 0.5 for values between 1 and 10, and increments of 5 for values above 10. The move suppression approach outperforms the schedule change suppression approach in every metric. It also has a wider range of tuning parameter values under which a stable result may be achieved.

To further examine the stability margin of the move suppression and schedule change suppression approaches, a Nyquist analysis was performed on simulation data of length 2^{17}. To further enhance the resolution in the low frequency portions of the curve, the Blackman-Tukey spectral analysis with frequency-dependent resolution was employed. The planning policies were run with the performance-optimal tunings as noted in Table 1. Figure 9 shows the result for the schedule change suppression approach. Note that in this case, the analysis suggests that the policy may in fact be closed-loop unstable. It is clear that the schedule change suppression approach encircles the critical point -1. Figure 10 shows the Nyquist analysis for the move suppression approach. As shown the move suppression approach has substantially more stability margin than the schedule change suppression approach. One can conclude from the data presented in Table 1 and the Nyquist analysis that not only does move suppression provides a more performance-optimal solution, but it also does so with a greater stability margin.

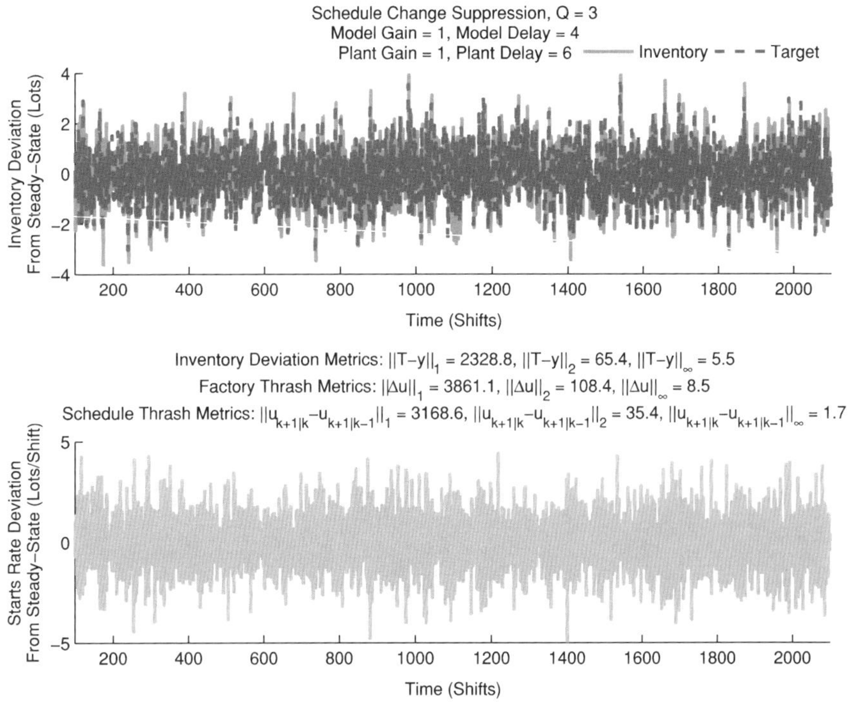

Fig. 8 Time series of the schedule change suppression policy with delay plant-model mismatch of +50%

Table 1 Delay plant-model mismatch: move suppression versus schedule change suppression metrics

Method	Stable?	Q	$\|y-r\|_\infty$	$\|\Delta u\|_\infty$	$\|\delta u\|_\infty$
Move suppression	Boundary	1.5	5.6	2.5	1.8
	Yes	30	3.1	0.1	0.4
	Boundary	765	4.4	0.1	0.8
Schedule change suppression	Boundary	1.5	7.1	12.4	5.7
	Yes	3	5.5	8.5	1.7
	Boundary	12	8.4	15.0	11.1

4.2 Gain Plant-Model Mismatch

Consider a plant-model mismatch in the gain parameter. For these results, a gain of one is used in the planning policies, and the plant model in the simulator will be configured with a value of two (i.e. $H_p = 2 \cdot z^{-2}/(1 - z^{-1})$). In this way, a gain plant-model mismatch of -50 % is introduced into the closed-loop response.

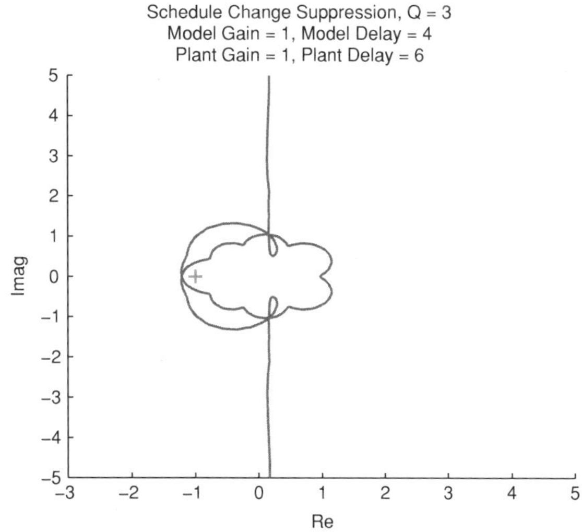

Fig. 9 Nyquist analysis for a schedule change suppression policy with delay plant-model mismatch of +50%

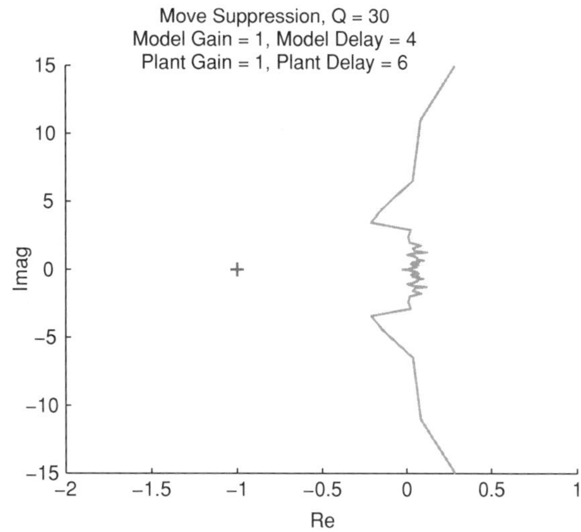

Fig. 10 Nyquist analysis for a move suppression policy with delay plant-model mismatch of +50%

In this scenario, the frozen horizon approach produces an unstable closed-loop system, however the extent of instability as shown in Fig. 11 is much more gradual compared to the result observed in the case of delay mismatch. The inventory level grows in an underdamped, oscillatory manner until it exceeds the inventory target variance by approximately a factor of 50. The oscillatory behavior is also observed in the factory starts signal as well.

The move suppression approach provides stability for this system, with the performance-optimal result shown in Fig. 12. As before, this result shows a sluggish response in inventory to changes in the inventory target, however the result

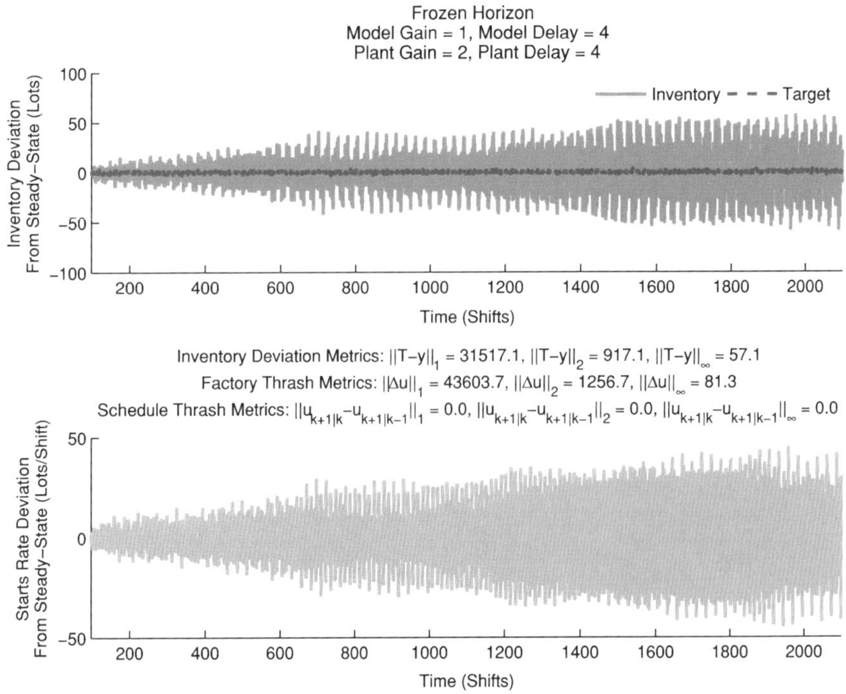

Fig. 11 Time series of the frozen horizon policy with gain plant-model mismatch of −50%

remains stable, and tracks the low frequency trend of the inventory target variation. As expected, the response in the factory starts is smooth.

The schedule change suppression approach also is capable of producing a stable result. Figure 13 illustrates how the schedule change suppression approach dramatically overshoots the inventory target due to the gain mismatch, and the values at times appear out of phase with the inventory target variation.

Table 2 provides a summary of the metrics for each approach. This time the metrics for the frozen horizon approach are included for comparison. The performance-optimal tuning (measured via the ∞-norm of the inventory deviation) and the range of stable tuning values were obtained via an exhaustive search. Tuning values were tested at increments of 0.1 for values less than 1, increments of 0.5 for values between 1 and 10, and increments of 5 for values above 10. Again the move suppression approach is able to achieve better metrics than the other two approaches. What is particularly surprising is the ability of the move suppression approach to maintain a lower schedule change metric given the fact that this is not explicitly part of the optimization criteria for this policy.

What makes this particular example challenging is that the configured gain in the planning policies is smaller than the actual gain. In general, the planning policy will think that it needs to make larger adjustments than is actually needed in order to bring

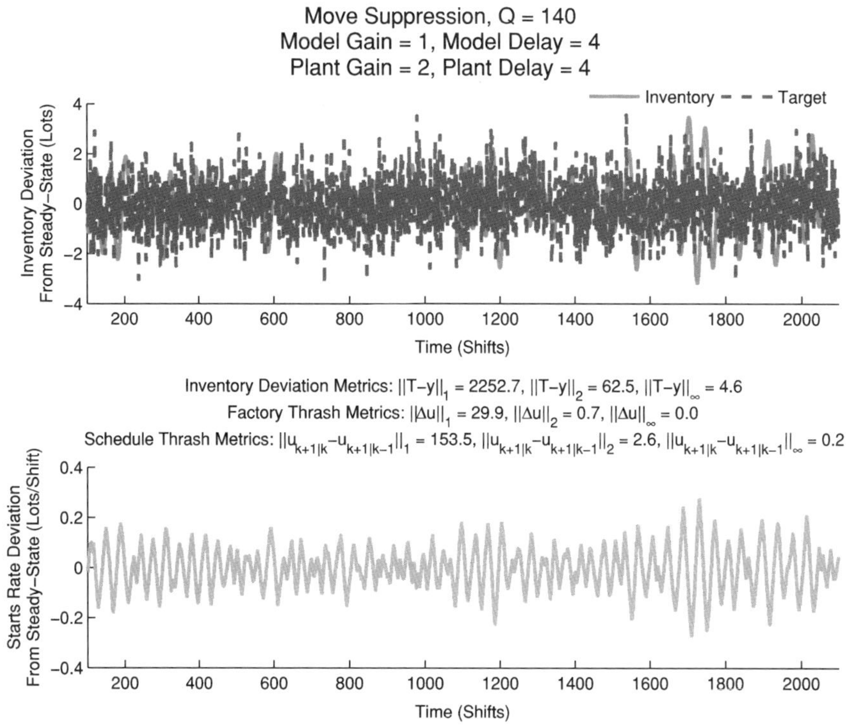

Fig. 12 Time series of the move suppression policy with gain plant-model mismatch of −50%

the inventory to target. The system responds with substantially more material. Then, the policy may over react again as is demonstrated in the frozen horizon results. By having some filtering action in place, instability is avoided as demonstrated by the other two approaches.

To provide some additional perspective on the robust stability properties of these three approaches, Nyquist stability analysis is again employed. For these results the ETFE analysis is used for the Frozen Horizon approach (Fig. 14), and the Blackman-Tukey with Frequency resolution is employed for the schedule change suppression and move suppression policies (Figs. 15 and 16). In this scenario, the instability of the frozen horizon approach is confirmed by the encirclement of the critical point at $(-1, 0)$. What is interesting is that the Nyquist analysis suggests that the stability margin, as measured by the distance from the critical point, is greater for the schedule suppression policy as opposed to the move suppression approach. This is inconsistent with the previous result and inconsistent with the fact that the move suppression has a much wider range of stable tunings as shown in Table 2. This warrants additional analysis not contained in this chapter.

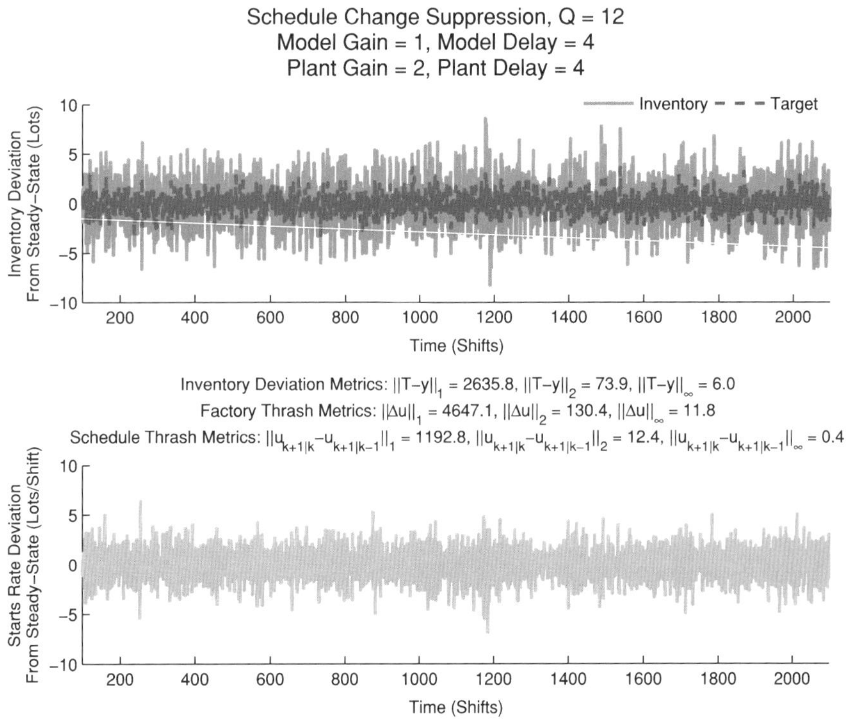

Fig. 13 Time series of the schedule suppression policy with gain plant-model mismatch of −50%

Table 2 Gain plant-model mismatch: frozen horizon versus move suppression versus schedule change suppression metrics

Method	Stable?	Q	$\|y - r\|_\infty$	$\|\Delta u\|_\infty$	$\|\delta u\|_\infty$
Frozen horizon	No	·	∞	∞	0
Move suppression	Boundary	25	6.4	0.2	0.5
	Yes	140	4.6	0.0	0.2
	Boundary	735	9.0	0.2	0.2
Schedule change suppression	Boundary	0.5	13.5	10.9	4.6
	Yes	12	6.0	11.8	0.4
	Boundary	90	15.6	24.1	0.2

5 Multiple-Input-Multiple-Output (MIMO) Case Study

5.1 System Topology

For the purposes of this study, the bill of materials (BOM), and system properties described in [12] are used. Consider the block diagram in Fig. 17. The discrete-time

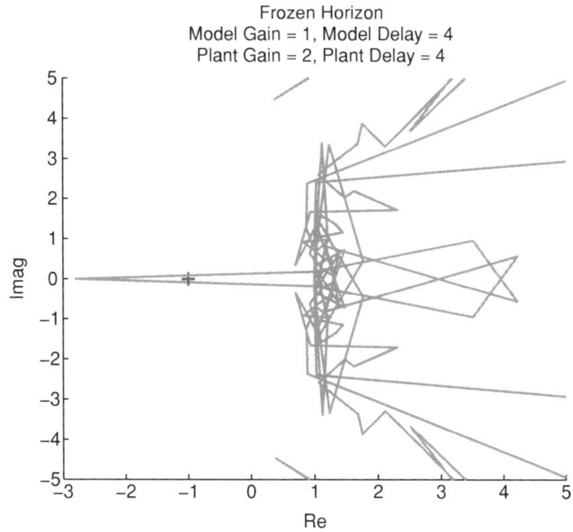

Fig. 14 Nyquist analysis for a frozen horizon policy with gain plant-model mismatch of −50%

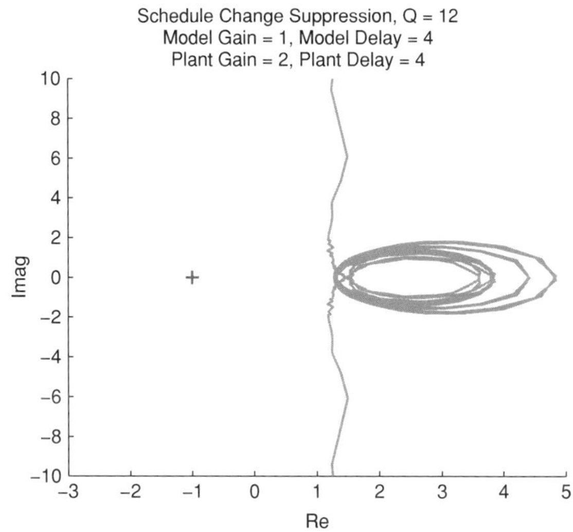

Fig. 15 Nyquist analysis for a schedule change suppression policy with gain plant-model mismatch of −50%

transfer functions can be written as

$$i_1(k) = \frac{z^{-3}}{1 - z^{-1}} f_1(k) - \frac{z^{-1}}{1 - z^{-1}} \min(\lfloor f_5(k) \rfloor, \lfloor 0.5 f_4(k) \rfloor) \tag{20}$$

$$i_2(k) = \frac{z^{-2}}{1 - z^{-1}} f_2(k) - \frac{z^{-1}}{1 - z^{-1}} \min(\lfloor f_5(k) \rfloor, \lfloor 0.5 f_4(k) \rfloor) \tag{21}$$

$$i_3(k) = \frac{z^{-2}}{1 - z^{-1}} f_3(k) - \frac{z^{-1}}{1 - z^{-1}} f_2(k) \tag{22}$$

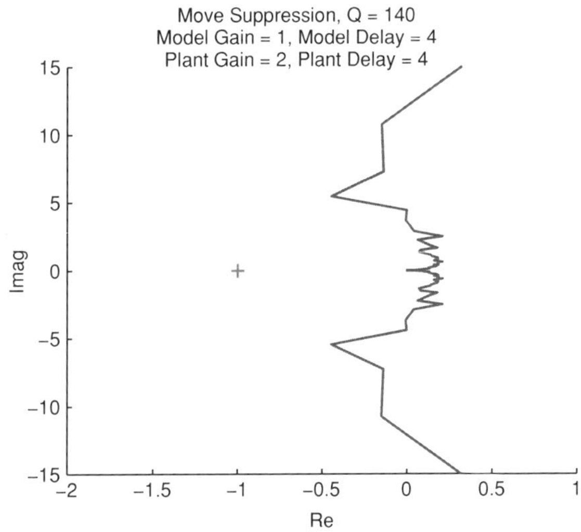

Fig. 16 Nyquist analysis for a move suppression policy with gain plant-model mismatch of −50%

$$i_4(k) = -\frac{z^{-1}}{1-z^{-1}} f_6(k) + \frac{z^{-2}}{1-z^{-1}} \min(\lfloor f_5(k) \rfloor, \lfloor 0.5 f_4(k) \rfloor) \tag{23}$$

Note that the stoichiometry is two f_4 for one f_5 to produce one i_4. *min* refers to the minimum of either argument; $\lfloor \cdot \rfloor$ refers to the *floor* function.

While production-inventory simulations can be useful for obtaining insight into the advantages and disadvantages of the proposed policies, it is desirable to evaluate their efficacy against realistic supply chain scenarios. In the following subsections the policies are simulated and subjected to varying levels of gain and delay plant-model mismatch.

5.2 Delay Plant-Model Mismatch

In the following simulations the supply-chain operates under steady-state conditions and no plant-model mismatch until shift 100. At this time the throughput time of Factory u_1 is increased by 1 shift, but the internal model is not updated to reflect this change. The result is one time unit of delay mismatch. The process is repeated at shifts 400 and 700, allowing one to see how the policy performs under increasing levels of delay mismatch. The demand for this system is randomly varied uniformly between zero and 100 units. The policies have visibility into the demand forecasts. The policies are tuned with the performance-optimal tunings from the prior section, to present a realistic scenario where stale settings are used.

The frozen horizon policy exhibits steady-state offsets from the inventory targets, particularly inventories y_1 and y_2 (Fig. 18). With the move suppression policy, the

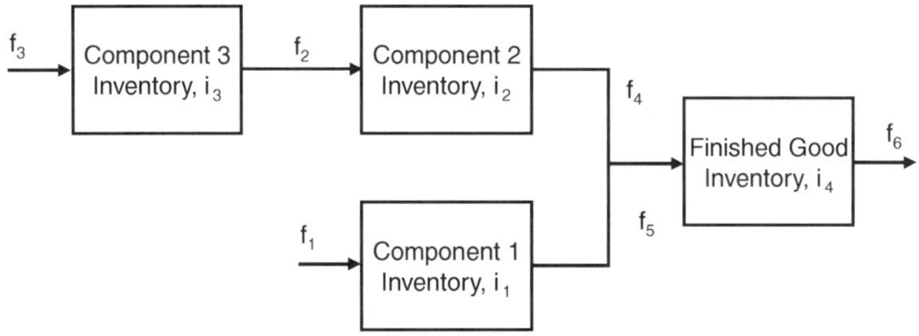

Fig. 17 Multi-item, multi-level BOM from Dolgui and Prodhon [12]

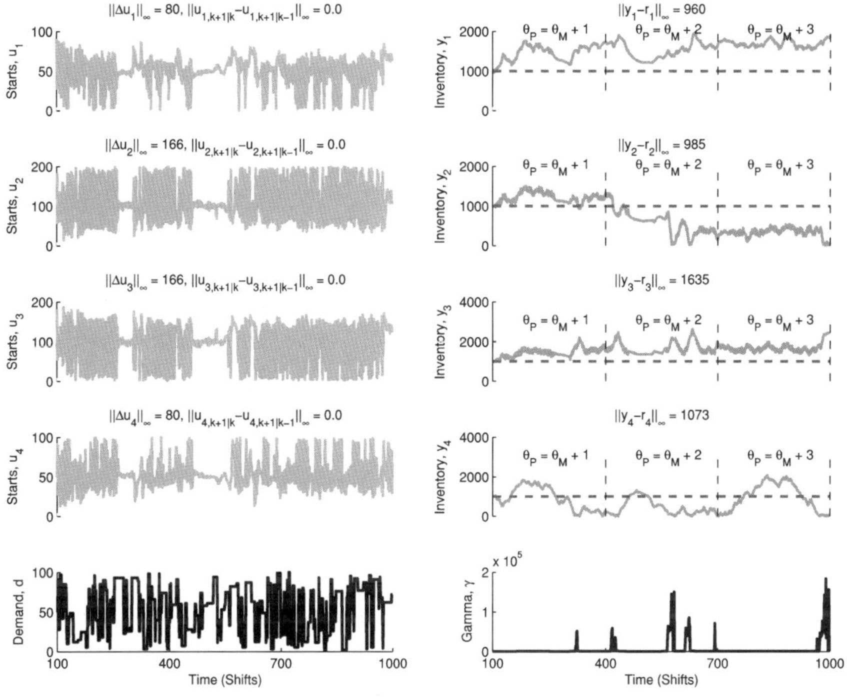

Fig. 18 Time series for a MIMO frozen horizon policy with varying levels of delay plant-model mismatch

inventories are on target average and the variance remains relatively constant throughout the simulation (Fig. 19). The move suppression provides less schedule change, and less move change as compared to the schedule change suppression approach, however the deviation from target is substantially more than the schedule change suppression approach (Fig. 20).

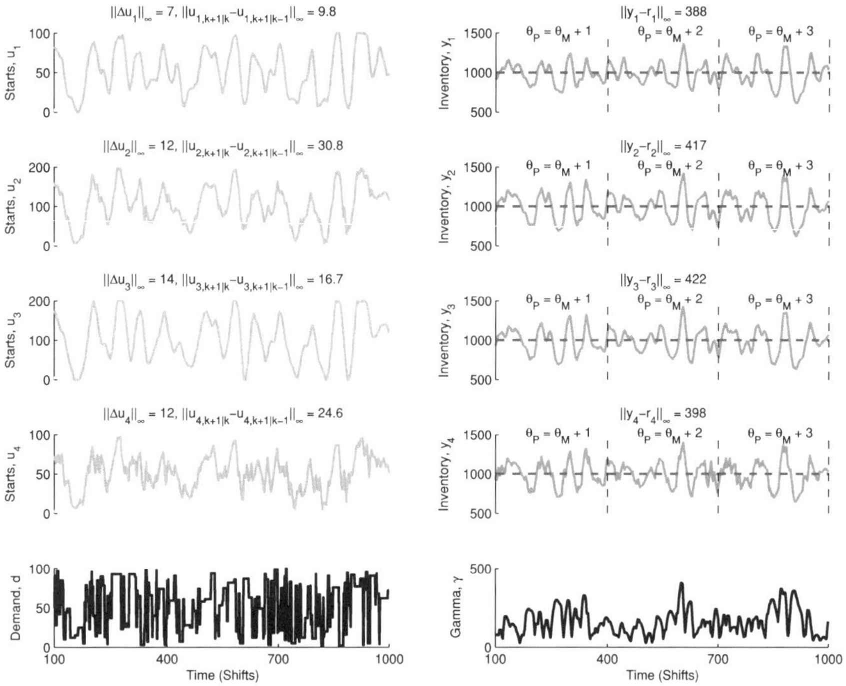

Fig. 19 Time series for a MIMO move suppression policy with varying levels of delay plant-model mismatch. $Q_{\Delta u} = 30$

5.3 Gain Plant-Model Mismatch

In this section, the multi-item BOM topology is simulated with increasing gain mismatch to illustrate an increasingly stale yield estimate. The simulations begin with the supply-chain operating under steady-state conditions and no plant-model mismatch until shift 100. At this time the gain of Factory u_1 is increased by 0.5, but the internal model is not updated to reflect this change. The result is a gain mismatch of -33% . The process is repeated at shifts 400 and 700, allowing one to see how the policy performs under increasing levels of gain mismatch.

In Fig. 21, the frozen horizon policy remains stable throughout the entire simulation, however the inventories are not kept to target. Inventories y_1 and y_3 increase substantially through interval 700 and level out. Inventory y_2 drops substantially to almost zero. Figures 22 and 23 show the results for the move suppression and schedule change suppression approaches, respectively. As in the prior subsection the schedule change suppression provides better performance in keeping the inventory to target since its tuning is less conservative. y_1 in this case does exhibit a steady-state offset in both the move suppression and schedule change suppression results, particularly toward the end of the simulation. The move suppression approach provides substantially reduced schedule change and move change metrics.

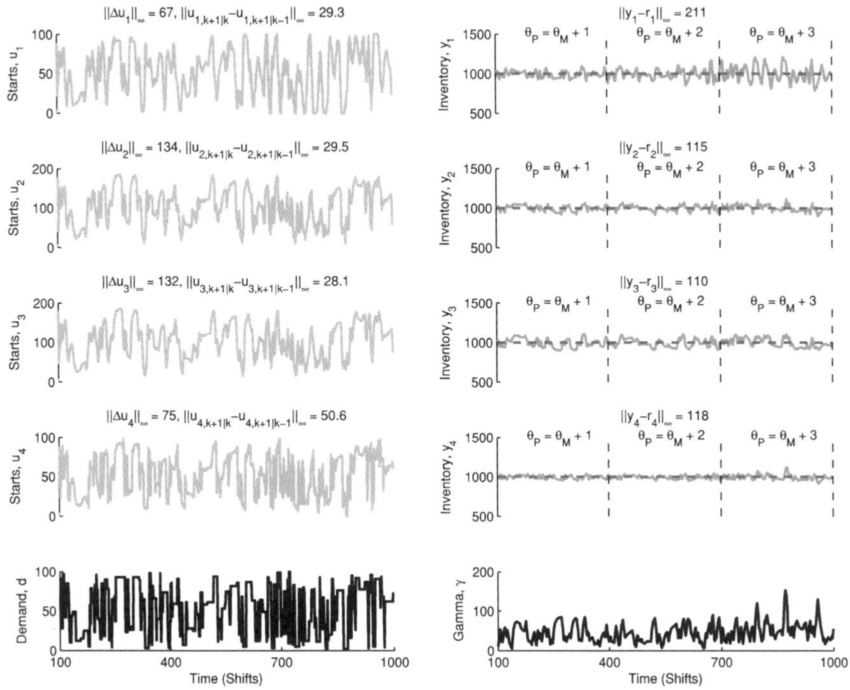

Fig. 20 Time series for a MIMO schedule change suppression policy with varying levels of delay plant-model mismatch. $Q_{\delta u} = 3$

This example is interesting because it demonstrates the performances of the proposed approaches in a more complex BOM scenario, along with stale tunings, and increasing plant-model mismatch. This is not unlike what one might experience in an industrial setting. Certainly the move suppression approach could be retuned to be more aggressive, but given the shallow region of convergence for the schedule change suppression approach it is unlikely the schedule nervousness could be mitigated with a more conservative tuning.

6 Conclusions

Industrial practitioners, especially supply chain managers and planners, who are seeking to reduce the costs of frequent and/or large schedule changes must consider the advantages and disadvantages of the three decision techniques presented in this chapter: frozen horizon, move suppression, and schedule change suppression. The frozen horizon technique does have utility as a technique to explicitly freeze the schedule and therefore eliminate changes to the WIP which might increase tool changeover or personnel costs. However, it is not able to provide filtering against the

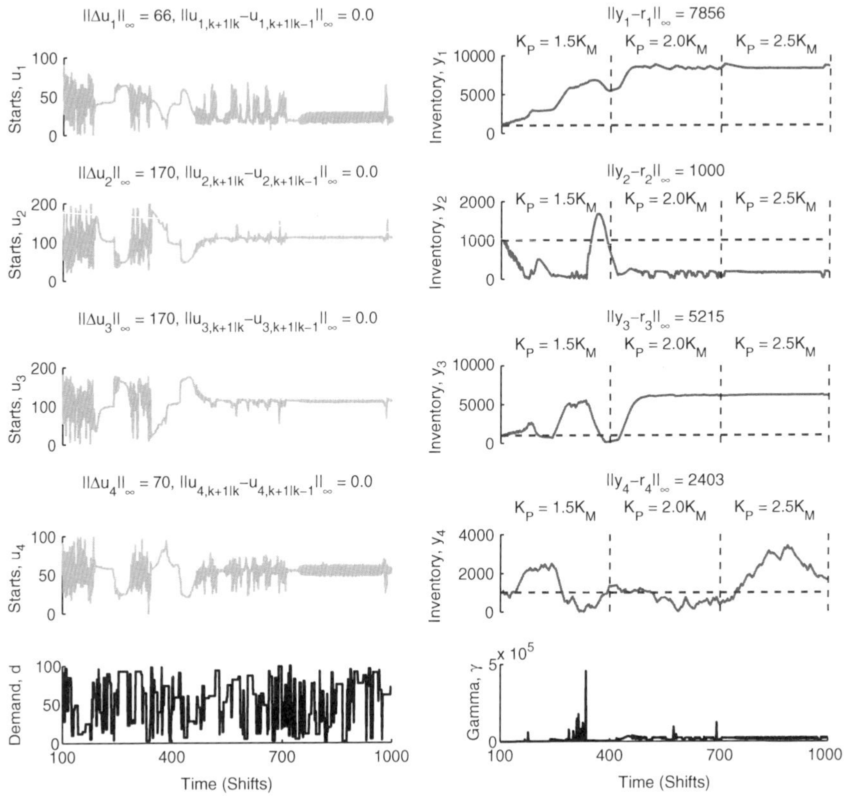

Fig. 21 Time series for a MIMO frozen horizon policy with varying levels of gain plant-model mismatch

effects of model or forecast inaccuracies. Left unchecked the uncertainty results in unnecessary swings in product starts, and in the worst case the result is an unstable inventory system. Move suppression as traditionally employed in process control systems is able to filter the effects of uncertainty and maintain system stability to a much higher degree. There is a cost in system agility. However this is the classic, fundamental tradeoff between agility and robustness of a system operating under uncertainty, that is only truly addressed through better models, forecasts, and reduced system throughput times. Lastly, the approach of schedule change suppression was examined. This approach provides a very modest ability to filter against the effects of uncertainty, and certainly does not fully freeze the schedule to eliminate changeover or personnel costs. Only in the last discussion of the results section was it able to outperform the move suppression case in inventory targeting metrics, but that was due to the particulars of this stale tuning example. In reality, with auto-tuning systems it would be possible for the move suppression policy to re-tune the policy, and provide better performance with a better stability margin.

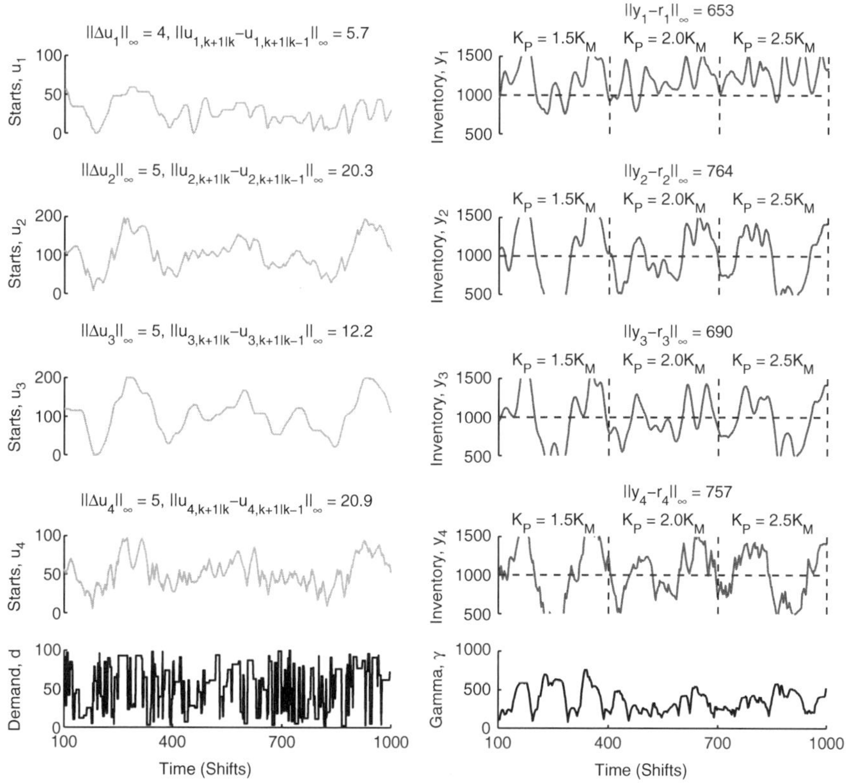

Fig. 22 Time series for a MIMO move suppression policy with varying levels of gain plant-model mismatch $Q_{\Delta u} = 140$

This chapter also presented a novel approach for evaluating the stability of tactical decision policies for supply chain management. Non-parametric empirical transfer functions were computed from simulated data and Nyquist stability analysis was performed to confirm the stability observations made from the time series data collected in the simulations. While the analysis was qualitatively helpful at best, it is fundamentally challenging to compute precise stability bounds with this analysis for these closed-loop systems under LP-based decision policies. Practitioners seeking to perform stability analysis might consider this as a useful addition to a suite of tools.

6.1 Future Work

In addition to gain and delay mismatch, manufacturing systems experience a wide variety of sources of uncertainty. Additional sources to examine include data

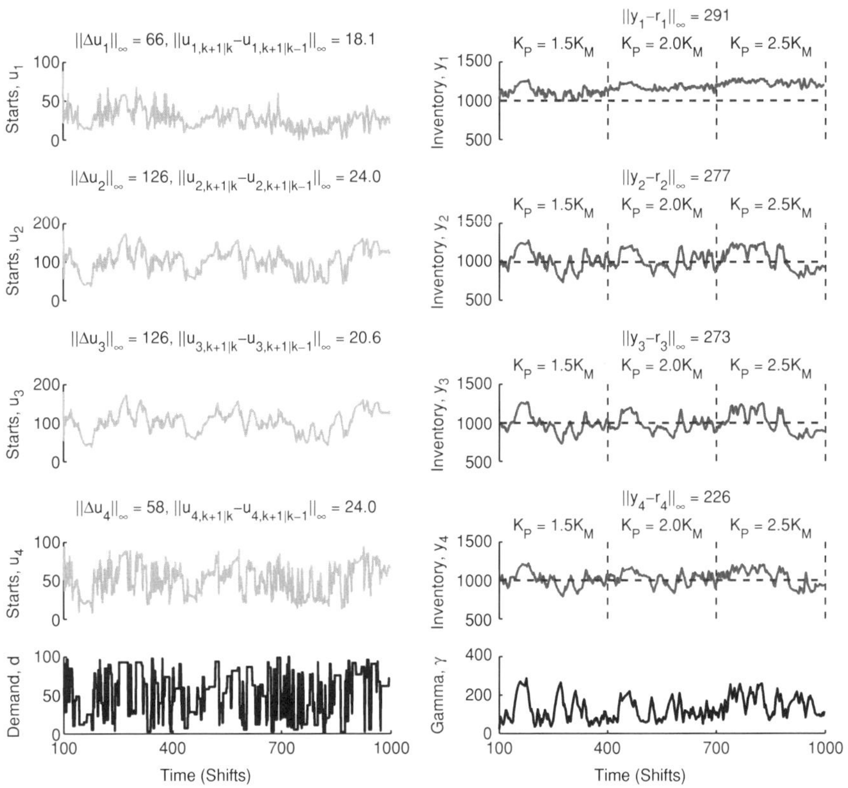

Fig. 23 Time series for a MIMO schedule change suppression policy with varying levels of gain plant-model mismatch $Q_{\delta u} = 12$

system integration mismatch (i.e. the data systems providing data for one part of the system topology may be out-of-sync with the other), demand or supply forecast error, and lastly impartial data. Further evaluation of the proposed methods would enable the practitioner to make a more informed choice of what methods to use for their particular scheduling problem.

Upon reflection, two additional methods readily become apparent for evaluation and may be examined in the future. One being the use of hard constraints on the changes in schedule from one time epoch to the next (a.k.a. "linearity constraints"). The other being a hybrid of the frozen horizon approach and the move suppression approach. This idea involves using the frozen horizon approach, but augmenting the logic to include move suppression in the remaining portion of the receding horizon. In this way it may be possible to meet the need for manufacturing to work from a fixed schedule in the WIP horizon, yet still provide stabilizing schedule changes.

References

1. Stadtler H (2005) Supply chain management and advanced planning—basics overview and challenges. Eur J Oper Res 163:575–588
2. Law KMY, Gunasekaran A (2009) A comparative study of schedule nervousness among high-tech manufacturers across the Straits. Int J Prod Res 20:31–39
3. Sridharan V, Berry WL, Udayabhanu V (1987) Freezing the master production schedule under rolling planning horizons. Manag Sci 33(9):1137–1149
4. Zhao X, Lee TS (1993) Freezing the master production schedule for materials requirements planning systems under demand uncertainty. J Oper Manag 11:185–205
5. Tang O, Grubbström RW (2002) Planning and replanning the master production schedule under demand uncertainty. Int J Prod Econ 78:323–334
6. Saffer DR, Doyle FJ (2004) Analysis of linear programming in model predictive control. Comput Chem Eng 28:2749–2763
7. Campo PJ, Morari M (1986) ∞-Norm formulation of model predictive control problems. American control conference, June 1986, pp 339–343
8. Pujawan IN (2004) Schedule nervousness in a manufacturing system: A case study. Prod Planning In: Control 15(5):515–524
9. Sahin F, Robinson EP, Gao L (2008) Master production scheduling policy and rolling schedules in a two-stage make-to-order supply chain. Int J Prod Econ 115:528–541
10. Ljung L (1999) System identification: theory for the user, 2nd edn. Prentice-Hall, Upper Saddle River
11. Blackman RW, Tukey JW (1958) The measurement of power spectra. Dover Publications, New York
12. Dolgui A, Prodhon C (2007) Supply planning under uncertainties in MRP environments: A state of the art. Ann Rev Control 31:269–279
13. Haugen F (2005) Discrete-time signals and systems. Tutorial, http://techteach.no/adm/fh/
14. Plummer AR, Ling CS (2000) Stability and robustness of discrete-time systems with control signal saturation. Proc Inst Mech Eng part I: J Syst Control Eng 214(1):65–76

Chance-Constraint-Based Heuristics for Production Planning in the Face of Stochastic Demand and Workload-Dependent Lead Times

Tarik Aouam and Reha Uzsoy

Abstract While the problem of planning production in the face of uncertain demand has been studied in various forms for decades, there is still no completely satisfactory solution approach. In this chapter we propose several heuristics based on chance-constrained models for a simple single stage single product system with workload-dependent lead times, which we compare to two-stage and multi-stage stochastic programing formulations. Exploratory computational experiments show promising performance for the heuristics, and raise a number of interesting issues that arise in comparing solutions obtained by the different approaches.

1 Introduction

In today's global supply chains, effective coordination of operations across space and time is vital to capital-intensive industries like semiconductor manufacturing with short product life cycles and rapidly changing market conditions. However, despite the fact that problems related to the planning of production and inventories have been the stock in trade of industrial engineering and operations research for the last five decades, a comprehensive solution to the problem as faced in industry is still unavailable [65]. Current research has followed the basic paradigms of deterministic mathematical programing and stochastic inventory models, resulting in highly compartmentalized streams of research that each focus on certain aspects of the

T. Aouam
School of Business Administration, Al Akhawayn University,
P.O. Box 104, 53000 Ifrane, Morocco
e-mail: t.aouam@aui.ma

R. Uzsoy (✉)
Edward P. Fitts Department of Industrial and Systems Engineering,
North Carolina State University, Campus Box 7906,
Raleigh, NC 27695-7906, USA
e-mail: ruzsoy@ncsu.edu

D. Armbruster and K. G. Kempf (eds.), *Decision Policies for Production Networks*,
DOI: 10.1007/978-0-85729-644-3_8, © Springer-Verlag London 2012

problem at the expense of others. In particular, the problem of planning production releases and allocating production capacity among different products has ignored the nonlinear congestion effects induced by capacitated resources subject to queueing, and has been treated in isolation from the problem of maintaining service levels in the face of stochastic demand.

This work is motivated by two basic limitations of the mathematical programing models used for production planning in both industrial practice and academia. The first of these is that the vast majority of these models fail to capture the nonlinear relationships between work in process inventories (WIP), cycle times, and work releases. Queueing models of production systems [17, 52] show that cycle times increase nonlinearly with resource utilization, which in turn is determined by the release plan produced by the planning system. Capital-intensive industries such as semiconductor manufacturing, with long, complex production processes, must run at high utilization to be profitable. Under these conditions small fluctuations in utilization may cause large changes in cycle times, rendering the effects of this dependence important to effective planning.

In addition to this nonlinear dependence, uncertain demand is a fact of life in most supply chains, requiring the deployment of safety stocks to ensure desired customer service levels. The production of these safety stocks, in turn, requires the release of additional work, affecting cycle times, and hence the work release and capacity allocation decisions made by planning models. It is thus notable that the planning of safety stocks [43, 107] has largely been addressed separately from capacity allocation, presumably due to the motivation for much inventory research arising from retail and distribution applications.

The large size and stochastic nature of industrial supply chain planning problems renders their exact solution computationally prohibitive. Thus industrial practice requires efficient approximations with reliable solution quality. However, when approximations are proposed, assessing the quality of their solutions is fraught with all the difficulties encountered in evaluating the quality of heuristic solutions for deterministic optimization problems [92]. There is thus a need to develop exact solution methods to provide insight into the structure of optimal solutions, as well as benchmarks against which different approximation methods can be compared and assessed.

This, in turn, presents additional complications. Problems of production planning and control in the face of stochastic demand admit several different formulations that often have quite different assumptions, advantages and drawbacks. Inventory and queueing models [107], for instance, tend to produce optimal solutions under steady-state conditions, but have difficulty in addressing transient solutions. Conventional mathematical programing models [103] solve a deterministic approximation to the actual stochastic problem, sometimes with inventory targets based on off-line analysis included as constraints. Stochastic dynamic programming models, including Markov decision processes [89], give state-based reactive decision rules that do not directly consider information about future demand that may be available. Stochastic programing [12] and chance-constrained models [87] make different assumptions about recourse actions that can result in subtle theoretical and practical difficulties.

A conclusive, unifying solution to these complex issues is clearly a long way in the future. Our objective in this chapter is more modest and exploratory in nature. We consider a simple single-stage single product production-inventory system subject to workload-dependent lead times and stochastic demand. We then develop a number of alternative formulations for this system, including two different chance-constrained models, a two-stage stochastic programing model, and a multi-stage stochastic programing approach. The multistage stochastic programing model is the only one of these that has potential to yield an exact solution, and that conditional upon the choice of scenarios; the other three are heuristics. We compare the solutions obtained from these different models by subjecting them to a simulation of uncertain demand realizations. Our exploratory computational experiments suggest that when parameters are appropriately chosen, heuristics based on chance constrained models may provide near-optimal solutions that are competitive with those from much larger stochastic programing models, although the stochastic programing models consider a very limited number of scenarios. Our results suggest a number of directions for future work on improving the heuristics, and further experimentation aimed at elucidating the strengths and limitations of the chance constraint-based heuristics.

2 Previous Related Work

A comprehensive review of the literature on production planning under uncertainty is clearly beyond the scope of this chapter. Instead, we briefly review the literature most relevant to this paper. Overviews of the production planning domain are given by de Kok and Fransoo [27], Voss and Woodruff [103] and Missbauer and Uzsoy [78].

Most deterministic production planning models establish optimal production, inventory and release levels over a given finite planning horizon to meet the total demand [16, 45, 50]. The planning horizon is divided into discrete periods during which production and demand rates are assumed to be constant; the capacity of the system is represented by the number of hours available on key resources in a planning period; and the production, inventory, WIP and demand associated with a period are treated as continuous quantities. These models allocate capacity to products to optimize a specified objective and satisfy aggregate constraints representing system capacity and dynamics. However, models of this type are subject to the utilization-lead time dependence discussed in Sect. 1. The estimates of cycle times used in planning models are referred to as *lead times*.

The most common approximation in both the research literature and industrial practice is to treat lead times as a fixed, exogenous quantity independent of resource load. The Material Requirements Planning (MRP) approach [82] uses fixed lead times in its backward scheduling step to determine job releases. Several authors have suggested ways of adapting MRP to uncertain demand. Meal [73] and Grubbstrom [39] derive component plans with safety stocks in the MRP records. Miller [75] proposes hedging of the master schedule to provide safety stocks within the system. However, all these approaches assume fixed exogenous lead times.

Another common approach to production planning under fixed lead times and deterministic demand is the use of linear(LP) and integer programing(IP) models, of which a wide variety exist [42, 56, 103]. These represent capacity as a fixed upper bound on the number of hours available at the resource in a period, and model input and output time lags between stages. However, these time lags are independent of workload.

Several authors have proposed enhanced models that address the dependency between lead times and resource utilization to some degree. Lautenschlager and Stadtler [69] suggest a model where the production in a given period becomes available over several future periods. Voss and Woodruff [103] propose a nonlinear model where the function linking lead time to workload is approximated as a piecewise linear function. Kekre et al. [63] and Ettl et al. [31] take a similar approach, adding a convex term representing the cost of carrying WIP as a function of workload to the objective function. Graves [37], Karmarkar [61], Missbauer [76], Anli et al. [2] and Asmundsson et al. [5, 4] use nonlinear clearing functions to model the dependency between workload and lead times. Several related models are proposed in the recent book by Hackman [41]. Pahl et al. [83] and Missbauer and Uzsoy [78] review production planning models with load-dependent lead times. We shall discuss clearing functions, which are used in the models in this chapter, more extensively in the next section.

Another approach to modeling the operational dynamics of the system has been the use of detailed simulation or scheduling models in the planning process. Dauzere-Peres and Lasserre [26] use a scheduling model to check whether the plans their IP model develops are feasible. Other approaches use simulation models in the same manner, e.g., Pritsker and Snyder [88]. The use of simulation or scheduling models captures the operational dynamics of the system correctly. However, this approach does not scale well, since simulation models of large systems are time-consuming to run and analyze. An innovative approach to integrating simulation and LP is that of Hung and Leachman [53]. Given initial lead-time estimates, an LP model for production planning is formulated and solved. The resulting plan is fed into a simulation model to estimate the lead-times the plan would impose on a real system. If these lead-times do not agree with those used in the LP, the LP is updated with the new lead-time estimates and resolved. This iteration is repeated until convergence. Similar models have been proposed by others [6, 18, 19, 66, 95]. However, the convergence of these methods is not well understood [55, 57]. The computational burden of the simulation runs required is also a significant disadvantage for large systems such as those encountered in semiconductor manufacturing.

Stochastic inventory models seek an optimal inventory policy (when to order, and how much to order) for individual items in the face of different environmental conditions (e.g. demand patterns, modes of shipment from suppliers) and constraints (e.g. supply restrictions, budget limitations, and desired customer service levels). Much of the work in this area [54, 59, 101, 102] is in the context of ordering from suppliers, modeling demand carefully but treating supply as known and unlimited, generally with a fixed lead time. Many subsequent papers have addressed variations of this basic problem [46, 47, 107]. However, the vast majority assume that a supplier can supply any amount of material within the specified lead time, i.e., has unlimited capacity.

Federgruen and Zipkin [32, 33] consider the capacitated inventory problem with uncertain demand and explore the optimality of "modified" base stock policies when the cost for the single period is convex in the base stock level. Tayur [99] extends this work by discussing the computation of the optimal base stock level. However, these models use simple capacity constraints that ignore the dependency between load and lead times. Ciarallo et al. [23] describe the structure of optimal policies for problems with uncertain production capacity and a time-stationary demand distribution. Anupindi et al. [3] provide bounds and heuristics for the problem with nonstationary demand and stochastic lead times, where the lead time distribution is stationary over time.

The idea of combining inventory and queueing models has attracted attention from many researchers [17, 52, 91]. Zipkin [106] develops a queueing framework to analyze supply chains facing a stationary demand distribution and where a (Q, r) policy is used to release units onto the shop floor. Ettl et al. [31] develop an optimization model combining queueing and inventory models to set base-stock levels for a multi-item batch production system facing non-stationary demands. Liu et al. [71] extend this approach.

One of the most popular frameworks for planning under uncertainty is stochastic programing [12, 58, 87]. Uncertainty is represented by using a number of discrete scenarios to represent possible future states, which allows stochastic linear programs to be modeled as large linear programing problems. Constraints are formulated requiring that an optimal solution be feasible for all scenarios, and the objective function is usually to minimize the expected value of the specified objective function. A number of authors have formulated production planning problems as multi-stage stochastic linear programs (M-SLPs) [48, 85], but the approach presents challenges.

A significant difficulty of M-SLPs is that the problem size tends to grow exponentially with the number of possible realizations (scenarios) of uncertain parameters, requiring solution methods that exploit their special structure. The scenario-based structure of M-SLPs makes decomposition methods attractive. Most decomposition methods exploit convexity of the recourse function to use outer linearization. Commonly used methods include Dantzig-Wolfe decomposition (inner linearization) and Benders decomposition (outer linearization), which decompose the large-scale problem into a master problem and several independent subproblems. Dantzig-Wolfe decomposition adds new columns to the master problem based on the suproblem solutions [25]. Benders decomposition, on the other hand, proceeds by adding new constraints (supporting hyperplanes known as optimality cuts) that are computed using dual solutions to the subproblems (e.g., Lasdon [68]).

Van Slyke and Wets [100] extended Benders' decomposition to solve two-stage stochastic linear programs (2-SLPs) via the L-Shaped Method. M-SLPs are much more challenging computationally than 2-SLPs. An extension of the L-shaped method to more than two stages, called nested decomposition, was first proposed by Louveaux [72] for multi-stage quadratic programs and by Birge [11] for multi-stage linear programs. The algorithm generates cuts for an ancestor scenario problem that has feasible completion in all descendant scenarios. As in the L-shaped method, nested decomposition achieves outer linearization by generating feasibility and optimality cuts until it converges to an optimal solution. A number of different strategies have been used to select the next subproblem for deterministic problems. Numerical experiments by Gassmann [35] found that the fast-forward-fast-back procedure of Wittrock [105] outperforms other strategies.

There have been recent attempts to model production planning problems using robust optimization approaches [7, 10]. Leung et al. [70] develop a robust optimization model to solve the aggregate production planning problem. Raa and el-Aghezzaf [90] use robust optimization to obtain a dynamic planning strategy for the stochastic lot-sizing problem.

Chance constrained programing dates back to the work of Charnes and Cooper [20, 21, 22]. A more recent overview of these methods is given by Prékopa [87]. In chance constrained programing, constraints can be violated with a specified probability, which is quite useful to model, for instance, service levels in supply chain problems [40]. Continuous probability distributions are often assumed on the uncertain parameters. This approach achieves a substantial decrease in the size of the model, and avoids the problem of defining the penalty function. However, it fails to capture the cost consequences of constraint violations, which can result in anomalous behavior [14].

Given that exact solutions to stochastic optimization problem are computationally challenging, a number of approaches to obtain solutions via decision rules have been proposed. These approaches classify decision variables according to whether they are implemented before (first stage decisions), or after (second stage decisions) an outcome of the random variable(s) is observed. However, in the decision rule-based approach, the second stage recourse decisions are determined by a rule that incorporates both the first stage decisions and the observed outcomes. A commonly encountered example of such a rule that is in fact optimal in form is the well-known base stock policy for inventory systems with unlimited capacity, deterministic replenishment lead time and linear holding and backorder costs. However, as pointed out by Garstka and Wets [34], the decision rule approach assumes a specific form for the optimal solution to the stochastic program. Since very few multistage stochastic programs yield a closed-form characterization of the optimal solution, solutions obtained assuming decision rules cannot be guaranteed to be optimal in the vast majority of cases.

A well-known family of decision rules are the Linear Decision Rules, where the second stage recourse decision is a linear function of the first stage decision variables and the observed outcomes. The pioneering Linear Decision Rule (LDR) was developed by Holt, Modigliani, Muth and Simon (HMMS) in the mid 1950s [49, 51]. Extensions to this rule have been proposed by several authors [8, 28, 36, 44, 84]. While the HMMS model and its variations incorporate demand uncertainty, these models treat capacity, specifically workforce levels, as a decision variable that can be varied continuously, which avoids the problem of workload-dependent lead times encountered under fixed capacity limits. In addition, the specific quadratic form of the objective function adopted allows the construction of a deterministic equivalent that simply replaces each random variable with its expectation. However, it is well known that this approach does not yield optimal solutions in general.

In summary, a variety of models have been proposed that address the issues of workload-dependent lead times and demand uncertainty separately at best, and in many cases do not address either. The LP and MRP approaches do not address workload-dependent lead times, and generally ignore stochastic demand. Most inventory models focus on modeling demand, with simple models of replenishment that do not consider workload-dependent lead times. The combined queueing-inventory models capture the interaction between workload-dependent lead times and inventory levels correctly, but assume specific inventory policies of the order up to type, and make different assumptions about the representation of a production unit. Stochastic programing approaches are hampered by their exponentially growing computational burden as the number of products and planning periods (stages) increase. Our heuristics, in contrast, consider non-stationary demand distributions to provide production plans over a finite planning horizon, taking available information about future demand into account. The work in this paper is an initial step in assessing the performance of this approach.

In the next section we present an overview of the clearing function concept that we use to develop a LP model that addresses the load dependent lead time and demand uncertainty aspects simultaneously for a single-product supply chain.

3 Clearing Function Basics

Clearing functions (CF) [37, 61, 78, 98], express the expected throughput of a capacitated resource over a given period of time as a function of some measure of WIP level at the resource over that period, which in turn, is determined by the average resource utilization over the period. We shall use the term "WIP" and the generic variable *W* to denote any reasonable measure of WIP level over a planning period.

To motivate the use of a nonlinear CF, it is helpful to begin with a single resource that can be modeled as a *G/G/1* queueing system in steady state. The expected number in system (i.e., expected WIP) for a single server is given by Medhi [74] as:

$$W = \frac{(c_a^2 + c_s^2)}{2}\frac{\rho^2}{(1-\rho)} + \rho$$

where c_a and c_s denote the coefficients of variation of service and interarrival times, respectively and ρ the utilization of the server. Setting $c = (c_a^2 + c_s^2)/2$ and rearranging (1) we obtain a quadratic in *W* whose positive root yields the desired ρ value. Solving for ρ with $c > 1$, we obtain

$$\rho = \frac{\sqrt{(W+1)^2 + 4W(c^2-1)} - (W+1)}{2(c^2-1)}$$

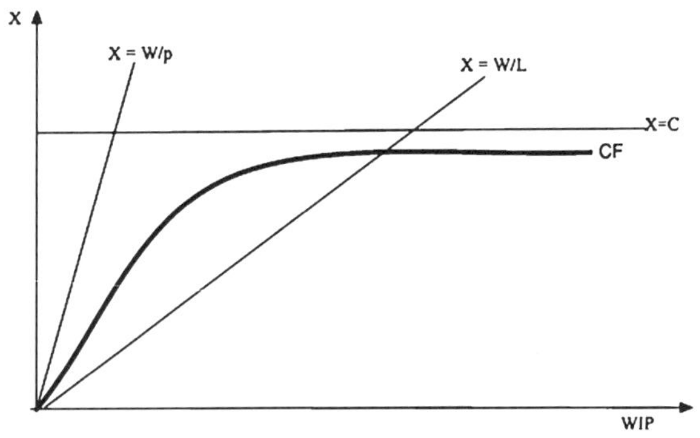

Fig. 1 Examples of CFs (Karmarkar [61])

which has the desired concave form. When $0 \leq c < 1$, the other root of the quadratic will always give positive values for ρ. When $c = 1$, the expression simplifies to yield $\rho = W/(1+W)$, again of the desired concave form. If we use utilization as a surrogate for output, we see that for a fixed c value, utilization, and hence throughput, increase with WIP but at a declining rate. Utilization, and hence output, is decreasing in c due to variability in service and arrival rates.

Figure 1, derived from Karmarkar [61], depicts several examples of CFs considered in the literature, where X denotes the expected throughput in a planning period. The horizontal line $X = C$ represents a fixed upper bound on output over the period, but without a lead-time constraint it implies that production can occur without any WIP in the system if work input and production are synchronized. This approach is implemented in most LP models but is supplemented with a fixed lead time as described above. The linear CF of Graves [37] is represented by the $X = W/L$ line, which implies a lead time of L periods that is maintained independently of the WIP level. If a fixed lead time is maintained up to a certain maximum output, we have $X = \min\{W/L, C\}$. When the parameters of the Graves CF are set such that the lead time is equal to the average processing time, with no queueing delays at all, we obtain the line $X = W/p$, where p denotes the average processing time. Assuming that lead time is equal to the average processing time up to a maximum output level gives the "Best Case" model $X = \min\{W/p, C\}$ of Hopp and Spearman [52]. The workload-independent fixed lead time in most LP models differs from the linear CF of Graves in that the former does not link output to WIP, while the latter does [81]. The CF always lies below the $X = W/p$ and $X = C$ lines. For most capacitated production resources subject to congestion, limited capacity leads to a saturating (concave) shape of the CF, for which Asmundsson et al. [4] and Selçuk et al. [96] provide analytical support.

Several authors discuss the relationship between throughput and WIP levels in the context of queueing analysis, focusing on the long-run steady-state expected throughput and WIP levels. Agnew [1] studies this behavior in the context of optimal control policies. Spearman [97] presents an analytic congestion model for a family of closed production systems that describes the relationship between throughput and WIP. Srinivasan et al. [98] derives the CF for a closed queueing network with a product form solution. Asmundsson et al. [4, 5] and Missbauer [77] study the problem of estimating CFs from experimental data, obtained either from industry or simulation models. Missbauer and Uzsoy [78] review the state of the art in this area.

An important advantage of CFs for our purposes is their ability to reflect different sources of variability in the production process. In queueing terms, this is accomplished by basing the CF on the effective processing time at the resources, which includes the effects of detractors such as uncertain yield, machine failures and setups, as discussed in Chap. 8 of Hopp and Spearman [52]. The manner in which these effects change the shape of the CF is described in Asmundsson et al. [4]. When the CFs are estimated from empirical data, the effects of the variability induced by detractors are present in the data to which the CF is fit, again capturing their effects.

Hence, given the current research on the derivation of CFs using both analytical and empirical approaches, in this chapter we shall proceed on the assumption that adequate methods of estimating CFs for different production systems will emerge from ongoing work. We focus on using CFs to develop production models that consider stochastic demand and the nonlinear relationship between workload and cycle in an integrated manner. We introduce our approach in the next section.

4 A Deterministic Model Based on Clearing Functions

In this section we develop a LP model for aggregate planning under the effects of congestion and demand uncertainty. We begin with a basic formulation prevalent in the literature, discuss its weaknesses, and use these to motivate our formulations, drawing heavily on the exposition in Bookbinder and Tan [15]. While there are clearly many formulations in the literature that capture additional aspects such as multiple stages, alternative production paths, etc., our focus is to find computationally tractable formulations that allow us to treat both the nonlinear dynamics of utilization and lead times and the stochastic nature of the demand as endogenous to the model. Hence to isolate these aspects of the problem for study, we focus on a single-stage single product system. The quantity of raw material released into the system in each time period is the key decision variable in our models. These releases are then converted into output according to different mechanisms defined by the models considered, which will be discussed as we proceed.

Consider the production planning problem for such a single stage production system producing a single product. The planning horizon is divided into T discrete periods of equal length. Demand in each period is assumed to be stochastic with known cumulative distribution function (CDF), and independent of demand in other periods. Service level requirements to be met are prespecified, and are thus treated as a constraint. We consider the simple objective of minimizing the sum of expected costs of holding finished goods inventory (FGI) and work in process (WIP) over the planning horizon. Following the literature, we do not consider the cost of stockouts in the objective function because we assume that the service level requirements are sufficiently high that the cost of stockouts is negligible. This assumption will be relaxed in our computational experiments. Clearly far more elaborate objective functions are possible, but our emphasis is on representation of production capacity and demand uncertainty.

To describe the models used in this paper we use three different classes of variables:

Decision or Control Variables: These variables represent the primary management decisions in a plan. In order to be implementable, a plan must specify either specific values for these variables, or specific rules by which they can be computed with the information available at the time a decision must be made.

State Variables: These variables define the behavior of the system, and their values are determined by the constraints determining the operational dynamics of the system and the values of the decision variables. These variables may be either deterministic or random.

Parameters: These are external inputs to the system and are prespecified in the model. We will assume these are always deterministic.

The notation used in the formulations is given below. We use a bold font, e.g., $\mathbf{X}$, for a random variable and a normal font, e.g., X, for a deterministic variable.

R_t : Planned quantity of product released into the system during period t

X_t : Planned production quantity during period t

$\mathbf{I_t}$: Inventory on hand at the end of period t. The initial inventory on hand at the start of period 1 will be denoted by I_0.

h_t : Unit inventory holding cost for period t

C_t : Capacity, e.g., total number of machine hours available, in period t

α : Specified service level

$G_{[t,t+k]}$: CDF of cumulative demand from period t to period $t+k$

$\mathbf{D_t}$: Demand during period t. Throughout this paper we shall assume the demand in each period t to be normally distributed with known mean μ_t and standard deviation σ_t. In our experiments we will assume demands are independent by time periods. However, the models presented remain valid for correlated demands as long as the variance-covariance matrix is known, or can be estimated with reasonable accuracy.

L_t : Average lead time in period t. For simplicity of exposition in presenting the models in this section we shall assume these are integer multiples of the planning period length. Fractional L_t values can be accommodated in a straightforward manner.

4.1 Basic Formulation

Most chance-constrained production planning models in the literature are similar to that of Bookbinder and Tan [15] given below; a slightly different version is given in Johnson and Montgomery [56]. Our model incorporates the following constraints:

• *Releases*

Since the lead time is L_t in period t, whatever is released into the system in period t is converted to output and available for consumption in period $t + L_t$. Hence the relationship between release quantities and output is given by

$$R_t = X_{t+L_t}, \text{ for all } t = 1, \ldots, T - L_t$$

The primary decision variable is the amount of work R_t released in period t, which must be specified at the start of the planning horizon. Hence both releases and production are deterministic. Note that the X_t and R_t variables are redundant, and the formulation can be written with only one of these two sets of variables.

In this model, work that is released into the production system at time t is in WIP for L_t periods until it emerges as finished product. Most LP models do not explicitly represent this quantity, or assign it a cost in the objective function, but it can easily be estimated for any period t as the difference between the cumulative releases and output up to a given period t.

• *Inventory balance*

The finished goods inventory on hand at the end of period t, $\mathbf{I_t}$, is a random variable for which the relationship

$$\mathbf{I_t} = \mathbf{I_{t-1}} + X_t - \mathbf{D_t}$$

holds for each time period t. Taking the expectation and repetitive substitution yields

$$\begin{aligned} E[\mathbf{I_t}] = E[\mathbf{I_{t-1}}] + X_t - E[\mathbf{D_t}] &= I_0 + \sum_{i=1}^{t} X_i - \sum_{i=1}^{t} E[\mathbf{D}_i] \\ &= I_0 + \sum_{i=1}^{t} X_i - \sum_{i=1}^{t} \mu_i, \text{ for all } t = 1, \ldots, T. \end{aligned}$$

All terms in this expression are now deterministic.

• *Capacity*

$$X_t \leq C_t, \text{ for all } t = 1, \ldots, T.$$

Service level: This constraint requires that the service level, defined by the probability of $\mathbf{I_t} < 0$, i.e., a stockout occurring, be less than $(1 - \alpha)$, implying

$$P\{\mathbf{I_t} \geq 0\} \geq \alpha \Rightarrow P\left\{ I_0 + \sum_{i=1}^{t} X_i \geq \sum_{i=1}^{t} \mathbf{D_i} \right\} \geq \alpha, \text{ for all } t = 1, \ldots, T.$$

The service level measure fits the chance constraint approach well, since the latter allows constraints to be violated with a certain probability. However, it does not

Table 1 Basic formulation

$\min \sum_{t=1}^{T} h_t\{I_0 + \sum_{i=1}^{t} X_i - \sum_{i=1}^{t} \mu_i\}$	subject to	
$P\{I_0 + \sum_{i=1}^{t} X_i \geq \sum_{i=1}^{t} \mathbf{D_i}\} \geq \alpha$	for all $t = 1, \ldots, T$	(SERVICE LEVEL)
$X_t \leq C_t$	for all $t = 1, \ldots, T$	(CAPACITY)
$X_t \geq 0$	for all $t = 1, \ldots, T$	(NONNEGATIVITY)

capture the degree to which the constraint is violated. Hence a production plan that has stockouts in many periods, but falls short by a very small fraction of the demand in each period, will appear to have a poor service level. To this end, we use the fill rate, the fraction of total period demand met from inventory, as another performance measure in our computational experiments. The basic formulation is summarized in Table 1, using the production variables X_t.

While the basic formulation is intuitive, it suffers from the following disadvantages:

- It ignores the effects of loading on WIP and lead times in a capacitated system [4, 5, 37, 60] by considering lead times to be fixed exogenous values.
- It assumes that safety stock must be held in finished goods inventory, based on the demand for each individual period. This is adequate when the lead times of the production system, which correspond to the replenishment time of the finished goods inventory, do not exceed one period, as in the models of Bookbinder and Tan [15] and Johnson and Montgomery [56]. However, if lead times span multiple periods, this becomes problematic. It is well known in the inventory literature [24] that in the presence of nonzero lead times the optimal policy in many cases, and a good heuristic in many more, is to set the inventory position, the sum of on-hand inventory and outstanding orders, to the desired percentile of the demand over the lead time (e.g., [29]). Hence this formulation fails to recognize that WIP can serve some of the function of safety stock [38], and hence might hold more finished goods inventory than required to maintain a given service level. We shall assume a replenishment policy of this form, which is not optimal for the production system we consider, in developing our heuristics.

  In the production-inventory context of this paper, outstanding orders are represented by material that has been released into the production line but has not yet emerged as finished goods, i.e., WIP [38]. The inventory position, which will be an important quantity for our development in the rest of this paper, will be defined in more detail in the following section.
- The model makes all decisions for the entire planning horizon at the beginning of the horizon, before any of the demands become known, and does not provide a way to use information as it becomes available. In other words, there is no recourse action.

In the rest of this section we extend this formulation to address these issues.

4.2 Development of Integrated Model

An elegant way of capturing the effect of capacity loading on WIP and lead times in production planning models is the use of clearing functions (CFs) as discussed in Sect. 3. Recall that up to this point all variables are deterministic except the inventory levels $\mathbf{I_t}$. Hence, incorporating CFs in the model requires:

- Introduction of WIP balance equations. If W_t is defined to be the WIP at a given time t, then the WIP balance equations are given as $W_t = W_{t-1} + R_t - X_t$ for all periods $t = 1, \ldots, T$. We treat R_t as a deterministic decision variable that is specified at the start of the planning horizon by solution of the planning model, and cannot be modified as uncertainty is realized.
- Replacing the original capacity constraint with a set of linear inequalities that represent the outer linearization of the original CF [4, 5]. The set of inequalities representing the CF is given by $X_t \leq a_k W_{t-1} + b_k$, for all periods $t = 1, \ldots, T$ and line segments $k = 1, \ldots, n$ used to outer linearize the CF.

The use of CFs to represent the capacity of the production system takes a more complex view of the relationship between the planned release quantity R_t in period t and the planned output X_t of the system in that period. The releases in a period determine the planned WIP level W_t at the end of the period, together with the linearized CF represented by the constraints above, determines the planned system output X_{t+1} in the next period The release variables R_t are defined such that releases are made at the end of period t, and hence cannot contribute to output during period t. This is necessary because in later models, our linear decision rule observes the realization of the random demand $\mathbf{D_t}$ in period t to determine the releases R_t at the end of period t. This definition, together with the definition of the CF and the WIP balance equations, is thus internally consistent.

In inventory theory an optimal or near-optimal policy, when there is no fixed ordering cost and shortage and holding costs are linear, is to maintain the inventory position, the sum of on-hand and on-order inventory, at a critical fractile of the demand over the replenishment lead time [24]. Hence if $\mathbf{IP_t}$ denotes the inventory position at the end of period t, we have $\mathbf{IP_t} = W_t + \mathbf{I_t}$, where W_t represents orders that have been released to production but not yet completed. This analogy with inventory models suggests a service level constraint requiring a probability α that $\mathbf{IP_t}$ is at least as great as the demand over the replenishment lead time [38]. Assuming this replenishment lead time, corresponding to the cycle time of the production system under study, is known to be L_t periods in period t, we have

$$P\left\{\mathbf{IP_t} \geq \sum_{i=t+1}^{t+L_t} \mathbf{D}_i\right\} \geq \alpha \Rightarrow P\left\{\mathbf{I_t} + W_t \geq \sum_{i=t+1}^{t+L_t} \mathbf{D}_i\right\} \geq \alpha .$$

The L_t parameters on the right hand sides of our chance constraints define the distribution of the lead time demand that will be used to set safety stock levels. Noting that

$$\mathbf{I_t} = I_0 + \sum_{i=1}^{t} X_i - \sum_{i=1}^{t} \mathbf{D}_i \text{ and}$$

$$W_t = W_0 + \sum_{i=1}^{t} R_i - \sum_{i=1}^{t} X_i$$

we obtain

$$\mathbf{IP}_t = \mathbf{I_t} + W_t = (I_0 + W_0) + \sum_{i=1}^{t} R_i - \sum_{i=1}^{t} \mathbf{D}_i.$$

The chance constraint is now of the form

$$P\{\mathbf{IP}_t \geq 0\} \geq \alpha \Rightarrow P\left\{I_0 + W_0 + \sum_{i=1}^{t} R_i - \sum_{i=t+1}^{t+L_t} \mathbf{D}_i \geq \sum_{i=1}^{t} \mathbf{D}_i\right\} \geq \alpha$$
$$\Rightarrow P\{I_0 + W_0 + \sum_{i=1}^{t} R_i \geq \sum_{i=1}^{t+L_t} \mathbf{D}_i\} \geq \alpha\,.$$

Following the approach of Charnes and Cooper [22] the deterministic equivalent of the service level constraint can be written as

$$I_0 + W_0 + \sum_{i=1}^{T} R_i \geq G^{-1}_{[1,t+L_t]}(\alpha), \text{ for all } t = 1, \ldots, T.$$

where $G_{[1,\,t]}(\cdot)$ denotes the cumulative distribution function (CDF) of the cumulative demand random from periods 1 to t,

Replacing the probabilistic service level constraint with its deterministic equivalent yields the Zero-Order Inventory Position (ZOIP) formulation shown in Table 2. This formulation embodies a service level constraint on inventory position and a zero order decision rule where all decision variables are specified irrevocably at the start of the planning horizon.

It is important to note that there are two different lead times at work in the ZOIP model. The first of these is the estimated replenishment lead time L_t used to establish the inventory position required to approximately achieve the desired service levels. The second lead time in question is that realized in the production system, the time required for work released into the system to become available as finished product. The workload-dependent nature of this realized lead time is explicitly represented by the clearing function, whose effectiveness for this purpose we have demonstrated in prior work [4, 5]. Ideally, the two lead times should be equal, with the replenishment lead time used for setting inventory targets matching that realized by the production

Table 2 ZOIP formulation

$\min \sum_{t=1}^{T} h_t\{I_0 + \sum_{i=1}^{t} X_i - \sum_{i=1}^{t} \mu_i + W_t\}$	subject to	
$W_t = W_{t-1} - R_t - X_t$	for all $t = 1, \ldots, T$	(WIP BALANCE)
$I_0 + W_0 + \sum_{i=1}^{T} R_i \geq G^{-1}_{[1,t+L_t]}(\alpha)$,	for all $t = 1, \ldots, T$	(SERVICE LEVEL)
$X_t \leq a_k W_{t-1} + b_k$	for all $t = 1, \ldots, T; k = 1, \ldots n$	(CAPACITY)
$X_t, R_t, W_t \geq 0$	for all $t = 1, \ldots, T$	

system in the face of the release schedule recommended by the model. In other words, in an ideal situation the L_t should be an output of the model. This would require us to estimate L_t using Little's Law as $(W_t + W_{t-1})/2X_t$ assuming the planning periods are long enough for the law to apply; for the shorter periods some transient version of Little's Law such as those discussed by Bertsimas and Mourtzinou [9], Whitt [104] and Riaño [95] would be required. Even the use of the classical stationary version of Little's Law yields a highly nonlinear constraint. Hence for the sake of tractability we treat the replenishment lead time L_t on the right hand side of the chance constraints as an exogenous parameter, which reduces the right hand sides to constants that can be precomputed easily. Our model thus captures workload-dependent lead times correctly in defining the relationship between releases R_t, planned WIP level W_{t-1}, and expected output X_t, but uses an exogenous parameter to approximate the distribution of the lead time demand, which will be used to set the safety stocks. Computational experiments indicate that the realized lead time may deviate somewhat from the exogenously assumed value used to establish the chance constraints when used in this manner, but results are still favourable over base stock type models that do not consider clearing functions [94].

A full resolution of this issue appears to be challenging, and must be left for future research. A promising approach is to use an iterative scheme, where we solve the ZOIP model using an initial set of lead time estimates to obtain a release plan, i.e., a set of R_t values. These R_t values are then used to compute the resulting state variables X_t, W_t, and I_t, from which a new set of L_t values can be estimated as $L_t = W_t/X_t$. These new L_t values are then substituted into the model and the process is repeated until convergence is, hopefully, achieved. Orcun et al. [79] have implemented this procedure with favourable results, but formal analysis of its convergence remains for future work.

Up to this point we have developed a formulation that combines the modeling of congestion and lead times in the production system with the explicit representation of random demand using chance constraints. We now move on to adding flexibility to the decision mechanism by utilizing information as it becomes available.

4.3 A Linear Decision Rule

So far our formulations have zero order static decision rules, where the values of all decision variables are determined at the beginning of the time horizon and there is no recourse action after the outcomes are observed. We now follow Charnes and Cooper [22] and propose a linear decision rule to introduce flexibility in the decision mechanism, recalling that this approach does not yield an optimal solution. Since the releases are the decision variables in the CF formulations, the decision rule is based on releases.

We use a simple rule closely following that described in Johnson and Montgomery [56] that allows the releases to be modified as uncertain demand is observed, rendering them random variables. We define auxiliary variables Y_t that represent the change in planned inventory position from period $t-1$ to period t, implying that $\mathbf{R_t} = Y_t + \mathbf{D_t}$. Thus Y_t represents the amount of work released in period t over and above that necessary to replenish the inventory position after that period's demand has been withdrawn; note that it may be negative, if demand is decreasing in a given time interval. This decision rule thus represents a base stock policy, and it is straightforward to show that $Y_t = IP_t - IP_{t-1}$ for specified values of $\mathbf{IP_t}$ and $\mathbf{IP_{t-1}}$. Our heuristic establishes chance constraints that set the planned inventory position at the end of period t, $\mathrm{IP_t}$, to a percentile of the lead time demand distribution as described below. The releases $\mathbf{R_t}$ are now random variables derived from the Y_t and the realized demand $\mathbf{D_t}$. Thus the WIP variables $\mathbf{W_t}$ are now also random. Since the production $\mathbf{X_t}$ in a given period depends on the realized WIP level $\mathbf{W_{t-1}}$ at the start of that period, $\mathbf{X_t}$ is also a random variable. We have the relation

$$\begin{aligned} E[W_t] &= E[\mathbf{W}_{t-1} + \mathbf{R}_t - \mathbf{X}_t] = E[\mathbf{W}_{t-1} + Y_t + \mathbf{D}_t - \mathbf{X}_t] \\ &= W_0 + \sum_{i=1}^{t} (Y_i + \mu_i - \mathbf{X}_i) \end{aligned}$$

Since the release quantities are now random variables, there exists a possibility that they may be negative. To prevent this, we use the chance constraint

$$P\{\mathbf{R_t} \geq 0\} \approx 1 \Rightarrow P\{Y_t + \mathbf{D_t} \geq 0\} \approx 1 \Rightarrow Y_t + D_t^{\min} \geq 0,$$

where $D_t^{\min}$ is a value of demand in period t such that the probability of demand falling below this level is deemed by management to be extremely small. This is clearly an approximation when demand follows a distribution with unbounded support, like the normal distribution we assume, and is unlikely to be binding except when there is a very sudden, large decline in demand from one period to another.

We again define the event of a stockout as the event that the total lead time demand exceeds the inventory position $\mathbf{IP_t} = \mathbf{W_t} + \mathbf{I_t}$, yielding the chance constraint

$$W_0 + I_0 + \sum_{i=1}^{t} Y_i \geq G^{-1}_{[t+1,t+L_t]}(\alpha).$$

Table 3 DYNIP formulation

$\min \sum_{t=1}^{T} h_t\{I_0 + \sum_{i=1}^{t} X_i - \sum_{i=1}^{t} \mu_i +$		
$W_0 + \sum_{i=1}^{t} (Y_i + \mu_i - X_i)\}$	subject to	
$W_0 + I_0 + \sum_{i=1}^{t} Y_i \geq G^{-1}_{[t+1,t+L_t]}(\alpha)$	for all $t = 1, \ldots, T$	(SERVICE LEVEL)
$X_t \leq a_k(W_0 + I_0 + \sum_{i=1}^{t-1} Y_i) + b_k$	for all $t = 1, \ldots, T; k = 1, \ldots n;$	(CAPACITY)
$Y_t + \mathrm{D}_t^{\min} \geq 0$	for all $t = 1, \ldots, T$	(REL. NON-NEG.)
$X_t, Y_t \geq 0$	for all $t = 1, \ldots, T$	

Since releases, WIP and production are all interrelated, all decision variables are now random variables except the Y_t, creating difficulties in establishing a tractable formulation. Hence for tractability in the solution procedure, we will assume that the production variables X_t and the auxiliary variables Y_t are determined at the start of the planning horizon, with the $\mathbf{W_t}, \mathbf{I_t},$ and $\mathbf{D_t}$ remaining as random variables. The assumption here is that when a production target X_t is in danger of not being met, the system will take extraordinary measures to meet it, such as running an extra shift or buying from an outside source. The cost of this is not captured in the models, but is, of course, considered in our computational experiments, where we assume the production system has no outside recourse when planned production levels cannot be achieved. This ensures that all models are treated similarly in the computational experiments. Incorporating this rule in the ZOIP formulation gives us our final Dynamic Inventory Position (DYNIP) formulation summarized in Table 3.

The models presented above have been analyzed by Ravindran et al. [93]. They compare the performance of the ZOIP and DYNIP models with a static base stock policy and find that DYNIP performs significantly better in terms of backorders. They also analyze the structure of optimal solutions to the model under the linear clearing function of Graves. These results indicate that the ZOIP model will overstock consistently, while DYNIP will not.

5 Stochastic Programing Models

For comparison with the chance constrained models, we develop two different stochastic programing models along with their implementation strategies. We first present a two-stage stochastic programing model. A multi-stage stochastic programing formulation is also presented along with static and dynamic implementation strategies.

5.1 The Two-Stage Model (2-SP)

As in the rest of the paper, we assume the primary source of uncertainty is the demand in each period, and consider the simple objective of minimizing the sum of expected WIP holding, FGI holding and backorder costs over the planning horizon of T periods. We assume that the demand evolves as a discrete time stochastic process with a finite probability space. This information structure can be interpreted as a scenario tree, where the nodes in stage t of the tree constitute the states of the world that can be distinguished by information available up to period t. The size of the scenario tree is clearly exponential in the number of periods T, and depends on the number of possible demand realizations considered at each stage.

The computational burden of any model based on scenario trees will rapidly become impractical. Therefore even for relatively small problem instances used to benchmark our heuristics, some means of reducing the size of the scenario tree must be devised. To this end, we shall follow Escudero et al. [30] and consider a two-stage formulation which consists of specifying a number of scenarios ξ composed of demand realizations for all periods. The first-stage problem involves deciding the production, release, and planned WIP levels for all periods, regardless of the state of the world. The second stage determines the FGI and backlog levels at the end of each period subject to the realized state. Thus the X_t, R_t, and W_t variables are only indexed by time periods (since they do not change with the realized state) while FGI variables I_t^{ξ} and backorder B_t^{ξ} at the end of period are indexed by the scenario ξ. The model can be stated as follows:

$$\text{Min} \sum_{t=1}^{T} h_t W_t + E_{\xi}[Q(X_t, \xi)]$$

subject to

$$W_t = W_{t-1} + R_t - X_t \quad \text{for all } t = 1, \ldots, T$$

$$X_t \leq f(W_t), \text{ for all } t = 1, \ldots, T$$

$$X_t, R_t, W_t \geq 0 \text{ for all } t = 1, \ldots, T$$

where $Q(X_t, \xi)$ denotes the recourse function which is defined as

$$Q(X_t, \xi) = \min \sum_{t=1}^{T} (h_t I_t^{\xi} + b_t B_t^{\xi})$$

subject to

$$I_t^{\xi} - B_t^{\xi} = I_{t-1}^{\xi} - B_{t-1}^{\xi} + X_t - D_t^{\xi}, \text{ for all } t = 1, \ldots, T$$

$$I_t^{\xi} - B_t^{\xi} \geq 0, \text{ for all } t = 1, \ldots, T$$

Unlike DYNIP, this model assumes no recourse for the R_t variables. In fact under the two-stage model the first stage decision variables X_t, R_t, and W_t are determined at the beginning of the planning horizon, while the second stage problem simply computes the realized FGIs and backorders after demands are realized. This model has the advantage that the size of the model grows linearly with the number of scenarios considered, and that it has complete recourse, in that all first-stage decisions are feasible for the second stage. The disadvantage is that it does not allow recourse action to be taken as demand is realized, placing it on a par with the ZOIP model in this regard.

In order to determine a 2-SP production planning strategy, one has to generate multiple scenarios, each consisting of demand realizations for periods $1, \ldots, T$. The 2-SP model is then solved and the optimal decisions $(R_t^*, X_t^*, W_t^*), t = 1, \ldots, T$ yield a production plan that is completely defined at the beginning of the planning horizon.

5.2 *The Multi-Stage Model (M-SP)*

A natural extension of the two-stage model is to allow recourse actions as demand is observed. This is accomplished by representing the demand process $\{D_t\}$ as a scenario tree. Each node n in the tree represents a demand realization in the corresponding period $t(n)$ with a probability q_n. The root node $(n=1)$ of the tree represents the current demand, i.e. D_1. Node *a(n)* is the direct ancestor of node n. The direct descendants of node n are called the children of node n. The subtree with root node n is denoted by $T(n)$. A path from the root node to a node n describes one realization of the stochastic process from the present (period 1) to period $t(n)$. The set of all the nodes on this path is denoted as $P(n)$. A full evolution of the demand process over the entire planning horizon, i.e., the path from the root node to a leaf node, is called a scenario.

The scenario tree representation of the demand process is an approximation of the actual demand distribution due to its use of a finite number of possible demand outcomes in each period. Also, generally the size of scenario tree increases exponentially with increasing time horizon. The cumulative demand, production, and releases for the partial realization of the demands represented by a path from the root node 1 to a node n in the tree are given by

$$D(1,n) = \sum\nolimits_{m \in P(n)} D_m$$

$$X(1,n) = \sum\nolimits_{m \in P(n)} X_m$$

$$R(1,n) = \sum\nolimits_{m \in P(n)} R_m$$

The stochastic programing formulation of the production planning problem with congestion is given by the following model:
(MSP):

$$\begin{array}{lll} \min & \sum_{n\in T(1)} q_n[h_{t(n)}(I_n + W_n) + b_{t(n)}B_n] & \\ \text{s.t.} & W_n = W_0 - X(1,n) + R(1,n) & \\ & I_n = I_0 + X(1,n) - D(1,n) + B_n & \forall n \\ & X_n \leq f(W_{a(n)}) & \forall n \\ & R_n,\ X_n,\ I_n,\ W_n,\ B_n \geq 0 & \forall n \end{array}$$

The objective in the M-SP model is to minimize the expected cost over the planning horizon, which includes the present cost determined by the root node decisions and the expected future cost. In any given period t, the release, WIP, and production can be determined before the knowledge of demand, and are hence called first-stage decisions. On the other hand inventory and backorder are recourse decisions because they depend on the first-stage decisions as well as the realization of the uncertain parameter (demand). In our implementation, the constraints related to the clearing function are piecewise linearized as in Asmundsson et al. [4] for computational convenience.

The MSP model has considerable similarities to the Model Predictive Control approach deployed in the engineering disciplines. The similarities between control theoretic and mathematical programing approaches were noted early on by Kleindorfer et al. [67] and their application to supply chain management problems has been discussed by Kempf [64].

5.3 Implementation Strategies for the M-SP Model

Based on the multi-stage stochastic programming model (M-SP) we develop two production planning strategies to satisfy future demand over the planning horizon. The first strategy is a static strategy (MSP) and the second is a dynamic strategy (MSP-DYN).

5.3.1 A Static Strategy (MSP)

MSP is a static strategy, which specifies completely the release, production, and WIP decisions for all future periods at the beginning of the planning horizon. Once demands are realized, the FGI and backorders can be determined and the performance of the solution evaluated, in a manner similar to that used for ZOIP. The primary difference between ZOIP and MSP lies in the manner they model the uncertainty in the demand process. ZOIP assumes a known demand distribution in each period, and establishes constraints that may be violated with a prespecified probability. MSP, on the other hand, captures the uncertainty of demand through a limited number of demand values in each period. Another important difference between ZOIP and MSP is that ZOIP assumes no recourse action is possible as uncertain demand is revealed.

In order to determine the MSP strategy, i.e., the production planning decisions for all periods, we follow the procedure below. Note that all the steps are performed at the beginning of the planning horizon.

For $t = 1$, we construct a scenario tree $T(1)$, set the first period demand to the current demand and the initial inventories to some preset initial values, then solve the MSP model. We obtain the optimal production decisions for all the nodes in the tree, i.e., $(R_n, X_n, W_n)^*$. However we only save the root node decisions, which correspond to the decisions to be implemented in period 1, $(R_1, X_1, W_1)^*$, under the MSP strategy. Also, $(I_1, W_1)^*$ serve as initial inventories for the next period.

For $t = 2$, we construct a scenario tree $T(2)$ over the periods $2, \ldots, T$, set the root node demand to μ_2 and solve the M-SP. Here μ_2 is the forecast of period 2 demand available to us in the beginning of the planning horizon.

The root node optimal decisions are recorded as $(R_2, X_2, W_2)^*$ and will be implemented in the second period under the MSP strategy. Repeat the same for $t = 3, \ldots, T$. The optimal decisions $(R_t, X_t, W_t)^*$, $t = 1, \ldots, T$ constitute the MSP production plan that is completely defined at the beginning of the planning horizon.

5.3.2 A Dynamic Strategy (MSP-DYN)

As pointed out in Powell et al. [86], a model is dynamic if "it incorporates explicitly the interaction of activities over time". A model is applied dynamically if "the model is solved repeatedly as new information is received". Under this definition, DYNIP is a dynamic model, while MSP-DYN presented below is a model applied dynamically.

In the MSP-DYN, the multi-stage SP model is applied dynamically over the planning horizon and only the decisions of the first period are implemented. As new information about demand becomes available the model is resolved and the release, production, and WIP decisions are made. Therefore, at the beginning of the planning horizon only period 1 decisions are known and future decisions will only be determined once the corresponding demand is realized. More specifically, we proceed as follows:

For the current period, $t = 1$, we construct a scenario tree $T(1)$, set the first period demand to the current demand and the initial inventories to some pre-set initial values, then solve the MSP model. We obtain the optimal production decisions for the root node decisions to be implemented in period 1, $(R_1(D_1), X_1(D_1), W_1(D_1))^*$. $(I_1, W_1)^*$ serve as initial inventories for the next period.

The current period is $t = 2$, the demand of period 2 is now realized and corresponds to the root node demand in a scenario tree to be constructed for periods $2, \ldots, T$. The MSP model is solved and the root node decisions $(R_2(D_2), X_2(D_2), W_2(D_2))^*$ are implemented. This process is repeated for $t = 3, \ldots, T$. At the end of the planning horizon the values $(R_t(D_t), X_t(D_t), W_t(D_t))^*$, for $t = 1, \ldots, T$ constitute the MSP-DYN production plan.

6 Computational Experiments

In this section, we present a computational study where we compare the performance of the ZOIP, DYNIP, 2-SP, MSP, and MSP-DYN models considering various demand profiles and based on Fill Rate and Inventory Position. The former is a proxy for the level of customer service provided, while the latter serves as a proxy for the average inventory holding cost, considering both WIP and finished goods inventory levels. The models have been implemented in GAMS and solved using CPLEX 11.0. We begin by discussing the experimental design and then present the results and analysis.

Demand Profiles: Demand is forecasted over a horizon of three months, each consisting of four working weeks ($T = 12$ weeks). Demand in each period is independent and normally distributed. However, the means and variances of demand are allowed to vary across periods. Three possible levels of mean demand in a given month are considered: H (High = 140), M (Medium = 100), and L (Low = 60). Based on these levels, seven demand profiles are constructed by considering different levels for each month (i.e., four week subinterval): LLL, MMM, HHH, LMH, HML, LHL, and HLH. For example, demand profile LMH represents an increasing monthly demand, where demand from week 1 to week 4 is 60, from week 5 to week 8 is 100, and from week 9 to week 12 is 140. These profiles show how the mean values of the demand distributions vary over the planning horizon. In all these profiles, we assume a constant coefficient of variation $\rho_t = \sigma_t / \mu_t = 0.25$ for the demand distributions in every period. This yields very small probability of negative demands; in the few cases in our experiments in which they arose, negative demands were set to zero.

In order to implement the stochastic programs 2-SP and M-SP, scenario trees based on the various demand profiles must be constructed. For the 2-SP model, three scenarios are considered, Low, Medium, and High, with demand in each period t is set to each of the values $\mu_t - \sigma_t$, μ_t, and $\mu_t + \sigma_t$, respectively. The probabilities of the three demand realizations are assumed to be 0.25, 0.5, and 0.25, respectively. This is clearly a limited representation of the demand uncertainty, and we shall return to this issue in our discussion of our computational results.

In the case of the M-SP model, successive stochastic programs (one in each period) have to be solved in order to obtain a production plan for the entire horizon. Therefore, in each period t a binary scenario tree starting from period t up to the end of the horizon is constructed. In each period we consider two possible demand realizations, Low Demand ($\mu_t - \sigma_t$) and High Demand ($\mu_t + \sigma_t$), with equal probabilities. Thus in any given period t, a M-SP is formulated and solved with a scenario tree containing $2^{T-t+1} - 1$ nodes and 2^{T-t} scenarios (number of leaf nodes).

The capacity of the production system is represented by a clearing function which captures the effect of congestion as discussed in Sect. 3. Following Karmarkar [61], we assume the form of the clearing function to be

$$f(W) = \frac{K_1 W}{K_2 + W},$$

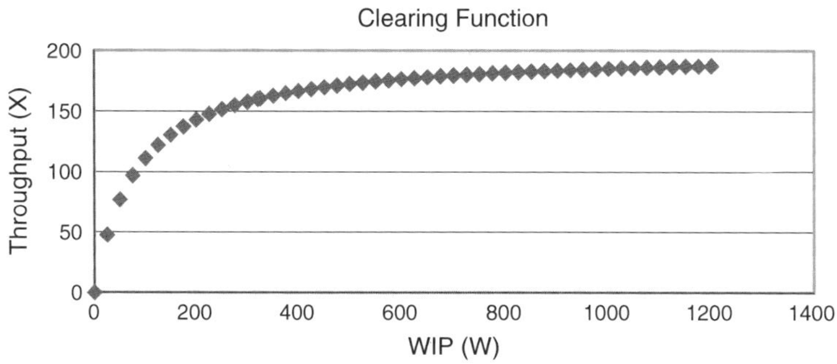

Fig. 2 Clearing function used in experiments

Table 4 Clearing function approximation

Segment	Intercept	Slope
1	0.0	0.5
2	136.0	0.069
3	154.8	0.036
4	161.8	0.023
5	180	0

where $K_1 = 200$ is the production capacity, and $K_2 = 80$ measures the curvature of the CF.
The resulting CF is shown in Fig. 2. Our piecewise linearization of this CF is given in Table 4.

There are clearly many specific issues involved in the estimation and piecewise linearization of CFs which are beyond the scope of this paper. These issues have been discussed extensively in Missbauer and Uzsoy [78]; specific approaches are illustrated in Asmundsson et al. [4], Missbauer [77], and Selcuk et al. [96], among others. Extensive experimentation in the course of this work has shown that the specific manner in which an appropriately fitted CF is piecewise linearized does not have much effect on the quality of the resulting production plans, although it does affect the estimates of the dual prices obtained for the associated constraints [62]. Since the primary purpose of this paper is to compare the solutions obtained from different formulations of the production planning problem with stochastic demand, all the models compared use the same piecewise linearized function. Hence the quality of the fit of the CF is not a factor in this study.

The values of L_t, i.e. the lead times in period $t = 1, .., T$ used in the formulations were chosen to be the same for all periods. This value, based on Little's Law, was chosen to be $L = W/\mu$, where μ is the average of all the demand means over the planning horizon and W the WIP value corresponding to a throughput of μ on the CF. This represents the behavior of a practitioner establishing a model based on historical data. The choice of values for I_0 and W_0 can be arbitrary, but the values we use are those recommended by Graves [38], setting $W_0 = L\mu$, and $I_0 = z_\alpha \sigma \sqrt{L}$.

To compare the performance of the production planning models, ZOIP, DYNIP, 2-SP, and M-SP (including the MSP and MSP-DYN strategies) we evaluate their optimal production plans in the face of simulated demand scenarios. For each demand profile, the evaluation procedure is as follows:

For ZOIP, 2-SP, and MSP

- Step 1 : Solve the four models for each demand profile and obtain the optimal values of the variables (R_t, X_t, W_t) for all periods to be specified at the beginning of the horizon before any actual demand has been observed, i.e. the first stage decision variables. These constitute the optimal production plan.
- Step 2 : Generate $N = 100$ demand scenarios from the normal distribution for each period and simulate the production plans for the models for each scenario. For each scenario a realization of inventories and backorders is obtained.
- Step 3 : Compute the performances for each model i.e., average and variance of backorders, fill rate, inventory position, and holding cost.

For DYNIP

- Step 1 : Solve the model for each demand profile and obtain the optimal values of the variable Y_t for all periods to be specified at the beginning of the horizon before any actual demand has been observed, i.e. the first stage decision variables.
- Step 2 : Generate $N = 100$ demand scenarios. For each scenario, once demand is realized in a given period, the corresponding (R, X, W) are determined and hence, the inventory and backlogs can be computed.
- Step 3 : Compute the performances for each model i.e., average and variance of backorders, fill rate, inventory position, and holding cost.

For MSP-DYN

- Step 1 : Generate 100 demand scenarios. For each scenario, once demand is realized in a given period t, solve a M-SP model for the periods $t, \ldots, T$ and implement the first period decisions, i.e., the (R, X, W) are determined as well as the ending inventory (I) and backlogs (B).
- Step 2 : Compute the performances for each model i.e., average and variance of backorders, fill rate, inventory position, and holding cost.

Since the chance constrained models ZOIP and DYNIP and the stochastic programing models 2SP, MSP and MSP-DYN use rather different modeling assumptions, care must be exercised when making comparisons. The chance constrained models assume a form for the demand distribution in each period, and do not consider shortage costs. However, it can be argued that an implicit judgement on the relative magnitude of holding and shortage costs is made in the specification of the required service level α, which also serves as the probability of constraint violation. The chance constrained models do not specify any particular recourse action when constraints are violated; in our computational experiments we assume any missed demands can be backlogged.

We thus consider three levels of the service level in our experiments: 90, 95 and 99.9%.

The stochastic programs, on the other hand, do not represent the demand distribution in a closed form. Instead, they use a discrete set of scenarios of outcomes to represent the uncertain nature of demand. Hence the effectiveness of these models is clearly linked to the number and degree of representativeness of the scenarios used to obtain the solutions. Another interesting issue is that stochastic programing models provide, by their nature, values for the decision variables corresponding to first stage decisions that must be made at the present time, as well as decision variables corresponding to each of the scenarios considered. However, since the scenarios considered in the model represent only a sample of possible realizations of the demand process, it is highly likely that in the future we will face a demand realization that does not match any of the scenarios used in obtaining them unless the stochastic program is solved on a rolling horizon basis. Since the size of the formulation to be solved for the stochastic programs is directly driven by the number of scenarios considered, this raises some interesting questions.

The performance of the stochastic programs (2-SP, MSP and MSP-DYN) is mainly affected by the magnitude of the backorder cost relative to the holding cost. We assume a unit production cost of $c = \$100$, and set the holding cost to $h = 0.2\,c$ and consider three levels for b the backorder cost: $0.5c$, c, and $4c$.

7 Results of Experiments

In order to facilitate a fair comparison between the different models, we have taken the approach of multiobjective optimization. The solution produced by any model represents a tradeoff between shortage and holding costs as that model perceives them, subject to the specific parameter settings used. The issue is further complicated by the different definitions of shortage that are possible. The chance constrained models require the specification of a maximum stockout probability. However, there is clearly a practical difference between a solution that stocks out by a large amount in one period, and one that stocks out by very small amounts in several.

We shall thus examine the issue in stages. We shall first consider the tradeoff between average inventory position, defined as the total finished goods and work in process inventory, and the fill rate, which is the fraction of demand in each period met from inventory. We shall then examine the difference between the planned and realized service levels in the chance constraint models, and also explore their sensitivity to errors in the estimation of the demand distributions used.

7.1 Inventory Position-Fill Rate Tradeoff

In order to examine the performance of the different models in terms of their tradeoff between inventory position and fill rate, we shall compute the scaled inventory

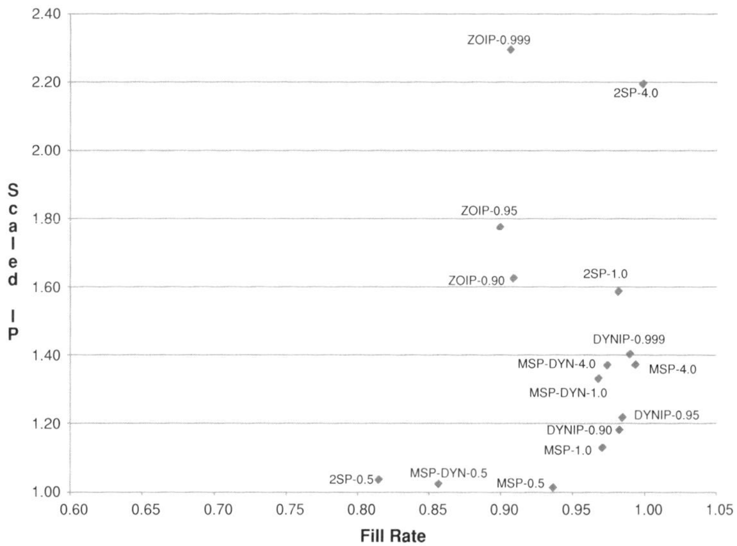

Fig. 3 Average performance of models over all demand configurations

position for each algorithm under each of our seven demand configurations. Let $IP(i,k)$ denote the average inventory position realized under model i under demand configuration k. Then we define the *Scaled* $IP(i,k) = IP(i,k)/\min_j\{IP(j,k)\}$. This quantity indicates the level of inventory position of a given model relative to the model with the lowest average inventory position obtained by any model for that demand configuration.

Figure 3 depicts the tradeoff between the models based on average performance across all demand configurations. The fill rate is plotted on the horizontal axis and the scaled inventory position on the vertical. Since we want fill rate to be high, and scaled IP to be low, the efficient frontier is to the bottom right of the plots.

Figure 3 yields a number of interesting insights. The ZOIP model is completely dominated, as we would expect. This is due to its complete lack of a recourse action, leaving it unable to react to the realized demand after it is observed. In particular, this leaves the model unable to react to demand that is lower than expected, causing it to overstock by a significant amount, as indicated by Ravindran et al. [93]. The two-stage stochastic program 2SP is also dominated. The efficient frontier is made up entirely of DYNIP and the static multistage model MSP, while the dynamic implementation of the M-SP, MSP-DYN, is also dominated.

Two salient features emerge from these results. The first and most encouraging from our perspective is the excellent performance of DYNIP. This model is highly competitive at service levels of 0.90 and 0.95, although it is dominated for a service level of 0.999. The relatively small difference in fill rate between the three service levels suggests that the model overstocks to some degree. There are two possible reasons for this behavior. One is that the assumptions of the chance constrained model are violated in the simulations we use, constituting an interesting direction for future work in understanding the sources of this behavior. Another possibility is that the lead time estimate used to set the safety stock levels is too high. The success of DYNIP over ZOIP is due to its incorporation of a dynamic recourse action—it can modify releases based on observed demand in the past, while ZOIP fixes all decisions at the start of the planning horizon; note that ZOIP and DYNIP use the same information about the demand process.

The comparison between DYNIP and MSP is more interesting. The results indicate that DYNIP obtains the same performance as MSP for a specific choice of service levels corresponding to a choice of parameters for MSP lying between $b = 100$ and $b = 400$. Given the very limited recourse action incorporated in DYNIP, this seems surprising at first sight; one would expect MSP to perform considerably better. However, we need to bear in mind that DYNIP is using a complete characterization of the demand distribution in each period, while MSP characterizes the demand uncertainty through the use of scenarios. Thus the number and choice of scenarios is critical for the MSP to obtain a good solution.

However, this is also where the size of the competing formulations needs to be taken into account. For a planning horizon of T periods, the DYNIP model requires $O(T)$ decision variables and constraints. Assuming two possible realizations for demand in each period as we do in this study, the scenario tree for MSP has $O(2^{T-1})$ nodes, implying that number of decision variables for what is a minimal amount of information on demand uncertainty. These results hold out the encouraging possibility that a minimal number of well-chosen scenarios may be sufficient for a stochastic program to make near-optimal decisions. However, the sheer size of the scenario trees required to model an industrial problem with multiple products, each with their own different demand processes, suggests that scaling conventional stochastic programing models up to solve industrial-sized problems poses substantial challenges.

Another interesting observation from Fig. 3 is the fact that the static MSP outperforms the dynamic version, MSP-DYN. The latter differs from the former in that the M-SP model is resolved at each period in the planning horizon, using the information from the realized demand in previous periods. Hence the recourse action taken at each period is to resolve the M-SP in the light of previously realized demand.

This result is particularly interesting since implementation on a rolling horizon or dynamic basis has been held up as a solution to the problem of uncertain demand in production planning for decades; the assumption is that only the decisions in the next period matter, and as long as we can revise decisions in the light of observed information we can obtain good results. However, some recent results suggest that our faith in this insight may be misplaced, at least under some circumstances. Orcun and Uzsoy [80] have shown that when the planning model does not accurately represent the behavior of the production system under study, rolling horizon implementations can result in undesirable oscillatory behavior similar to the nervousness discussed in the Material Requirements Planning (MRP) literature (e.g., [13]). What is striking in this case is that the extremely simple recourse action used in DYNIP yields just as good results as the far more sophisticated recourse action in MSP-DYN. This may well be due in part to the very limited demand information used in M-SP, as discussed above, which could potentially be remedied by including additional scenarios in the M-SP model. However, this would come at the cost of increasing the size of an already very large model. It is important to note that in the current experiments, the planning horizon T is fixed and does not recede into the future, which will cause ending effects to arise in decisions towards the end of the planning horizon. In particular, the limited planning horizon may cause the models to take decisions that are very good within the current horizon, but have very unfavorable consequences outside the current planning horizon. This issue clearly needs to be more carefully examined in future work.

7.2 Effect of Estimation Errors

In order to further explore the performance of DYNIP relative to MSP, we conducted two additional experiments in which the mean of the demand distribution used in the DYNIP models are perturbed by a random error uniformly distributed between 0 and 0.2 times the mean, representing a situation where demand is systematically overestimated. Our second case represents the case when demand is underestimated, represented by an error uniformly distributed between −0.2 and 0. The standard deviations are subjected to a random error uniformly distributed between −0.2 and 0.2. The purpose of this experiment is to examine the sensitivity of DYNIP to errors in the estimation of the demand distributions used.

The results of these experiments are shown in Fig. 4. The suffix "H" denotes the results for the case with overestimated demand, and "L" for the case with underestimated demand. The results for MSP are included for comparison. The results are quite intuitive. The impact of errors in demand estimation increases with the required service level. When $\alpha = 0.90$, the scaled IP varies between 1.12 and 1.21; for $\alpha = 0.95$, from 1.14 to 1.24; and for $\alpha = 0.999$, from 1.3 to 1.47. The changes in fill rate are all less than 1%. The MSP results are dominated except for MSP-4.0, which achieves a higher service level than DYNIP-0.999 and DYNIP-0.999-H with lower inventory position. These results together suggest that DYNIP is relatively robust to errors in demand estimation, while at the same time supporting the earlier evidence that it tends to overstock relative to the desired service level.

The tradeoff between fill rate and scaled IP for the individual demand configurations was also examined, although detailed results are not presented for brevity. Comparing the HHH, LLL and MMM results indicates that for LLL and MMM, DYNIP dominates MSP, while for HHH MSP enters the efficient frontier, obtaining slightly lower fill rates with substantially lower inventory position than MSP, although DYNIP-0.90 and DYNIP-0.95 remain on the efficient frontier. In all demand configurations except HML, DYNIP is represented in the efficient frontier; in that configuration MSP dominates all the DYNIP models, obtaining both higher fill rate and lower inventory position. Interestingly, the converse is true for the LHL configuration, where MSP is dominated by the DYNIP models.

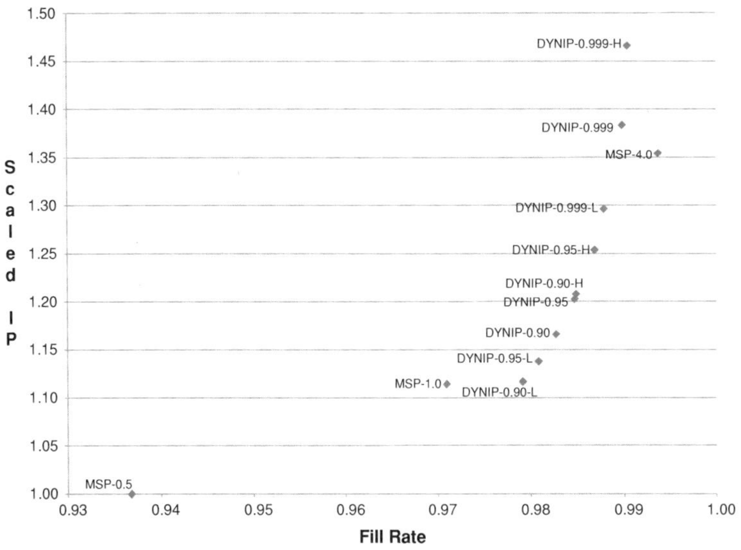

Fig. 4 Sensitivity of DYNIP to errors in demand estimation

Taken as a whole, these results suggest that DYNIP is at least a contender as a solution technique for the planning problem considered in this paper. While it exhibits some weaknesses in the face of high demand variability, its performance appears to be relatively robust to errors in the estimation of the demand distributions it uses, and it consistently achieves a position on the efficient frontier of the fill rate—inventory position tradeoff. It appears to have a tendency to overstock, which is likely due to the discrepancy between the assumptions of the model and the environment in which the simulations take place.

7.3 Service Level-Fill Rate Comparison

An interesting comparison that sheds some additional light on the behavior of the different models is to compare the average service levels and fill rates. The entries in Table 5 are computed by taking the average over all periods in each realization, and then taking the grand average of these over all realizations of a specific demand configuration. It is immediately apparent that the service levels realized by DYNIP are higher than the planned service levels, resulting in even higher fill rates. The reason for this behavior is very likely that the lead time being used to compute the inventory targets is higher than the average lead time that is realized in the simulations. Interestingly, even though the same lead time parameters are used in the ZOIP model, ZOIP's service level is markedly worse than that of DYNIP. ZOIP and DYNIP appear to perform better when the demand distribution is time-stationary

Table 5 Realized fill rates and service levels

		LLL	MMM	HHH	LMH	HML	LHL	HLH
2SP-0.5	SL	0.580	0.580	0.618	0.662	0.558	0.616	0.583
	FR	0.815	0.815	0.828	0.839	0.791	0.812	0.806
2SP-1.0	SL	0.935	0.944	0.948	0.954	0.967	0.966	0.948
	FR	0.978	0.979	0.979	0.978	0.991	0.990	0.978
2SP-4.0	SL	0.983	0.992	1.000	0.996	0.978	0.994	0.993
	FR	0.998	0.999	1.000	1.000	0.995	0.999	0.999
DYNIP-0.90	SL	1.000	0.992	0.958	0.950	0.833	0.983	0.892
	FR	1.000	1.000	0.993	0.987	0.931	0.999	0.969
DYNIP-0.95	SL	1.000	1.000	0.967	0.975	0.833	0.983	0.908
	FR	1.000	1.000	0.995	0.991	0.934	1.000	0.973
DYNIP-0.999	SL	1.000	1.000	0.983	0.983	0.875	1.000	0.925
	FR	1.000	1.000	0.998	0.999	0.952	1.000	0.980
MSP-0.5	SL	0.885	0.892	0.898	0.832	0.869	0.742	0.788
	FR	0.966	0.967	0.969	0.938	0.950	0.872	0.896
MSP-1.0	SL	0.917	0.926	0.951	0.883	0.903	0.884	0.884
	FR	0.979	0.980	0.985	0.965	0.971	0.957	0.960
MSP-4.0	SL	0.965	0.983	0.997	0.951	0.957	0.971	0.973
	FR	0.994	0.997	0.999	0.989	0.989	0.993	0.995
MSP-DYN-0.5	SL	0.846	0.846	0.713	0.667	0.775	0.633	0.658
	FR	0.939	0.937	0.879	0.837	0.897	0.728	0.782
MSP-DYN-1.0	SL	0.917	0.929	0.971	0.817	0.892	0.863	0.879
	FR	0.980	0.986	0.995	0.932	0.964	0.950	0.970
MSP-DYN-4.0	SL	0.917	0.929	0.983	0.842	0.887	0.896	0.921
	FR	0.980	0.986	0.996	0.937	0.965	0.970	0.986
ZOIP-0.90	SL	0.858	0.858	0.650	0.697	0.668	0.772	0.660
	FR	0.964	0.964	0.877	0.909	0.858	0.912	0.880
ZOIP-0.95	SL	0.801	0.737	0.655	0.713	0.682	0.795	0.666
	FR	0.925	0.910	0.881	0.915	0.863	0.921	0.883
ZOIP-0.999	SL	0.776	0.748	0.684	0.738	0.732	0.848	0.656
	FR	0.920	0.913	0.891	0.925	0.887	0.933	0.878

(demand configurations LLL, MMM, and HHH) than when it is not. In contrast, MSP maintains a consistent level of fill rate across all scenarios. The fact that the fill rate is consistently higher than the service level for the chance constrained models (ZOIP and DYNIP) suggests that even though stockouts occur, the amount of the stockout is quite modest in most cases.

8 Conclusions and Future Directions

Acknowledging at the outset the exploratory nature of this chapter, our results raise some interesting issues. Planning procedures with recourse (MSP, MSP-DYN and DYNIP) consistently outperform those without recourse (ZOIP and 2SP) as one would expect. However, the performance of DYNIP suggests that when appropriately parameterized it may be able to compete effectively, in terms of producing near-optimal solutions in reasonable CPU time, with far larger multi-stage stochastic programing models that employ a limited number of scenarios to capture demand uncertainty—at least under certain conditions. It also appears to be relatively robust to errors in estimation of the demand distribution used to construct the model. On the other hand, the MSP model appears to be able to produce good solutions with a minimal number of demand scenarios, considering only two possible values in each planning period. Even so, the MSP approach results in very large models relative to DYNIP. Finally, a dynamic, rolling horizon implementation of MSP yielded no apparent advantage over the static procedure. This finding is interesting in itself, since a rolling implementation is widely held to be the remedy for demand uncertainty.

Given the limited number of experiments carried out, these findings raise more questions than they answer, suggesting several directions for future work to clarify or confirm these findings. Clearly future work needs to focus on procedures with recourse, such as MSP and DYNIP. The reason why DYNIP appears to consistently overstock needs to be understood, and methods found to alleviate this issue if possible. It may be as simple as setting the lead time parameter used to compute the inventory targets more accurately, but it may also be related to the fact that the assumptions used in developing the model are violated in the experimental environment. If the latter is the case, careful mathematical analysis must be carried out to reveal the reason, and suggest an approach to correct the problem. The sensitivity of DYNIP to errors in estimating the demand distributions, and approaches for using it in the face of very limited demand information also need to be explored.

The MSP model used in this work highlights the primary issue with multistage stochastic programing when applied to production planning: the size of the scenario tree grows very rapidly, resulting in very large formulations even when a very limited number of different demand realizations are considered in each period. There needs to be a systematic investigation of how many scenarios need to be considered to provide a reasonably good solution (however that is to be defined, which is another complex issue), and possible solution methods that will allow a scaling up of these approaches to problems of industrial size.

Finally, as we have noted in our analysis, the chance constrained and stochastic programing models make quite different assumptions in formulating the models. The chance constrained models ignore shortage costs, and require a specified stockout probability. The stochastic programing models require a number of scenarios that describe the demand uncertainty and explicit holding and shortage costs. These different assumptions have been shown in the literature to lead to paradoxical behavior for the chance constrained models under certain circumstances, such as a negative value of the expected value of perfect information [14]. While the mathematical existence of such behavior is well documented, it may yet remain the case that chance constrained models, when appropriately formulated and parameterized, can provide effective heuristics for the problem of production planning under uncertain demand.

Acknowledgments The research of Reha Uzsoy was partially supported by the National Science Foundation under Grant No. CCM-1029706. The opinions in this paper are those of the authors, and do not necessarily reflect the position of NSF.

References

1. Agnew C (1976) Dynamic modeling and control of some congestion prone systems. Oper Res 24(3):400–419
2. Anli OM, Caramanis M, Paschalidis IC (2007) Tractable supply chain production planning modeling non-linear lead time and quality of service constraints. J Manuf Syst 26(2):116–134
3. Anupindi R, Morton TE, Pentico D (1996) The nonstationary stochastic lead-time inventory problem: near-myopic bounds, heuristics, and testing. Manag Sci 42(1):124–129
4. Asmundsson JM, Rardin RL, Turkseven CH, Uzsoy R (2009) Production planning models with resources subject to congestion. Naval Res Logist 56:142–157
5. Asmundsson JM, Rardin RL, Uzsoy R (2006) Tractable nonlinear production planning models for semiconductor wafer fabrication facilities. IEEE Trans Semicond Manuf 19:95–111
6. Bang JY, Kim YD (2010) Hierarchical production planning for semiconductor wafer fabrication based on linear programming and discrete-event simulation. IEEE Trans Autom Sci Eng 7(2):326–336
7. Ben-Tal A, Nemirovski A (2000) Robust solutions of linear programming problems contaminated with uncertain data. Math Program 88(3):411–424
8. Bergstrom GL, Smith BE (1970) Multi-Item production planning—an extension of the Hmms rules. Manag Sci 16(10):B614–B629
9. Bertsimas D, Mourtzinou G (1997) Transient laws of non-stationary queueing systems and their applications. Queueing Syst 25:115–155
10. Bertsimas D, Thiele A (2006) A robust optimization approach to inventory theory. Oper Res 54(1):150–168
11. Birge JR (1985) Decomposition and partitioning methods for multistage stochastic linear programs. Oper Res 33(5):989–1007
12. Birge JR, Louveaux F (1997) Introduction to stochastic programming. Springer, New York
13. Blackburn JD, Kropp DH, Millen RA (1986) A comparison of strategies to dampen nervousness in Mrp systems. Manag Sci 32(4):412–439
14. Blau RA (1974) Stochastic programming and decision analysis: an apparent dilemma. Manag Sci 21(3):271–276
15. Bookbinder JH, Tan JY (1988) Strategies for the probabilistic lot sizing problem with service level constraints. Manag Sci 34(9):1096–1108
16. Buffa ES, Taubert WH (1972) Production-inventory systems; planning and control. R.D. Irwin, Homewood Ill
17. Buzacott JA, Shanthikumar JG (1993) Stochastic models of manufacturing systems. Prentice-Hall, Englewood Cliffs
18. Byrne MD, Bakir MA (1999) Production planning using a hybrid simulation-analytical approach. Int J Prod Econ 59:305–311
19. Byrne MD, Hossain MM (2005) Production planning: an improved hybrid approach. Int J Prod Econ 93–94:225–229
20. Charnes A, Cooper WW (1959) Chance-constrained programming. Manag Sci 6(1):73–79

21. Charnes A, Cooper WW (1963) Deterministic equivalents for optimizing and satisficing under chance constraints. Oper Res 11:18–39
22. Charnes A, Cooper WW, Symonds GH (1958) Cost horizons and certainty equivalents: an approach to stochastic programming of heating oil. Manag Sci 4(3):235–263
23. Ciarallo FW, Akella R, Morton TE (1994) A periodic review, production planning-model with uncertain capacity and uncertain demand—optimality of extended myopic policies. Manag Sci 40(3):320–332
24. Clark AJ, Scarf H (1960) Optimal policies for a multi-echelon inventory problem. Manag Sci 6(4):475–490
25. Dantzig GB, Wolfe P (1960) Decomposition principle for linear programs. Oper Res 8(1): 101–111
26. Dauzere-Peres S, Lasserre JB (1994) An integrated approach in production planning and scheduling. Springer, Berlin
27. de Kok AG, Fransoo JC (2003) Planning supply chain operations: definition and comparison of planning concepts. In: Kok AG, Graves SC (eds) Or handbook on supply chain management, Elsevier, Amsterdam, pp 597–675
28. Deckro RF, Hebert JE (1984) Goal programming approaches to solving linear decision rule based aggregate production planning-models. IIE Trans 16(4):308–315
29. Eppen G, Martin RK (1988) Determining safety stock in the presence of stochastic lead times. Manag Sci 34:1380–1390
30. Escudero LF, Kamesan PV, King AJ, Wets JB (1993) Production planning via scenario modelling. Ann Oper Res 43:311–335
31. Ettl M, Feigin G, Lin GY, Yao DD (2000) A supply chain network model with base-stock control and service requirements. Oper Res 48:216–232
32. Federgruen A, Zipkin P (1986) An inventory model with limited production capacity and uncertain demands I: the average cost criterion. Math Oper Res 11(2):193–207
33. Federgruen A, Zipkin P (1986) An inventory model with limited production capacity and uncertain demands II: the discounted cost criterion. Math Oper Res 11(2):208–215
34. Garstka SJ, Wets RJB (1974) On decision rules in stochastic programming. Math Program 7(2):117–143
35. Gassmann HI (1990) Mslips: a computer code for the multistage stochastic linear programming problem. Math Program 47:407–423
36. Goodman DA (1974) Goal programming approach to aggregate planning of production and work force. Manag Sci Ser B Appl 20(12):1569–1575
37. Graves SC (1986) A tactical planning model for a job shop. Oper Res 34:552–533
38. Graves SC (1988) Safety stocks in manufacturing systems. J Manuf Oper Manag 1:67–101
39. Grubbstrom RW (1998) A net present value approach to safety stocks in planned production. Int J Prod Econ 56(57):213–229
40. Gupta A, Maranas CD (2003) Managing demand ncertainty in supply chain planning. Comput Chem Eng 27(8–9):1219–1227
41. Hackman S (2008) Production economics. Springer, Berlin
42. Hackman ST, Leachman RC (1989) A general framework for modeling production. Manag Sci 35:478–495
43. Hadley G, Whitin TM (1963) Analysis of inventory systems. Prentice-Hall, Englewood Cliffs
44. Hanssmann F, Hess SW (1960) A linear programming approach to production and employment scheduling. Manag Technol 1(1):46–51
45. Hax AC, Candea D (1984) Production and inventory management. Prentice-Hall, Englewood Cliffs
46. Heyman DP, Sobel MJ (1982) Stochastic models in operations research. McGraw-Hill, New York
47. Heyman DP, Sobel MJ (1990) Stochastic models. Elsevier Science Publishing Co., New York

48. Higle JL, Kempf KG (2010) Production planning under supply and demand uncertainty: a stochastic programming approach: stochastic programming: the state of the art. G. infanger. Springer, Berlin
49. Holt CC, Modigliani F, Muth JF (1956) Derivation of a linear rule for production and employment. Manag Sci 2(2):159–177
50. Holt CC, Modigliani F, Muth JF, Simon HA (1960) Planning production, inventories and work force. Prentice Hall, Englewood Cliffs
51. Holt CC, Modigliani F, Simon HA (1955) A linear decision rule for production and employment scheduling. Manag Sci 2(1):1–30
52. Hopp WJ, Spearman ML (2001) Factory physics: foundations of manufacturing management. Irwin/McGraw-Hill, Boston
53. Hung YF, Leachman RC (1996) A production planning methodology for semiconductor manufacturing based on iterative simulation and linear programming calculations. IEEE Trans Semicond Manuf 9(2):257–269
54. Iglehart DL, Karlin S (1962) Optimal policy for dynamic inventory process with nonstationary stochastic demands. Stanford University Press, Stanford Calif, pp 127–147
55. Irdem DF, Kacar NB, Uzsoy R (2010) An exploratory analysis of two iterative linear programming-simulation approaches for production planning. IEEE Trans Semicond Manuf 23:442–455
56. Johnson LA, Montgomery DC (1974) Operations research in production planning, scheduling and inventory control. Wiley, New York
57. Kacar NB, Irdem DF, Uzsoy R (2010) An experimental comparison of production planning using clearing functions and iterative linear programming-simulation algorithms. Research Report, Edward P. Fitts Department of Industrial and Systems Engineering, North Carolina State University
58. Kall P, Wallace SW (1994) Stochastic programming. Wiley, Chichester
59. Karlin S (1960) Dynamic inventory policy with varying stochastic demands. Manag Sci 6(3):231–258
60. Karmarkar US (1987) Lot sizes, lead times and in-process inventories. Manag Sci 33(3): 409–418
61. Karmarkar US (1989) Capacity loading and release planning with Work-in-Progress (WIP) and lead-times. J Manuf Oper Manag 2:105–123
62. Kefeli A, Uzsoy R, Fathi Y, Kay M (2011) Using a mathematical programming model to examine the marginal price of capacitated resources. Int J Prod Econ 131(1):383–391
63. Kekre S (1984) Some issues in job shop design. University of Rochester, Rochester NY
64. Kempf KG (2004) Control-oriented approaches to supply chain management in semiconductor manufacturing. In: Proceedings of the American control conference, Boston, MA, United States
65. Kempf KG, Keskinocak P, Uzsoy R (2010) Preface. In: Kempf KG, Keskinocak P, Uzsoy R (eds) Planning production and inventories in the extended enterprise: a state of the art handbook, Springer, Amsterdam, pp 1–20
66. Kim B, Kim S (2001) Extended model for a hybrid production planning approach. Int J Prod Econ 73:165–173
67. Kleindorfer PR, Kriebel CH, Thompson GL, Kleindorfer GB (1975) Discrete optimal control of production plans. Manag Sci 22(3):261–273
68. Lasdon LS (1970) Optimization theory for large systems. Macmillan, New York
69. Lautenschläger M, Stadtler H (1998) Modelling lead times depending on capacity utilization. Research Report, Technische Universitat Darmstadt
70. Leung SCH, Wu Y (2004) A robust optimization model for stochastic aggregate production planning. Prod Planning Control 15(5):502–514
71. Liu L, Liu X, Yao DD (2004) Analysis and optimization of multi-stage inventory queues. Manag Sci 50:365–380

72. Louveaux F (1980) A solution method for multistage stochastc programs with recourse with application to an energy investment problem. Oper Res 28(4):889–902
73. Meal H (1979) Safety stocks in Mrp systems. Operations Research Center, Massachusetts Institute of Technology, Cambridge MA
74. Medhi J (1991) Stochastic models in queuing theory. Academic Press, Boston
75. Miller JG (1979) Hedging the master schedule. Dissagregation problems in manufacturing and service organizations. LP Ritzman, Martinus Nijhoff, Boston MA
76. Missbauer H (2002) Aggregate order release planning for time-varying demand. Int J Prod Res 40:688–718
77. Missbauer H (2011) Order release planning with clearing functions: a queueing-theoretical analysis of the clearing function concept. Int J Prod Econ 131(1):399–406
78. Missbauer H, Uzsoy R (2010) Optimization models for production planning. In: Kempf KG, Keskinocak P, Uzsoy R (eds) Planning production and inventories in the extended enterprise: a state of the art handbook, Springer, New York, pp 437–508
79. Orcun S, Kempf KG, Uzsoy R (2009) An integrated production planning model with load-dependent lead times and safety stocks. Comput Chem Eng 32:2159–2136
80. Orcun S, Uzsoy R (2011) The effects of production planning on the dynamic behavior of a simple supply chain: an experimental study. In: Kempf KG, Keskinocak P, Uzsoy R (eds) Planning in the extended enterprise: a state of the art handbook, Springer, Berlin, pp 43–80
81. Orcun S, Uzsoy R, Kempf KG (2006) Using system dynamics simulations to compare capacity models for production planning. Winter Simulation Conference, Monterey
82. Orlicky J (1975) Material requirements planning: the new way of life in production and inventory management. McGraw-Hill, New York
83. Pahl J, Voss S, Woodruff DL (2005) Production planning with load dependent lead times. 4OR Q J Oper Res 3:257–302
84. Parlar M (1985) A stochastic production planning model with a dynamic chance constraint. Eur J Oper Res 20(2):255–260
85. Peters RJ, Boskma K, Kupper HAE (1977) Stochastic programming in production planning: a case with non-simple recourse. Statistica Neerlandica 31:113–126
86. Powell WB, Jaillet P, Odoni A (1995) Stochastic and dynamic networks and routing. In: Ball M, Magnanti T, Monma C (eds) Handbooks in operations research and the management sciences. Amsterdam, Elsevier, pp 141–295
87. Prékopa A (1995) Stochastic programming. Kluwer Academic Publishers, Boston
88. Pritsker AAB, Snyder K (1997) Production scheduling using factor. In: Artiba A, Elmaghraby SE (eds) The planning and scheduling of production systems. Chapman and Hall
89. Puterman ML (2005) Markov decision processes: discrete stochastic dynamic programming. Wiley, New York
90. Raa B, Aghezzaf EH (2005) A robust dynamic planning strategy for lot-sizing problems with stochastic demands. J Intell Manuf 16(2):207–213
91. Rao SS, Gunasekaran A, Goyal SK, Martikainen T (1998) Waiting line model applications in manufacturing. Int J Prod Econ 54(1):1–28
92. Rardin RL, Uzsoy R (2001) Experimental evaluation of heuristic optimization algorithms: a tutorial. J Heuristics 7:261–304
93. Ravindran A, Kempf KG, Uzsoy R (2008) Dynamic base stock models for production-inventory systems with nonstationary demand. Research Report, Edward P. Fitts Department of Industrial and Systems Engineering, North carolina State University
94. Ravindran A, Kempf KG, Uzsoy R (2011) Production planning with load-dependent lead times and safety stocks. Int J Plan Sched 1(1–2):58–89
95. Riaño G (2003) Transient behavior of stochastic networks: application to production planning with load-dependent lead times. School of industrial and systems engineering. Georgia Institute of Technology, Atlanta GA
96. Selçuk B, Fransoo JC, de Kok AG (2007) Work in process clearing in supply chain operations planning. IIE Trans 40:206–220

97. Spearman ML (1991) An Analytic congestion model for closed production systems with Ifr processing times. Manag Sci 37(8):1015–1029
98. Srinivasan A, Carey M, Morton TE (1988) Resource pricing and aggregate scheduling in manufacturing systems. Graduate School of Industrial Administration, Carnegie-Mellon University, Pittsburgh
99. Tayur SR (1993) Computing the optimal policy for capacitated inventory models. Commun Stat Stoch Models 9(4):585–598
100. Van Slyke RM, Wets JB (1969) L-shaped linear programs with applications to optimal control and stochastic programming. SiAM J Appl Math 17(4):638–663
101. Veinott AF (1965) Optimal policy for a multi-product, dynamic, nonstationary inventory problem. Manag Sci 12(3):206–222
102. Veinott AF (1965) Optimal policy in a dynamic single product nonstationary inventory model with several demand classes. Oper Res 13(5):761–778
103. Voss S, Woodruff DL (2003) Introduction to computational optimization models for production planning in a supply chain. Springer, Berlin
104. Whitt W (1991) A Review of $L = \Lambda w$ and Extensions. Queueing Syst 9:235–268
105. Wittrock RJ (1983) Advances in a nested decomposition algorithm for solving staircase linear programs. Technical Report SOL-83-2, Systems Optimization Laboratory
106. Zipkin PH (1986) Models for design and control of stochastic, multi-item batch production systems. Oper Res 34(1):91–104
107. Zipkin PH (2000) Foundations of inventory management. Irwin, Burr Ridge IL

Traffic Flow Models and Service Rules for Complex Production Systems

Christian Ringhofer

Abstract We present an overview over recent developments of traffic flow models for production networks. Particular emphasis is given to the implementation of service rules for complex systems, involving multiple product types and re-entrant loops. A rather general scheduling concept is introduced and demonstrated on some numerical experiments.

1 Introduction

This article gives an overview over recent developments in traffic flow type models for complex production systems. Traffic flow models represent, in some sense, an intermediate stage between simple rate equations (or so called fluid models) [1, 8, 22] and a detailed discrete event simulation (DES) [7]. They allow for a more detailed description of transient phenomena than rate equations, and represent the many body mean field limit of multi-agent models. The price to be paid for this detail is that they involve not only the solution of systems of ordinary differential equations, but also the solution of systems of (in general hyperbolic) conservation laws. While all these macroscopic approaches will never be able to recapture the detail provided by discrete event simulation or multi-agent models, they do have some significant advantages.

- They are scalable. That is, they compute a density of parts in the production system. A large number of parts does not require a larger number of agents, but simply

Work supported by NSF grants DMS-0604986 and DMS-0757309.

C. Ringhofer (✉)
School of Mathematical and Statistical Sciences,
Arizona State University, Tempe, AZ 85287-1804, USA
e-mail: ringhofer@asu.edu

D. Armbruster and K. G. Kempf (eds.), *Decision Policies for Production Networks*, 209
DOI: 10.1007/978-0-85729-644-3_9, © Springer-Verlag London 2012

results in higher densities. This means that the computational effort required is independent of the number of parts considered.
- Since they involve differential equations they readily render themselves to optimization and optimal control algorithms, where a function of the system is optimized, and the dynamics of the system enter as constraints [16, 19].
- Discrete event simulations usually suffer from a model maintenance problem. A complex production system is not a physical system that is governed by a few basic laws. So, by the time all the details of a complex system are entered in the discrete event simulator, they probably have changed already, negating the advantage of the detailed description. So we are dealing with transport in a constantly evolving medium.

In general, traffic flow models are much better able to predict the dynamic response of a production system than fluid models. So, they are more capable to describe non-equilibrium situations, such as temporary overloads or breakdowns. They are, in some sense, a generalization of fluid models, since they will in general reduce to fluid models (rate equations) if only one computational cell is used in the spatial direction.

The main disadvantage of macroscopic descriptions, such as fluid or traffic flow models, lies in their lack of versatility. While it is relatively simple to modify the rules of a discrete event simulation or multi-agent model, this requires usually a major modeling step in the macroscopic model. This paper tries to give an overview over recent approaches to address this problem.

In Sect. 2 we give a brief introduction to traffic flow models and to possible approaches to their derivation. These approaches generally fall into two categories. The first is the use of clearing functions [15]. That is, we use methods of steady state queueing theory to compute the actual dynamic response of a stochastic system. This necessarily results in a quasi-steady-state theory, with all the obvious limitations. The second approach is to use mean field theories, borrowed from many body physics. While this is in some sense much more rigorous (capturing the actual dynamics) it is limited to a certain set of relatively simple interaction mechanisms. (The same argument could be made, of course, for or against the clearing functions derived from queuing theory.)

One of the main obstacles to use macroscopic models, such as fluid or traffic flow models, is the inclusion of policies and service rules. A complex production system, producing more than one product type and exhibiting a re-entrant topology, will require rules which product to serve first at what stage of a re-entrant loop in the production cycle. While, again, the implementation of such rules is rather straight forward in multi-agent or DES models, the implementation in macroscopic models, based on differential equations, represents a challenge. In Sect. 3 we give a general outline—or a recipe—how to model more or less arbitrarily complex service rules in the context of traffic flow models. The key ingredient is to attach a, dynamically changing, vector-valued attribute to each type of part. We implement an arbitrarily general service rule by defining a priority function, which determines, depending on the attribute, which part is served first.

Section 4 is devoted to numerical examples, which demonstrate the use of this strategy, and which verify its accuracy against discrete event simulations.

2 Clearing Functions and Fluid Models

Fluid models or rate equations based on clearing functions, are a relatively inexpensive way to model the behavior of queueing systems. The basic idea of a clearing function [15] is to consider a simple system, consisting of a queue and a server. Parts arrive in random intervals at the end of the queue, and the server processes them at a (in general random) rate. The clearing function gives the expectation of the time τ it takes for a part to pass through the system in terms of the expectation of the Work in Progress (WIP) W, i.e. the number of parts in the queue plus the number of parts currently processed. So, it is of the form $\tau = \tau(W)$. The clearing function $\tau(W)$ is derived from steady state queuing theory, and its form depends obviously on the type of arrival and service processes under consideration [15]. An alternative for more complex systems with multiple servers and multiple queues, is to fit the clearing function to either observed data or detailed discrete event simulations (DESs) [4, 7]. Using Little's law [12], the outflux of the system is then given, in steady state, by $\phi = \frac{W}{\tau(W)}$. A fluid model [1] represents then a simple rate equation of the form

$$\frac{\mathrm{d}}{\mathrm{d}t} W(t) = \lambda(t) - \phi(t) = \lambda(t) - \frac{W}{\tau(W)}, \tag{1}$$

for the evolution of the expected WIP $W(t)$. Here $\lambda(t)$ denotes the mean arrival rate. Note, that λ and W are now time-dependent, whereas the derivation of of the outflux $\phi = \frac{W}{\tau}$ relies on what is essentially a steady state theory. Herein lies the basic problem and most of the limitation of clearing functions. We note, that the term fluid model—although generally used in the literature—is a bit unfortunate, since the simple rate Eq. (1) involves none of the properties usually associated with a physical fluid, other than simple mass conservation. However, because they are so simple, they allow for the efficient simulation of large coupled systems of individual queue—server molecules. They also render themselves readily to optimization algorithms and therefore allow the design of optimal policies for large systems.

The inconsistency of using a steady flux function in a time-dependent model becomes apparent in the following observation: A sudden change in the influx λ in (1) will produce an immediate response in the outflux $\phi = \frac{W}{\tau}$, i.e. the change in λ will instantaneously change the WIP W, and therefore instantaneously change the outflux. In the actual system a sudden change in the arrival rate will only have an impact on the outflux, once this change has worked itself through the queue. So the response will be time delayed. This time delay is neglected in the fluid model, since it is valid essentially only close to the equilibrium situation where the outflux f almost equals the influx λ and the mean WIP W only varies relatively slowly in time. A more accurate model, producing a time delayed response, would have to make the outflux ϕ dependent not only on the current WIP $W(t)$, but in some way on the history of the evolution of W. This would replace the fluid model (1) by a delay differential equation with a distributed delay, and the question arises how to construct such an equation. One simple way to produce a delayed response in the system is to replace the ordinary differential equation (1) by a conservation law for a density. In this approach, the WIP $W(t)$ is replaced by a part density $\rho(x,t)$, where the additional independent variable x is an artificial construct denoting the degree of completion or a continuous stage of the process. So a part enters the system at $x = 0$ and leaves the system at $x = S$. $\rho(x,t)$ denotes the density of parts per stage and the WIP $W(t)$ is related to $\rho(x,t)$ by

$$W(t) = \int_0^S \rho(x,t)\,\mathrm{d}x.$$

The density $\rho(x,t)$ satisfies the conservation law

$$\partial_t \rho(x,t) + \partial_x [v_\rho \rho] = 0. \tag{2}$$

Here, $v_\rho(x,t)$ denotes the velocity (measured in stages/time) with which parts move through the artificial stages (depending, in some functional form, on the density ρ), and integrating (2) from $x = 0$ to $x = S$ gives

$$\frac{\mathrm{d}}{\mathrm{d}t} W(t) = v_\rho(0,t)\rho(0,t) - v_\rho(S,t)\rho(S,t).$$

So, the arrival rate $\lambda(t)$ in (1) has to equal the term $v_\rho(0,t)\rho(0,t)$, giving a boundary condition for the conservation law (2) at stage $x = 0$. The outflux $\phi(t)$ is now given by $\phi(t) = v_\rho(S,t)\rho(S,t)$, i.e. not by a simple functional relation in terms of the WIP $W(t)$, but in terms of the density ρ, which encodes the history of the WIP W. An instantaneous change in the arrival rate $\lambda = v_\rho(0,t)\rho(0,t)$ will now produce a time delayed change in the outflux, because any information in (2) is transported with a velocity $v_\rho(x,t)$. The cardinal question is, of course, how to construct the velocity v_ρ.

2.1 Traffic Flow Models Based on Clearing Functions

A First Order Model

The simplest, somewhat heuristic way to construct the velocity $v_\rho(x,t)$ is to essentially linearly interpolate the cycle time, given by the clearing function $\tau(W)$. To this end, we split $\tau(W)$ into two parts, setting $\tau(W) = \tau_q(W) + \tau(0)$, $\tau_q(W) = \tau(W) - \tau(0)$. Here, $\tau(0)$ denotes the pure processing time of a part arriving at an empty system, and $\tau_q(W)$ denotes the time the part has to wait until being processed. Breaking down the system into an infinite number of identical parts gives for a part at stage x, who still has to cover a distance $S - x$ in order to reach the end, the formula

$$\tau(x,t) = \tau_q\left(\int_x^S \rho(z,t)\,dz\right) + \frac{S-x}{S}\tau(0) = \tau\left(\int_x^S \rho(z,t)\,dz\right) - \frac{x}{S}\tau(0).$$

Here the term $\int_x^S \rho(z,t)\,dz$ denotes the number of parts in front of the part at stage x, and the pure processing time $\tau(0)$ is equidistributed over the interval $[0, S]$. This gives for the velocity $v_\rho(x,t)$ in (2) the formula

$$v_\rho(x,t) = \frac{S-x}{\tau(x,t)} = \frac{S-x}{\tau(\int_x^S \rho(z,t)\,dz) - \frac{x}{S}\tau(0)}$$

$$v_\rho(S,t) = \frac{S}{\tau(0) + S\tau'(0)\rho(S,t)}$$

Lagrangian Methods Based on Clearing Functions

The natural way of implementing the clearing function into the conservation law model is to start from the Lagrangian picture of fluid dynamics, where the coordinates of individual particles (parts) are used as the primary variables, instead of the spatial density ρ. We start from the following picture:

- A part arrives in the system at time $t = a$.
- We estimate the cycle time τ by using the clearing function. Thus, we determine a velocity $\eta = \frac{S}{\tau}$ at the time the part enters the system, which we keep constant for this particular part, and move the part with the velocity η from $x = 0$ to $x = S$.

Numbering the parts by a continuous index y, and denoting the position of part number y at time t by $x = \xi(y,t)$, this gives the model

$$\partial_t \xi(y,t) = \eta(y), \quad \xi(y, a(y)) = 0. \tag{3}$$

Here $a(y)$ denotes the arrival time of part number y in the system. The velocity $\eta(y)$ is determined at the time the part enters the system. So $\eta(y) = \tau(W(a(y)))$ holds (where we still have to relate the ensemble of coordinates ξ to the WIP W). The relation between the part coordinates $\xi(y,t)$ the WIP $W(t)$ and the density $\rho(x,t)$ is established by a simple counting argument. Counting all the parts in the interval $[0, S]$ gives

$$W(t) = \int [H(\xi(y,t)) - H(\xi(y,t) - S)]\,dy. \tag{4}$$

Here $H(x)$ denotes the usual Heaviside function, with $H(x) = 1$ for $x > 0$ and $H(x) = 0$ for $x < 0$. The function $H(\xi(y,t)) - H(\xi(y,t) - S)$ equals unity for $0 < \xi < S$ and zero else. So the formula (4) counts the number of parts in the interval $[0, S]$ (the number currently in the system). Similarly, if we count the number of parts in a given interval $[x, x + \Delta x]$, we have

$$\int_x^{x+\Delta x} \rho(z,t)\,dz = \int [H(\xi(y,t) - x) - H(\xi(y,t) - x - \Delta x)]\,dy.$$

Dividing by Δx and letting $\Delta x \to 0$, we obtain

$$\rho(x,t) = \int \delta(\xi(y,t) - x)\,\mathrm{d}y, \tag{5}$$

where δ denotes the Dirac δ-function (i.e. the derivative of the Heaviside function H). The conservation law (2) is then obtained in the following way: The velocity $v_\rho(x,t)$ of a part remains constant for this particular part for all times. So, we set

$$v_\rho(\xi(y,t),t) = \eta(y), \ \forall y, t. \tag{6}$$

Differentiating (6) with respect to time gives (because $\partial_t \xi = \eta$ holds)

$$\partial_t v_\rho(x,t) + \eta(y)\partial_x v_\rho(x,t)|_{x=\xi(y,t)} = 0,$$

and therefore we obtain the equation

$$\partial_t v_\rho(x,t) + v_\rho \partial_x v_\rho(x,t) = 0, \tag{7}$$

for the velocity v_ρ. Although the transport equation for v_ρ is independent of the density ρ, the velocity v_ρ still depends on ρ through the boundary conditions, since we have

$$v_\rho(0,a(y)) = v_\rho(\xi(y,a(y)),a(y)) = \eta(y) = \frac{S}{\tau(W(a(y)))},$$

giving the boundary condition

$$v_\rho(0,t) = \frac{S}{\tau(W(t))}, \quad W(t) = \int_0^S \rho(x,t)\,\mathrm{d}x. \tag{8}$$

This means, that the velocity v_ρ is chosen once at the entrance at $x = 0$, and then transported along the particle trajectories according to Eq. (7). To obtain the conservation law (2), we differentiate the definition (5) for $\rho(x,t)$ with respect to time, and obtain

$$\begin{aligned}\partial_t \rho(x,t) &= \int \delta'(\xi(y,t)-x)\partial_t\xi(y,t)\mathrm{d}y = -\partial_x\left[\int \delta(\xi(y,t)-x)\eta(y)\,\mathrm{d}y\right]\\ &= -\partial_x\left[\int \delta(\xi(y,t)-x)v(\xi(y,t),t)\,\mathrm{d}y\right]\\ &= -\partial_x\left[v(x,t)\int \delta(\xi(y,t)-x)\,\mathrm{d}y\right] = -\partial_x[v_\rho \rho(x,t)]\end{aligned}$$

or

$$\partial_t \rho + \partial_x[v_\rho \rho] = 0 \tag{9}$$

We still have to determine a boundary condition for the density $\rho(x,t)$ in terms of the influx $\lambda(t)$. Integrating the conservation law (9) from $x = 0$ to $x = S$ gives

$$\begin{aligned}\frac{\mathrm{d}}{\mathrm{d}t}W(t) &= \int_0^S \partial_t\rho(x,t)\,\mathrm{d}x = -\int_0^S \partial_x[v_\rho\rho(x,t)]\,\mathrm{d}x = v_\rho\rho(0,t) - v_\rho\rho(S,t)\\ &= \lambda(t) - \phi(t).\end{aligned}$$

So the influx $\lambda(t)$ enters the conservation law picture via the boundary conditions

$$v_\rho\rho(0,t) = \lambda(t) \tag{10}$$

and the outflux $\phi(t)$ in the fluid model is replaced by $\phi(t) = v_\rho\rho(S,t)$. We conclude by mentioning how the arrival distribution time $a(y)$ is related to the influx $\lambda(t)$. Given, that the parts move with a constant velocity η, the solution of (3) is given by

$$\xi(y,t) = \eta(y)(t - a(y)),$$

which implies

$$v_\rho\rho(x,t) = \int \eta\delta(\eta(t-a(y)) - x)\,\mathrm{d}y = \int \delta\left(t - a(y) - \frac{x}{\eta}\right)\mathrm{d}y.$$

So, in particular

$$\lambda(t) = v_\rho\rho(0,t) = \int \delta(t - a(y))\,\mathrm{d}y.$$

The term $\int \delta(t - a(y))\,\mathrm{d}y$ is nothing else but the derivative of the functional inverse of a. Given that $a(y)$ is a monotonically increasing function, there exists an inverse function $a^{-1}(t)$ satisfying $a(y) = t \iff y = a^{-1}(t)$. Substituting $y = a^{-1}(s)$ in the integral, we obtain

$$\lambda(t) = \int \delta(t-s)(a^{-1})'(s)\,\mathrm{d}s = (a^{-1})'(t).$$

So, in order to compute the arrival times $a(y)$ in the particle model (3), given an arrival rate λ, we first have to compute the cumulative arrivals $a^{-1}(t) = \int_0^t \lambda(s)\,\mathrm{d}s$, and form the inverse of this function.

Equations (7) and (9) constitute the equations of pressureless gas dynamics, and the clearing function enters the model through the boundary conditions (8) and (10).

2.2 Mean Field Models

An alternative to use steady state queueing theory to derive the flux function for a conservation law is to directly model the velocity of parts via mean field theory [6, 9]. This proceeds in general according to the following approach: Given a certain interaction rule between parts we first derive an evolution equation for the trajectories $\xi_1(t), \ldots, \xi_N(t)$ of N parts. This trajectory equations consist usually of rather simple ODE systems, and replicate more or less a discrete event simulation. One then derives immediately an transport equation for the probability density

$$\mathbf{f}(x_1, \ldots, x_N, t) = \frac{\mathrm{d}\mathscr{P}}{\mathrm{d}x_1, \ldots, x_N}[\xi_1 = x_1, \ldots, \xi_N = x_N]. \tag{11}$$

The density function in (11) is of course a rather complicated object, since it depends on N (the number of parts in the system) independent variables. So, in physics terms, we really treat the N—body problem. It is therefore not usable in direct computations. The basic idea of a mean field model is that, for a large number of parts, the correlation between two randomly chosen parts can be assumed to be small. Neglecting the correlation and assuming identical parts the ansatz

$$\mathbf{f}(x_1, \ldots, x_N, t) = \prod_{n=1}^{N} f(x_n, t)$$

is made for the probability density in (11). Integrating out the variables $x_2, \ldots, x_N$ yields a transport for the effective single part probability density $f(x, t)$ which then can—up to a scaling factor—be identified with the part density $\rho(x, t)$. It should be noted that this approach depends heavily on the type of part interaction, and the process of integrating out all but one variables is usually not an easy task. It is, however, the standard way to derive gas dynamics equations from microscopic physics models and, in this case, rigorously establishes the link between multi-agent models and macroscopic PDE models.

A Deterministic Automaton

We first start with a deterministic agent-based model with a rather simple rule:

- Each server is located in an interval of length Δx.
- Each server has a processing time of $\frac{\Delta x}{V}$ time units (where the velocity V is already measured in stages/time).
- Each server can handle p parts at the same time. So it accepts a new part every $\frac{\Delta x}{pV}$ time units, yielding a maximum capacity of $c = \frac{pV}{\Delta x}$ parts per time.

In [2, 5] it has been shown that—in the limit of a large number of servers and a large number of parts—the resulting limiting fluid model is of the form

$$\partial_t \rho + \partial_x \Phi = 0, \quad \Phi(x, t, \rho) = v_\rho \rho = \min\{c, V\rho\}. \tag{12}$$

Equation (12) models a purely deterministic behavior, i.e. an automaton. It simply states that the velocity is of a part moving through the stages is given by V as long as there is no buildup of queues. The total flux, however, can never exceed the capacity

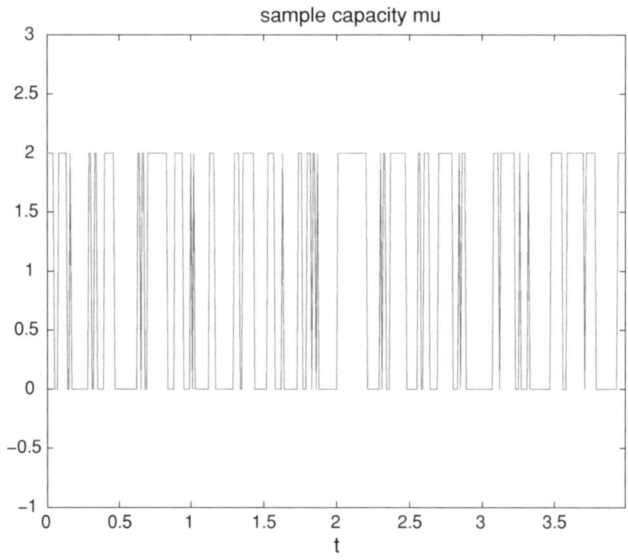

Fig. 1 Capacity of one server, generated from a Markov process for breakdowns and repairs. Average availability: 75%

of the servers given by c. If the servers are not identical, the V and c have to be made dependent of the stage variable x.

Random Breakdowns

We modify the above model by making the capacities c not only dependent on the stage variable x, but on time as well. Temporal fluctuations in the capacity model the breakdown of servers. These breakdowns occur in random intervals, making $c(x,t)$ not only time-dependent but also a random variable. In the simplest case we assume that each server has only two modes, namely up when $c(x,t) = \mu(x)$ (the on capacity) or down, when $c(x,t) = 0$. For each server we create an independent time series of capacities. Figure 1 shows the randomly generated capacity of one particular server which is down 25% of the time. In [11] it has been shown that—again under the assumption of many parts and many servers—the long time behavior of the system obeys the equation

$$\partial_t \rho + \partial_x \Phi = 0, \quad \Phi(x,t,\rho) = v_\rho \rho = a(x)\mu\left[1 - \exp\left(\frac{V\rho}{\mu}\right)\right]. \tag{13}$$

The flux function (13) is the generalization of the deterministic case in (12) in the following sense: For $\rho \to 0$ (in the case when the system is almost empty) we recuperate the flux $\Phi = V\rho$. In the same way for $\rho \to \infty$ (for full queues) we again have the flux given by the maximum 'on' capacity μ. In both cases the flux has to be multiplied by the availability a the ratio $\frac{T_{\text{up}}}{T_{\text{up}}+T_{\text{down}}}$ of the mean up and down times, since the breakdowns will affect the throughput regardless of the size of the density.

3 Policies and Service Rule

When considering a more complex system [13, 14] than just a simple queue, it becomes necessary to consider policies [3, 10]. The need for policies arises when one and the same server serves more than one product or, in the case of re-entrant systems, the same product at different stages of the production cycle. In this case service rules have to be established, determining in what order the parts are processed. One possibility is to simply give a type of part or part priority over the other: As a part arrives it is put into a buffer or queue serving only its type. The buffers are served in order of their priority, that is queue number two is served only when queue number one is empty and so on. From the point of implementation in the actual physical system, the simplest service rule is first in–first out, or first come–first serve (FIFO), where the parts are simply processed in the order of their arrival. Alternative rules, also sometimes employed, are first in system first out (FISFO) and last in system first out (LISFO), scheduling to due date, i.e. each part has a certain delivery date and parts with the closest delivery date are processed first. In practice, an arbitrarily complicated combination of these rules is possible.

While it is rather straight forward to implement these rules in a discrete event simulator, implementation in a fluid model, or in a model based on PDE conservation laws, as discussed in the previous section, is not so simple. In the following, we try to give a rather general recipe to implement what can be almost arbitrarily complex service rules into PDE conservation law models and rate equations.

We start with a linear chain of servers. We assume that we have decided on a conservation law model, using any of the approaches discussed in the previous section. We therefore have a model for the total flux of the form

$$\partial_t \rho + \partial_x [v_\rho \rho] = 0, \quad x \in [0, S], \ t \geq 0, \quad v_\rho \rho(0, t) = \lambda(t). \tag{14}$$

We note, that it might be necessary to solve an auxiliary differential equation (as in (7)) to actually compute the velocity v_ρ in (14). This has, however, no consequence for the following, and we omit the possible auxiliary equation for the velocity. The function $\lambda(t)$ in (14) denotes the total influx into the system.

3.1 Scheduling by Type

This is, from the point of conservation laws, the service rule which by far the easiest to implement. We assume that the total density ρ, and therefore the total WIP $W(t) = \int_0^S \rho(x, t)\, \mathrm{d}x$, consists of N different types of parts which are processed in the order of their importance. So we have

$$\rho(x, t) = \sum_{n=1}^{N} \rho_n(x, t), \quad \lambda(x, t) = \sum_{n=1}^{N} \lambda_n(t)$$

in (14). We number the individual densities ρ_n such that ρ_1 is the density of parts with the highest priority and ρ_N the one with the lowest priority. The challenge is now to model the velocities and fluxes of the individual components ρ_n. To this end, we define the flux $\Phi(x, t, \rho) = v_\rho \rho$. We will derive a system of conservation laws for the individual densities ρ_n of the form

$$\partial_t \rho_n + \partial_x F_n(x, t, \rho_1, \dots, \rho_N) = 0,$$

where the individual fluxes will have to satisfy

$$\sum_{n=1}^{N} F_n(x, t, \rho_1, \dots, \rho_N) = \Phi\left(x, t, \sum_{n=1}^{N} \rho_n\right) = v_\rho \rho$$

The basic principle is rather simple: The parts of type 1 are basically not aware of the other parts, since they are always served first, i.e. their flux will not be influenced by $\rho_2, \dots, \rho_N$. So, we set

$$F_1(x, t, \rho_1, \dots, \rho_n) = \Phi(x, t, \rho_1).$$

Now, the flux of the cumulative density $\rho_1 + \rho_2$ of the part types with the two highest priorities will again not be influenced by $\rho_3, \dots, \rho_N$, giving

$$\partial_t \rho_1 + \partial_x \Phi(x, t, \rho_1) = 0, \quad F_1(x, t, \rho_1, \dots, \rho_N) = \Phi(x, t, \rho_1),$$

$$\begin{aligned}\partial_t(\rho_1 + \rho_2) + \partial_x[\Phi(x, t, \rho_1 + \rho_2)] = 0, \quad &\Phi(x, t, \rho_1 + \rho_2)\\ &= F_1(x, t, \rho_1, \dots, \rho_N) + F_2(x, t, \rho_1, \dots, \rho_N).\end{aligned}$$

Repeating the argument gives the system

$$\partial_t \sum_{n=1}^{m} \rho_n + \partial_x \Phi\left(x, t, \sum_{n=1}^{m} \rho_n\right) = 0, \quad m = 1 : N$$

$$\Phi\left(x, t, \sum_{n=1}^{m} \rho_n\right) = \sum_{n=1}^{m} F_n(x, t, \rho_1, \dots, \rho_N), \quad m = 1 : N.$$

In a first-order model, where the velocity v_ρ and the flux F are simple local functions of ρ, the individual fluxes F_n can simply be computed by the recursion

$$F_n = \Phi\left(x, t, \sum_{k=1}^{n} \rho_k\right) - \sum_{k=1}^{n-1} F_k, \quad n = 1 : N.$$

In a second-order model, where the velocities are given by an auxiliary differential equation, it is necessary to solve this auxiliary equation for the cumulative velocities $v_{\rho_1 + \dots + \rho_n}$, $n = 1 : N$. In either case, the basic principle is, that the lower priority parts receive the components of the flux left over by the higher priority parts.

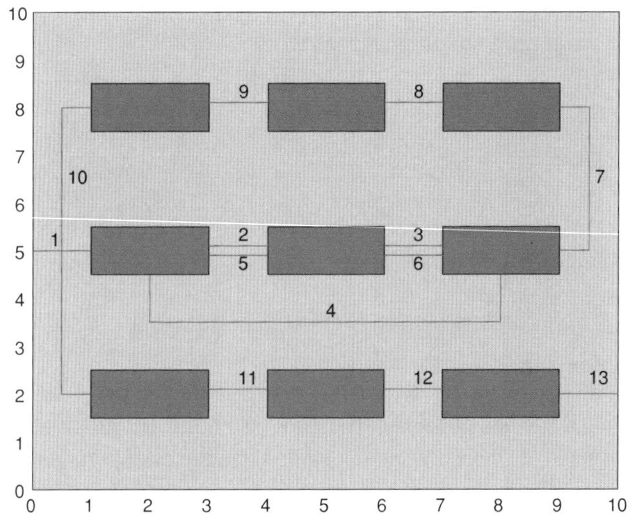

Fig. 2 Schematic diagram of a simple re-entrant system. Servers 1–3 are used in steps 1–3 and again in steps 4–6

3.2 Re-entrant Systems

In practice, the need for service rules arises not only when different product types have to be considered, but also when one and the same server is used at different stages of the production process. In Fig. 2 we schematically depict a process with a simple re-entrant loop. Servers 1–3 are used in steps 1–3, but then again in steps 4–6. The remaining six servers then just form a linear chain. Producing only a single product, the first three servers still require a rule on whether to treat parts in their first or their second pass preferentially.

There are basically two approaches to translating this scenario into a fluid model. The first one is to introduce artificial product types. That is we treat the loop, given by servers 1–3 separately, and solve a system of the form

$$\begin{aligned} &(a)\ \partial_t\rho_1+\partial_x F_1=0, \quad 0<x<S_1, \quad F_1(0,t,\rho_1,\rho_2)=\lambda \\ &(b)\ \partial_t\rho_2+\partial_x F_2=0, \quad 0<x<S_1, \quad F_2(0,t,\rho_1,\rho_2)=F_1(S_1,t,\rho_1,\rho_2) \\ &(c)\ \partial_t\rho_3+\partial_x F_3=0, \quad S_1<x<S, \quad F_3(S_1,t,\rho_3)=F_2(S_1,t,\rho_1,\rho_2). \end{aligned} \tag{15}$$

So, the interval $[0, S_1]$ corresponds to the first three servers in Fig. 2 and the interval $[S_1, S]$ to the last six servers. The density ρ_1 is the density of parts on their first pass through the loop and ρ_2 is the density on the second pass. The densities are linked via their boundary conditions. So the outflux of ρ_1 at $x = S_1$ gives the influx at $x = 0$ for ρ_2, and the outflux of ρ_2 at $x = S_1$ gives the influx for ρ_3. If parts in their the second pass receive preferential treatment, which establishes a so-called PULL policy, we have, given a flux function $F(x, t, \rho) = v_\rho \rho$,

$$F_2(x,t,\rho_2)=\Phi(x,t,\rho_2), \quad F_1(x,t,\rho_1,\rho_2)=\Phi(x,t,\rho_1+\rho_2)-\Phi(x,t,\rho_2),$$

$$F_3(x,t,\rho_3) = \Phi(x,t,\rho_3).$$

Conversely, if parts in their the first pass receive preferential treatment, which establishes a so called PUSH policy, we have

$$F_1(x,t,\rho_1) = \Phi(x,t,\rho_1), \quad F_2(x,t,\rho_1,\rho_2) = \Phi(x,t,\rho_1+\rho_2) - \Phi(x,t,\rho_1),$$
$$F_3(x,t,\rho_3) = \Phi(x,t,\rho_3).$$

The second approach, which is actually preferable in the case of a more complex topology of the system is to introduce virtual processors instead of artificial product types. We 'stratify' the graph in Fig. 2, and now the spatial variable x denotes the actual 12 stages of the process mapped into the interval $[0, S]$. We assign the first three stages to the interval $[0, S_1]$, the second three stages to the interval $[S_1, 2S_1]$, and the last six stages to the interval $[2S_1, S]$. There is now only one density equation of the form

$$\partial_t \rho + \partial_x F = 0, \quad F(x,t,\rho) = v_\rho \rho, \quad F(0,t,\rho) = \lambda(t).$$

However, we have to account for the fact that the interval $[S_1, 2S_1]$ actually corresponds to thc same physical servers as the interval $[0, S_1]$. So, the interval $[S_1, 2S_1]$ corresponds to virtual processors, and we have to make sure that the flux in the physical servers is given by the correct total flux. This makes the definition of the flux function nonlinear. Again, if, in a PUSH policy parts in their first pass receive preferential treatment, then the definition of the flux F is given by

$$F(x,t,\rho) = \begin{pmatrix} \Phi(x,t,\rho(x,t)) & \text{for } 0 < x < S_1 \\ \Phi(x,t,\rho(x-S_1,t)+\rho(x,t)) - \Phi(x,t,\rho(x-S_1,t)) & \text{for } S_1 < x < S_2 \\ \Phi(x,t,\rho(x,t)) & \text{for } S_2 < x < S \end{pmatrix}.$$

Conversely, if, in a PULL policy parts in their second pass receive preferential treatment, then the definition of the flux F is given by

$$F(x,t,\rho) = \begin{pmatrix} \Phi(x,t,\rho(x,t)+\rho(x+S_1,t)) - \Phi(x,t,\rho(x+S_1,t)) & \text{for } 0 < x < S_1 \\ \Phi(x,t,\rho(x,t)) & \text{for } S_1 < x < S_2 \\ \Phi(x,t,\rho(x,t)) & \text{for } S_2 < x < S \end{pmatrix}.$$

3.3 General Policies

We now generalize the concept from Sects. 3.1 and 3.2 to more general policies. Assume that each part has a certain number K of attributes which we collect in the vector $Y = (y_1, \ldots, y_K)$. We now serve the parts in a sequence determined by a priority function $p(x, Y)$. That is, we group the parts into N bins, according to their

Y–values. Let $Y(n),\ n = 1 : N$ denote the (in general vector valued) attributes of the parts in bin number n. Parts in bin number m have priority over parts in bin number n if $p(x, Y(m)) > p(x, Y(n))$ holds. We write $p_n = p(x, Y_n)$ for short. Combining this with the flux splitting idea of Sect. 3.1 gives the ordering

$$\Phi(x, t, \sum_{p_n < p_m} \rho_n(x, t)) = \sum_{p_n < p_m} F_n(x, t, \rho_1, \dots, \rho_N), \quad m = 1 : N.$$

Clearly, this reduces to the model in Sect. 3.1 if we consider only one ($K = 1$) discretely distributed attribute, namely the type $Y = y_1 \in \{1, \dots, N\}$. The policy function in this case would be given by $p(x, Y) = -y_1$, guaranteeing that type 1 has the highest priority.

For a more general policy function $p(x, Y)$ this gives rise to the following algorithm [21]:

Priority Algorithm (16)

- Given the bin—densities $\rho_1, \dots, \rho_n$ and the bin attributes $Y(1), \dots, Y(N)$.
- Compute the priorities $p_n = p(x, Y(n)),\ n = 1 : N$.
- Reorder the bins. That is find a permutation σ of the numbers $\{1, \dots, N\}$, such that $p_{\sigma(1)} > p_{\sigma(2)} > \dots > p_{\sigma(N)}$ holds.
- Compute the individual bin fluxes $F_n,\ n = 1 : N$ according to the rule

$$\Phi\left(x, t, \sum_{n=1}^{m} \rho_{\sigma(n)}(x, t)\right) = \sum_{n=1}^{m} F_{\sigma(n)}(x, t, \rho_1, \dots, \rho_N), \quad m = 1 : N,$$

or, equivalently, solve the recursion

$$F_{\sigma(n)}(x, t, \rho_1, \dots, \rho_N) = \Phi\left(x, t, \sum_{n=1}^{m} \rho_{\sigma(n)}(x, t)\right) - \sum_{n=1}^{m-1} F_{\sigma(n)}(x, t, \rho_1, \dots, \rho_N),$$
$$m = 2 : N$$
$$F_{\sigma(1)}(x, t, \rho_1, \dots, \rho_N) = \Phi(x, t, \rho_{\sigma(1)}(x, t))$$

The algorithm 16 does, in itself, not represent much of a generalization of the model in Sect. 3.1, except for the fact that we now can consider a more complicated attribute than just the type. The additional key ingredient is to be able to make the choice of policy function as well as the evolution of the attributes dynamic.

We let the attributes in a each of the bins evolve dynamically. So, for a part in bin number n with position $x = \xi_n(t)$, the attribute vector $Y(n)$ evolves according to the differential equation

$$\frac{\mathrm{d}Y(n)}{\mathrm{d}t} = E_n$$

Since the part in bin number n moves itself with a velocity $v_{\rho,n} = \frac{F_n}{\rho_n}$, this yields a spatially dependent attribute density $Y(n, x, t)$, which evolves according to

$$\frac{\mathrm{d}}{\mathrm{d}t} Y(n, \xi_n(t), t) = E_n$$

or

$$\partial_t Y(n, x, t) + v_{\rho,n} \partial_x Y(n, x, t) = E_n. \tag{17}$$

Equation (17) simply states, that the attribute vector Y is moved along with the part flow, i.e. for $E_n = 0$ a part retains its attribute when moving through the system. The form of E_n will dependent on the type of attribute considered. We can make E_n dependent on the density ρ itself if so desired, since this information is available in the system. In addition we might want to make the policy, i.e. the way we dynamically order the bins dependent on the stage of the process as well as on the density itself.

For numerical reasons it is best to replace (17) by an equation for the vector valued variables $Z(n, x, t) = \rho_n Y(n, x, t)$. So the vector Z denotes the density of parts with attribute Y [17]. The reason for this is, that the theory of numerical methods for first-order hyperbolic equations is developed largely for conservation laws, and the equation for Z can be written in conservative form. Given, that ρ_n satisfies the conservation law $\partial_t \rho_n + \partial_x F_n = 0$ with $F_n = v_{\rho,n} \rho_n$ we compute the combined evolution equation for the density ρ_n and the vector Z as

$$(a)\ \partial_t \rho_n + \partial_x F_n = 0 \tag{18}$$

$$(b)\ \partial_t Z(n, x, t) + \partial_x \left[\frac{F_n}{\rho_n} Z(n, x, t) \right] = \rho_n E_n.$$

In order to shed a little more light on this construction, let us revisit the example in Sect. 3.2 with a somewhat more complicated policy.

- Up to some point, we use a PUSH policy, giving priority to parts in the first pass of the loop.
- This policy has the disadvantage that, in the case of overload, there will be no output for prolonged periods of time. If the first three processors work at capacity in the first pass, there is no capacity left for the second pass, and there will be no outflux from the whole system until the the total influx λ sinks below capacity.
- We therefore use a PULL policy (priority for the second pass) for those parts which have spent more than a certain amount T of time in the system.
- This requires the monitoring of the time elapsed since parts have entered the system.

In the interval $[0, S]$ in (15) wetherefore have two bins and two attributes, one being the pass number, which does not change within the interval $[0, S]$ and can therefore being identified with the bin number. The other attribute is the time elapsed since the part entered the system. This attribute changes dynamically according to

$$\frac{\mathrm{d}}{\mathrm{d}t} y_2(\xi, t) = 1 = E_2$$

Equation (15) must therefore be augmented by

$$(a)\ \partial_t \rho_1 + \partial_x F_1 = 0, \quad 0 < x < S_1, \quad F_1(0, t, \rho_1, \rho_2) = \lambda \tag{19}$$

$$(b)\ \partial_t \rho_2 + \partial_x F_2 = 0, \quad 0 < x < S_1, \quad F_2(0, t, \rho_1, \rho_2) = F_1(S_1, t, \rho_1, \rho_2)$$

$$(c)\ \partial_t \rho_3 + \partial_x F_3 = 0, \quad S_1 < x < S, \quad F_3(S_1, t, \rho_3) = F_2(S_1, t, \rho_1, \rho_2).$$

$$(d)\ \partial_t z_2(n, x, t) + \partial_x \left[\frac{F_n}{\rho_n} z_2(n, x, t) \right] = \rho_n, \quad n = 1:2$$

$$(e)\ \frac{F_1}{\rho_1} z_2(1, 0, t) = 0, \quad (f)\ \frac{F_2}{\rho_2} z_2(2, 0, t) = \frac{F_1}{\rho_1} z_2(1, S, t)$$

The boundary condition (19)(e) starts the clock at $t = 0$ on influx and the boundary condition (19)(f) expresses the fact that the attribute y_2 is preserved in the transition form the first to the second pass. The policy function $p(y_1, y_2) = p(n, y_2)$ is now given by

$$p(n, y_2) = \begin{pmatrix} -n \text{ for } y_2 < T \\ n \text{ for } y_2 > T \end{pmatrix},$$

which just means that in the algorithm (16) the permutation σ is chosen as

$$\sigma = \begin{pmatrix} \{1, 2\} \text{ for } y_2 < T \\ \{2, 1\} \text{ for } y_2 > T \end{pmatrix},$$

and the fluxes are computed accordingly.

4 Numerical Experiments

In order to demonstrate the applicability and accuracy of the conservation law models in Sect. 2 and the models for the service rules in Sect. 3, perform some numerical experiments on a relatively simple test case. We verify the model discrete event simulations in the deterministic as well as the stochastic case, including random breakdown of individual nodes. Section 4.2 deals with a straight First In System First Out (FISFO) policy and Sect. 4.3 uses a more complex mixed policy.

4.1 The Basic Setup

The service rules considered below require the monitoring of three attributes, namely the time elapsed since a part enters the system, the time left to a certain delivery due date, and the type of the part. Thus, we transport a three dimensional attribute vector Y. The elapsed cycle time grows linearly in time, whereas the time to the due date decays linearly, giving for the vector E in (17)

$$E(x,t) = \begin{pmatrix} E_1 \\ E_2 \\ E_3 \end{pmatrix} = \begin{pmatrix} 1 \\ -1 \\ 0 \end{pmatrix}.$$

Since the third component (the type) does not change dynamically, it can be identified with the bin number n. So, for N types of parts, we solve, according to (17)–(18), a system of hyperbolic conservation laws of the form

$$(a)\ \partial_t \rho_n + \partial_x F_n = 0, \quad (b)\ \partial_t Y(n,x,t) + \frac{F_n}{\rho_n}\partial_x Y(n,x,t) = E_n = \begin{pmatrix} 1 \\ -1 \end{pmatrix}, \quad n = 1:N, \tag{20}$$

where the individual bin fluxes F_n have to be computed, according to a service rule, by the algorithm (16), using a priority function $p(x,Y,t)$, such that

$$\sum_{n=1}^{N} F_n = \Phi\left(x, t, \sum_{n=1}^{N} \rho_n\right)$$

holds, for a given total flux model using the flux function $\Phi(x,t,\rho)$. The boundary conditions for the Eq. (20) are of the form

$$F_n(x=0,t) = \lambda_n(t), \quad Y_n(0,t) = \begin{pmatrix} 0 \\ d_n \end{pmatrix}, \quad n = 1:N;$$

i.e. parts of type n enter according to the influx λ_n with zero elapsed cycle time and a certain time to due date d_n. Writing the system (20) into conservative form, setting $Z_n = \rho_n Y_n$, gives

$$(a)\ \partial_t \rho_n + \partial_x F_n = 0, \quad (b)\ \partial_t Z(n,x,t) + \partial_x\left[\frac{F_n}{\rho_n} Z(n,x,t)\right] = \rho_n E_n, \quad n = 1:N, \tag{21}$$

$$(c)\ F_n(x=0,t) = \lambda_n(t), \quad (d)\ Z_n(0,t) = \begin{pmatrix} 0 \\ \rho_n(0,t)d_n \end{pmatrix}, \quad n = 1:N.$$

The examples below deal with the case of two different types of parts ($N = 2$), and will use either the simplest deterministic flux model (for verification) or the flux model from Sect. 2 modeling servers which break down randomly according to a Markov process. So we have

$$\Phi(x,t,\rho) = \min\{c(x,t), V\rho\} \quad \text{or} \quad \Phi(x,t,\rho) = a(x,t)\left[1 - \exp\left(-\frac{V\rho}{\mu}\right)\right]$$

Numerically, the system (21) is solved by a standard Lax–Wendroff scheme [20].

4.2 Verification for a FISFO Policy

We first verify the the multi-phase approximation of Sect. 3 against a discrete event simulation, i.e. a stochastic automaton. The discrete event simulation is briefly described as follows:

- Parts move linearly through a chain of M identical servers.
- Each server has a capacity μ, meaning it can serve μ parts at the same time. So, it accepts a new part every $\frac{1}{\mu}$ time units.
- Each server takes a processing time $\frac{\Delta x}{V}$ time units to process a part, with $\Delta x = \frac{S}{M}$.
- Servers break down and are repaired according to a Markov process. So, once up, they remain running for a time interval T_{up}, exponentially distributed according to $\mathrm{d}\mathscr{P}[T_{\text{up}} = t] = \frac{1}{\tau_{\text{up}}}\exp(-\frac{t}{\tau_{\text{up}}})\,\mathrm{d}t$, and once down, they remain down for a time interval T_{down}, distributed according to $\mathrm{d}\mathscr{P}[T_{\text{down}} = t] = \frac{1}{\tau_{\text{down}}}\exp\left(-\frac{t}{\tau_{\text{down}}}\right)\,\mathrm{d}t$, with τ_{up} and τ_{down} the respective means of the up- and down times.

It has been shown [3] that in the deterministic case ($\tau_{\text{down}} = 0$, the servers are always running), the system can, in the limit for the number of parts to ∞, be described by the conservation law

$$\partial_t\rho + \nabla_x\Phi = 0, \quad \Phi(x,t,\rho) = \min\{\mu, V\rho\} \tag{22}$$

for the total density of parts. In the stochastic case ($\tau_{\text{down}} > 0$) it can be shown [11] that the expectation $\mathbf{E}_\rho$ of the part density satisfies the limiting equation

$$\partial_t\mathbf{E}_\rho + \nabla_x\Phi = 0, \quad \Phi(x,t,\mathbf{E}_\rho) = \frac{\tau_{\text{up}}\mu}{\tau_{\text{up}} + \tau_{\text{down}}}\left[1 - \exp\left(-\frac{V\mathbf{E}_\rho}{\mu}\right)\right] \tag{23}$$

The System

We consider a chain of forty processors ($M = 40$, $\Delta x = \frac{S}{M}$), processing $N = 2$ different types of parts. Each processor has a throughput time $\frac{\Delta x}{V} = 1$, except for processors 11:20, which have a throughput time $\frac{\Delta x}{V} = 2$. Each processor can handle 30 parts at the same time. This yields a bottleneck capacity of $\mu(x) = 15\frac{\text{parts}}{\text{time}}$ for processors 11:20 and a capacity $\mu(x) = 30\frac{\text{parts}}{\text{time}}$ for the rest. Figure 3 shows the influx for both species. So, the total influx (the sum of both curves in Fig. 3) exceeds the bottleneck capacity in the interval $40 < t < 80$. In the case of a pure FISFO policy the priority is given just by the elapsed cycle time, and therefore we choose the priority function as

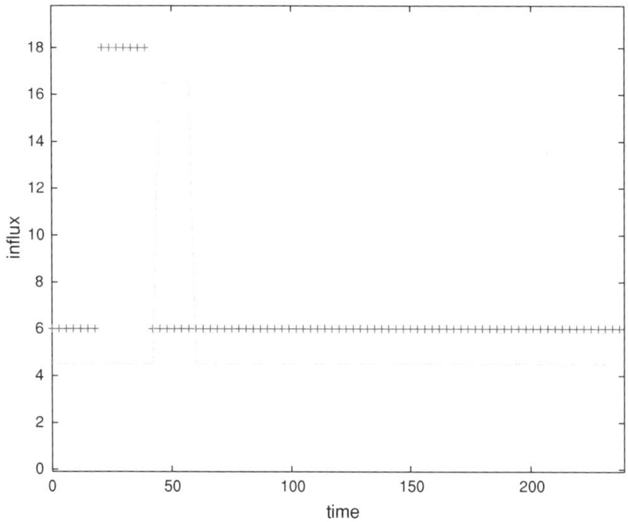

Fig. 3 *Influx*, $+$ = species 1, *solid* = species 2

$$p(x, Y, t) = Y_1.$$

The Deterministic Case

We first test the deterministic case. So, we set the mean down times τ_{down} in the discrete event simulator equal to zero (or choose $T_{\text{down}} = 0$ deterministically, which reduces the discrete event simulation to a simple automaton). We compare this to the multi-phase solution, using the deterministic flux function from (16.3) and the FISFO priority function $p(x, Y, t) = Y_1$. Figure 4 shows the quantity $\Delta x\rho$, i.e. the number of parts corresponding to each processor (those in the queue and the processor itself), for one of the species. To give a more quantitative comparison, we plot the density over time at various stages. Figure 5 shows the densities of the two species in processor 5 (before the bottleneck) processor 11 (the first bottleneck stage) and stage 21 (after the bottleneck). The right panel shows the densities for the sum of both species. We see a temporary buildup of density in the bottleneck stage at server 11, since the total influx temporarily exceeds the bottleneck capacity. The buildup is higher for species 2, since the parts of type 1 have arrived earlier (see Fig. 3), and have therefore a higher priority at the bottleneck. The purpose of Fig. 5 is to demonstrate the accuracy of the policy model. The agreement of the total densities in the right panel is perfect, as has to be expected. The agreement of the individual species densities in the left two panels at the bottleneck at server 11 is reasonably good. The moderate error is explained by the fact, that, in the discrete event simulator individual parts arrive with individual arrival times, while in the conservation law

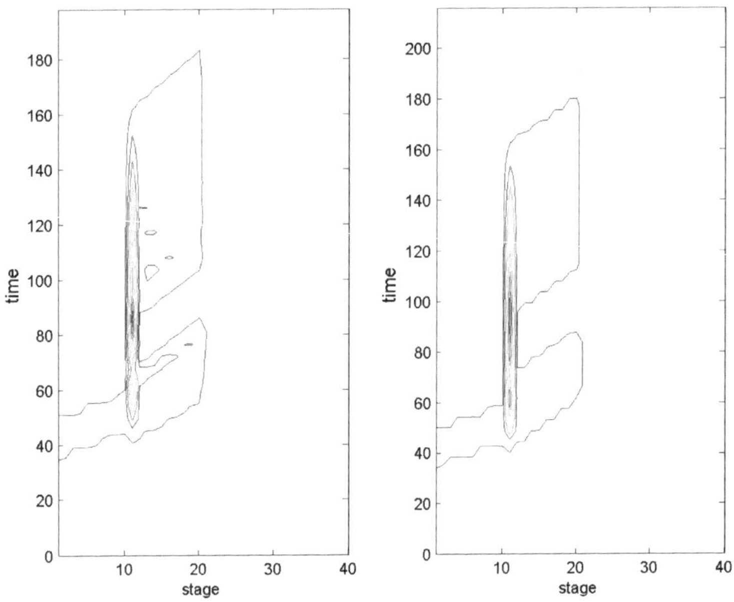

Fig. 4 Parts per processor comparison between the deterministic automaton and the deterministic conservation law for species 2. *Left panel* automaton (contour plot). *Right panel* traffic flow model

model they are lumped together and all parts arriving within the same time step receive the same arrival time.

The Random Case

Next we test the multi-phase solution on the stochastic system. So, in the discrete event simulator, we generate time series for the capacities from a Markov process using the means $\langle \tau_{\text{up}} \rangle = \frac{3\Delta x}{V}$ and $\langle \tau_{\text{down}} \rangle = \frac{\Delta x}{V}$. So each processor runs on average for three cycle times, and then shuts down for the next cycle time, giving an average availability of $a = 0.75$. We compute 300 realizations of the discrete event simulation, compute the means, and compare this to the mean field multi-phase solution, using the flux function from (23). As in the deterministic case, we compare the quantity $\Delta x \rho$, for one of the species in Fig. 6. To give a more quantitative comparison again, we plot the density over time at stages 5, 11, 21 in Fig. 7. At first glance, it is surprising that the present approach is able to accurately simulate the effects of the FISFO policy on the *expectation* of the density, since we apparently have interchanged the evaluation of the expectation operator with the evolution of a nonlinear stochastic dynamical system. However, as shown in [11], we are using the correct conservation law for the expectation of the density ρ of the whole ensemble, and the process of re-ordering the parts according to the elapsed cycle time can apparently be commuted with the expectation operator.

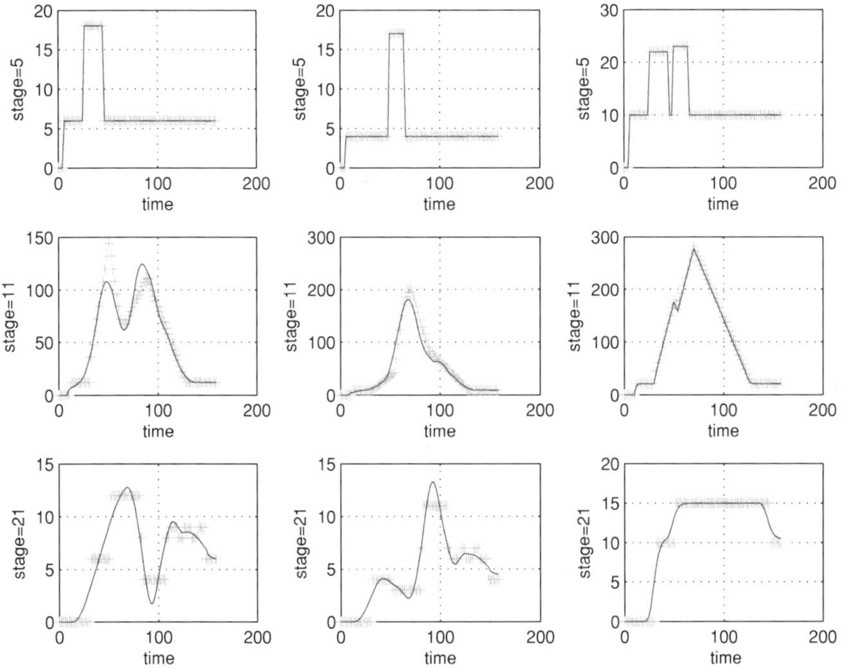

Fig. 5 Parts per processor comparison between the deterministic automaton and the deterministic conservation law for stages 5, 11 and 21. *Left panel* species 1. *Middle panel* species 2. *Right panel* both species. *solid line* = traffic flow model. + = automaton

4.3 A More Complex Policy

We conclude the numerical experiments with a numerical study of the influence of different policies. We consider the same system as in Sect. 4.2. Each species now has a 'due-date' [18], i.e. a certain limit on the cycle time, after which it is delivered late. This models essentially the production of a perishable good which is spoiled and worthless after spending too much time in the system. We compare the FISFO policy from Sect. 4.2, where the priority is set to $p(Y) = Y_1$, to a more complex policy with a priority function of the form

$$p(Y_n) = Y_1 H\left(Y_2 - \frac{1}{2}d(n)\right) - Y_2 H\left(\frac{1}{2}d(n) - Y_2\right). \tag{24}$$

The policy given by the priority function (24) is interpreted as follows:

- We schedule according to FISFO, using Y_1, until the time to due date Y_2 has reached half its limiting value $d(n)$, where d is dependent on the species number n. From this point on we switch policies, prioritizing the parts according to the time until the part becomes worthless.

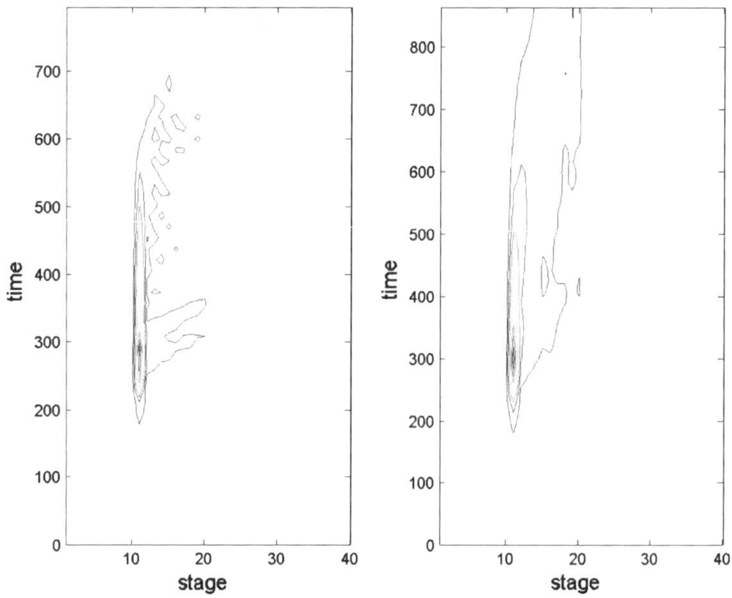

Fig. 6 Parts per processor comparison between the average over 300 realizations of the stochastic automaton and the mean field conservation law for species 2. *Left panel* the mean field traffic flow solution (contour plot). *Right panel* average over 300 DES realizations

- We choose as the as maximal acceptable cycle times $d(1) = 75,\ d(2) = 200$. So species one 'spoils' after 150% of the raw throughput time, while species number two can spend 400% of the raw throughput time in the system before spoiling.

The left panel in Fig. 8 shows result for the FISFO policy, and the right panel for the policy corresponding to the priority function (24). The top row shows the cycle time $Y_k^{(1)}(x = 40, t),\ k = 1:2$ at exit, i.e. the total time parts have spent in the system. The middle row shows the cycle time $Y_k^{(2)}(x = 40, t),\ k = 1:2$ at exit, i.e. the time to due date for each system at the exit.

- The cycle times at exit for FISFO (top, left panel in Fig. 8) are identical for both species, as they should be, using a pure FISFO policy.
- Using just FISFO, there is a significant amount of spoiled parts of species 1 at the exit. That is the curve in the middle left panel of Fig. 8 for species 1 dips significantly below 0. Using the policy given by (24), essentially all parts of both species can be delivered on time, as seen in the middle right panel of Fig. 8.
- Note, that the attributes vanish for certain periods of time. This is an artifact of the conservative discretization (21)(b) of the attribute equations. The primary variable used in the code is $Z_n = \rho_n Y_n$ and the attributes in the top two panels of Fig. 8 are computed as $\frac{\rho_n Y_n}{\rho_n}$. So, if there are no parts ($\rho_n = 0$,) the attribute Y_n is meaningless.

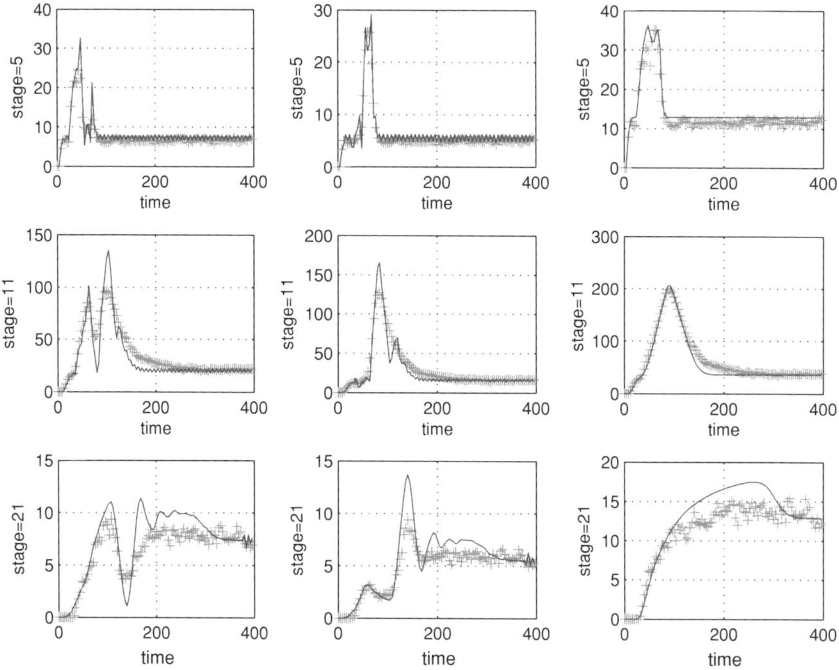

Fig. 7 Parts per processor comparison between 300 realizations of the stochastic automaton and the mean field conservation law for stages 5, 11 and 21. *Left panel* species 1. *Middle panel* species 2. *Right panel* both species. *solid line* = mean field traffic flow solution. + = average over 300 DES realizations

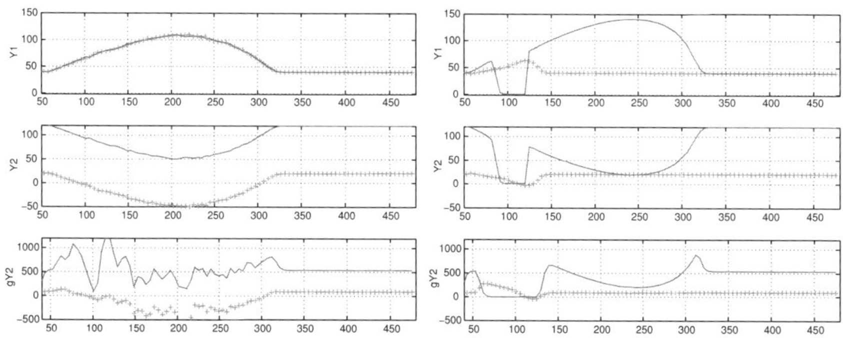

Fig. 8 Attribute comparison. *Left panel* FISFO. *Right panel* service rule for perishable goods, defined by the priority function (24). *Top* Cycle time at exit. *Middle* Time to due date at exit. *Bottom* Attribute density $\rho_n Y_n^{(2)}$ of the time to due date. + = species1, *solid* = species 2

- The bottom panel of Fig. 8 shows the average attribute $\rho_n Y_n$ for each species, for the FISFO policy and the policy given by (24). This represents a measure of the cost.

5 Conclusions

The evolution of parts in a supply chain, governed by a quite general class of service rules based on prioritizing attributes, can be modeled by a set of hyperbolic conservation laws. These conservation laws yield an exact solution to the underlying kinetic equations, as long as the level sets of parts of equal priority form a set of measure zero in a sufficiently high dimensional attribute space. This situation can be created by artificially inflating the attribute space, essentially breaking down parts of equal priority into subgroups. The resulting macroscopic model can even be used to model stochastic systems as long as the correct flux function for the evolution of the expectation of the whole ensemble is known.

References

1. Anderson EJ (1981) A new continuous model for job shop scheduling. Intern J Syst Sci 12:1469–1475
2. Armbruster D, Degond P, Ringhofer C (2006) A model for the dynamics of large queuing networks and supply chains. SIAM J Appl Math 66:896–920
3. Armbruster D, Degond P, Ringhofer C (2007) Kinetic and fluid models for supply chains supporting policy attibutes. Bull Inst Math Acad Sinica 2:433–460
4. Armbruster D, Ringhofer C (2005) Thermalized kinetic and fluid models for re-entrant supply chains. SIAM J Multiscale Model Simul 3:782–800
5. Armbruster D, Marthaler D, Ringhofer C (2004) Kinetic and fluid model hierarchies for supply chains. SIAM J Multiscale Model Simul 2(1):43–61
6. Aw A, Rascle M (2000) Resurrection of "second order" models of traffic flow. SIAM J Appl Math 60:916–938
7. Banks J, Carson J II, Nelson B (1999) Discrete event system simulation. Prentice–Hall, Upper Saddle River
8. Billings R, Hasenbein J (2001) Applications of fluid models to semiconductor fab operations. In Proceedings of the 2001 international conference on semiconductor manufacturing operational modeling and simulations, Seattle, WA
9. Daganzo C (1995) Requiem for second order fluid approximations of traffic flow. Transp Res B 29:277–286
10. Dai JG, Weiss G (1996) Stability and instability of fluid models for re-entrant lines. Math Oper Res 21:115–135
11. Degond P, Ringhofer C (2007) Stochastic dynamics of long supply chains with random breakdowns. SIAM J Appl Math 68(1):59–79
12. El-Taha M, Stidham S (1999) Sample path analysis of queuing systems. Internat Ser Oper Res Management Sci 11. Kluwer Academic Publishers, Boston
13. Fügenschuh A, Göttlich S, Herty M (2007) A new modeling approach for an integrated simulation and optimization of production networks. In Gunther H-O, Mattfeld D, Suhl L (eds.) Management logistischer netzwerke. Physica-Verlag, Heidelberg, pp 45–60
14. Göttlich S, Herty M, Klar A (2005) Network models for supply chains. Commun Math Sci 3(4):545–559
15. Graves WHSC, Kletter DB (1998) A dynamic model for requirements planning with application to supply chain optimization. Oper Res 46(3):35–49
16. Herty M, Ringhofer C (2007) Optimization for supply chain models with policies. Physica A 380:651–664

17. Jin S, Li X (2003) Multi-phase computations of the semiclassical limit of the Schrodinger equation and related problems: Whitham vs. Wigner. Physica D 182:46–85
18. Kumar PR, Lu SH (1991) Distributive scheduling based on due dates and buffer priorities. IEEE Trans Autom Control 36:1410–1416
19. La Marca M, Armbruster D, Herty M, Ringhofer C (2010) Control of continuum models of production systems. IEEE Trans Autom Control 55(11):2511–2526
20. LeVeque R (1992) Numerical methods for conservation laws. Birkhäuser, Basel
21. Ringhofer C (2010) A level set approach to modeling general service rules in supply chains. Commun Math Sci 8(4):909–930
22. Simchi-Levi D, Kaminsky P, Simchi-Levi E (2003) Designing and managing the supply chain. 2nd edn. McGraw-Hill Irwin, New York

Autonomous Decision Policies for Networks of Production Systems

Bernd Scholz-Reiter, Sergey Dashkowskiy, Michael Görges, Thomas Jagalski and Lars Naujok

Abstract Modern production and logistic systems are facing increasing market dynamics: customers demand highly individualized goods, the adherence to due dates becomes critical and stipulated delivery times are decreasing. Particularly logistic networks, e.g. production networks or supply chains, are strongly affected by this trend. On the other hand, production networks have to deal with inherent internal dynamics, which are caused by e.g. machine breakdowns or rush orders. The concept of autonomous control, coming from the theory of self-organization, offers decentralized autonomous decision policies (ADPs), which enable logistic objects to make and execute decision on their own. Due to this kind of decision making, autonomous control aims at a distributed coping with dynamic complexity and, at the same time, at an improvement of the logistic performance. This contribution addresses the concept of autonomous control and the underlying autonomous decision policies as a novel concept for the control of the material flows in networks of coupled production facilities. Moreover, it shows different concepts of modeling and analysis of autonomously controlled networks. To achieve this goal, a dual approach including both, mathematical methods as well as simulation models, is presented. Subsequently, the possibilities to analyze the dynamic behavior of the autonomous logistic system are discussed, i.e., the system's stability and its logistic performance. Finally, this contribution presents an exemplary case of a production

B. Scholz-Reiter (✉) · M. Görges · T. Jagalski
Bremen Institute of Production and Logistics (BIBA),
University of Bremen, Hochschulring 20, 28359 Bremen, Germany
e-mail: bsr@biba.uni-bremen.de

M. Görges
e-mail: goe@biba.uni-bremen.de

S. Dashkowskiy · L. Naujok
Centre of Industrial Mathematics,
University of Bremen, Bibliothekstraße 1, 28334 Bremen, Germany

D. Armbruster and K. G. Kempf (eds.), *Decision Policies for Production Networks*,
DOI: 10.1007/978-0-85729-644-3_10, © Springer-Verlag London 2012

network to demonstrate the practicability of the approach of modeling and analysis of autonomous control for production networks.

1 Introduction

Modern logistic systems are exposed to various dynamically changing parameters in its internal and external environment. Especially logistic networks, e.g., production networks or whole supply chains, are affected by dynamical changes [53, 56]. For example, these dynamics are caused by customers' increasing desires for individualized goods or the demand of decreasing delivery times and a strict adherence to due dates. Moreover, internal factors can cause unfavorable dynamic behavior of logistic networks, e.g., interdependencies between transportation and production processes or machine breakdowns. Manufacturing enterprises have to adapt to these changes rapidly. On the one hand, companies concentrate on their core competencies to sustain competitiveness. On the other hand they establish close cooperations with each other in order to satisfy the demand of their customers. In this context, several cooperation concepts for interconnected logistic networks were developed in the past. These concepts, for example virtual enterprises [7, 26] or production networks [57], aim at enabling companies to react promptly to dynamics. Related to this, several planning tasks for operating such networks occur in addition to classical production planning and control (PPC) functions. Comprehensible examples of these new tasks are the assignment of orders to production plants or the temporal coordination between transport and production processes. Especially the temporal coordination in geographically dispersed production networks gains importance [15, 40]. A lack of reconciliation between production and transport processes can lead to increasing throughput times, increasing tardiness of orders or underutilization of resources [28, 37]. Thus the integrated planning of transport and production processes has to ensure that an adequate quantity of raw material is supplied to the particular production plant at the right time. Furthermore, a high work-in-process (WIP) level should be avoided. A high level of WIP is unfavorable due to the resulting capital lockup. However, in highly dynamic and volatile situations centralized planning approaches, which solve the total planning problem incrementally, are not able to cope with occurring dynamics and unforeseen disturbances [21, 24]. Decentralized approaches, e.g., autonomous cooperating logistic processes, seem to be a suitable counterpart to classical centralized planning methods. This concept aims at enabling single logistic entities to make and execute operational decisions on their own. According to this idea, intelligent logistic objects (e.g., parts, machines or trucks) apply autonomous decision policies, in order to pursue their own logistic targets [59]. Due to the use of modern information and communication technologies (e.g., RFID, GSM, GPS, etc.) these objects are able to interact with others. Based on these interactions, logistic objects collect information about current local system states and use this information for decentralized decision making. Autonomous cooperating logistic processes aim at increasing the system's robustness and its performance, due

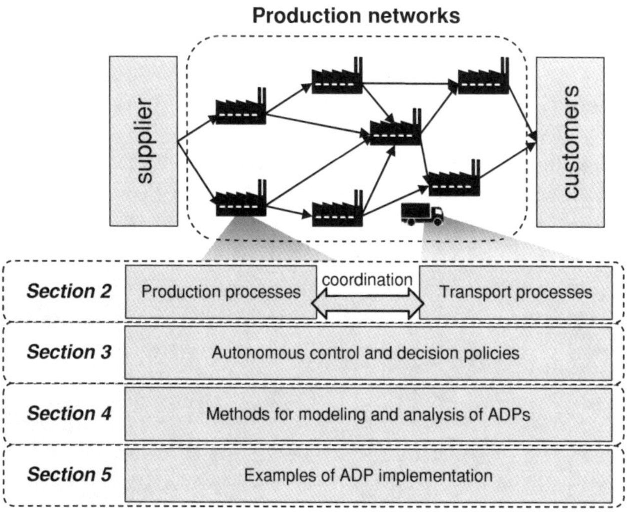

Fig. 1 Structure: ADPs for networks of production systems

to autonomous and distributed decision making of intelligent logistic objects. The implementation of autonomous decision policies (ADPs) in production systems and entire production networks already showed promising results, in terms of an increased logistic target achievement and the robustness against disturbances [45, 46].

However, this kind of autonomous decision making causes a decentralized system behavior, which may affect the total logistic performance negatively or even leads to instability of the system [34, 60]. Roughly speaking, stability means that the state of a plant remains bounded over time, whereas instability of a network leads to infinite states. A network with increasingly growing WIP can be called unstable.

This contribution aims at explaining the idea of autonomous cooperating logistic processes and the fundamental concepts of ADPs in large-scale logistic networks to practitioners. In the beginning the theoretical background will be outlined briefly. The general focus of this contribution is set on describing how to implement ADPs in production networks and furthermore how to determine key-indicators of the systems using ADPs, such as logistic performance and stability. Therefore, this contribution is structured as follows (Fig. 1): In Sect. 2 a general definition of production networks is given, while Sect. 3 addresses operative planning problems of production networks. It discusses classical central approaches in this context. The concept of autonomously cooperating logistic processes and the underlying ADPs are presented in Sect. 4.

Concrete approaches for modeling, simulating and analyzing the performance and the stability of ADPs in production networks are presented in Sect. 5. Subsequently, the application of these modeling and analysis approaches is presented in Sect. 6 in two examples of production network scenarios. Finally, Sect. 7 includes a summary and an outlook.

2 Production Networks

Relevant literature provides several definitions concerning networks of coupled interconnected production systems. In the context of this contribution the term production network is defined according to Wiendahl and Lutz [57], based on the orientation at the integrated planning of logistic processes: Production networks are company or cross-company owned networks of geographically dispersed production facilities. They focus on the mutual use of common resources and integrated planning of value adding processes in the network [57]. This allows achieving economies of scale through the joint planning and the common use of production resources. These types of networks may react promptly to internal or external disturbances due to redundancies of resources. An integrated view on production planning and transport planning requires additional tasks: Companies have to generate concepts for identifying new network partners, the network design and adjusting the PPC according to the network's purpose [53]. However, this creates complex interdependencies between PPC of plants and coordination of transports, e.g., allocation problems between plants or planning of transport schedules and transport capacity [2, 40, 52]. Besides these operational planning problems, which concern a short-term time horizon, there are also several planning problems on the tactical and strategic level. The supply chain planning matrix, introduced by Meyr et al. [27], comprises all relevant planning problems for short-, mid- and long-term time horizons. It covers all dimensions of corporate logistics: procurement, production, distribution and sales. A classical problem of the long-term time horizon is the strategic network planning, with tasks like selection of strategic partnerships or localization of production plants [39]. In the mid-term time horizon, the so called master planning describes tasks of coordinating all procurement production, distribution and sales activities, which are necessary to fulfill the customers' demands. The short-term time horizon concerns the operational level and contains classical tasks such as production planning and control. Furthermore, the operational level addresses the distribution and transport planning, as well as purchasing activities and material requirements planning. It is assumed that these tasks are implemented in software modules, which cover all of these planning problems [30]. Some authors argue that an integrated planning method, which solves all problems in an incremental way, may be challenged toughly by the occurrence of dynamics and unforeseen disturbances [21, 24]. Moreover, the structural complexity of large logistic networks is another limiting factor for the application of centralized optimization methods in the context of planning and operating such networks. At least single problems of production and transport logistics are NP hard [17]. Accordingly, optimal solutions can only be found for very small and simple instances in appropriate computational time. Thus, heuristics are commonly used for these kinds of scheduling problems. The following sections give a brief overview about different planning problems in large-scale logistic networks with a special focus on operative planning problems. These problem classes cover production logistics aspects, transport-related problems as well as integrated problem formulations.

3 Planning Problems in Production Networks

Sauer [40] describe planning tasks in supply or production networks as multi-site scheduling problems. Multi-site scheduling problems are an integrated formulation of production and transport problems, in terms of determining quantities and schedules for particular production facilities as well as determining transport schedules [9]. These approaches address three planning problems: the scheduling of the shop-floor, planning of transport operations and their coordination on the network level. The coordination in production networks comprises tasks of information updating of successors and predecessors. Dunbar and Desa [14] investigate in this context a distributed model adaptive control approach and compare it to a nominal feedback policy. They point out that this approach outperforms the nominal policy in situations with reliable demand forecasts. The MUST-architecture introduced by Sauer [40] describes an approach based on a central coordination instance, which creates a global schedule on the basis of locally generated schedules for all plants and the corresponding transportation activities. This global schedule takes the solutions of the sub-scheduling problems into account. Guinet [19] presents another centralized approach, which divides the total planning problem into sub-problems on the network level and on the shop-floor level. Shop-floor and transport problems will be characterized and described in the following.

3.1 Shop-Floor Problems

Shop-floor scheduling problems are a well-known problem class in operations research. The corresponding literature provides several comprehensive textbooks (e.g., [31] or [13, 35]. Hence, this section aims at giving a brief overview about different problem classes. Especially, the flexible flow shop problem will be discussed in detail, due to its realistic assumptions and its widespread application in analysis of autonomous controlled production systems.

General classification characteristics concerning shop-floor problems are: the machine types/the arrangement of machines, characteristic of jobs and objective functions. The machine types and the arrangement can be differentiated according to three main classes: single machine problems, multiple identical parallel machine problems and unrelated parallel machine problems [1, 36]. In contrast to single machine problems the class of multiple machine problems addresses the assignment of a job set to a set of multiple machines on one or more production stages. As a specification of parallel machine problems, unrelated parallel machines offer different processing times and setup times for different job types. As mentioned above, the flexible flow shop (FFS) problem is a special problem formulation of a shop-floor scheduling problem [22]. The FFS comprises a variable number of production stages, which contain a variable number of unrelated parallel machines per stage. Jobs running through the system have to pass each stage once. Due to the unrelatedness, the machines offer different process and setup times to the jobs. Algorithms for solving this problem type depend on the chosen logistic target system. Often the makespan is chosen as objective function. This means the timespan between the first order release time of the first job and the completion of the last job. Jungwattanakit et al. [22] propose multiple coupled algorithms which construct primarily a sequence for jobs on the first stage. Afterwards greedy algorithms assign the jobs to the machines on a stage. The greedy algorithm is repeated, until all jobs are assigned to stages. The contribution of Jungwattanakit et al. [22] shows that a combination of these algorithms with a genetic algorithm improves the optimization result.

The assumptions (different job types, unrelated parallel machines, variable number of resources, etc.) in the FFS problem formulation can be considered to be realistic and near to practice [1]. Thus, the FFS is often used for analyzing different autonomous decision policies in the production logistic context.

3.2 Transport Problems

The planning of transports in geographically dispersed networks is a complex task. Transport operation can be generally classified in short haul and long haul operations. Short haul operations describe the aggregation of different transport orders, which do not fully utilize the capacity of a transport carrier, to tours or round trips. Popular planning problems related to this area are the traveling sales man problem (TSP), the vehicle routing problem (VRP) or the pick-up and delivery problem (PDP) [54]. These problems and their derivatives focus on determining round tours starting and ending in one point (depot) for one or more transport carriers (trucks) to deliver a certain amount of goods to costumers.

Long haul planning addresses the delivery of goods over long distances with less nodes. Usually, in long haul transports line operations are implemented [16]. Thus, the particular transport route gets already fixed in advance and the transport operation takes place according to predefined policies. This type of transport initiation is commonly used in production networks. This planning problem can be divided in two sub-problems. The mid-term task of service network design includes the choice of a transport carrier (road, rail, sea, etc) and the circulation of the transport carriers [10]. The short-term planning aims at aggregating and assigning orders to loads. The triggering of transports in a long haul operation can be done by several policies. Usually, these transports are initiated in fixed frequencies according to a predefined schedule [18]. Another type is the so-called "go-when-full" policy. This policy implies that a truck starts a transport process, when a predefined loading quantity is reached [5]. In real word practice a mix-form of both can be found. This means, transports are initiated with predefined time windows, but within these time windows there is the possibility to operate with a go-when-full policy. The advantage of a go-when-full policy is an efficient utilization of the load carriers [10].

Their capacity is fully used in this case. A drawback in this kind of policy is the construction of loose schedules.

4 Autonomous Cooperating Logistic Processes

The idea of autonomous cooperating logistic processes is inspired by the theory of self organization. This section presents the definition of autonomous control and elaborates on autonomous decision policies.

4.1 Definition

According to the collaborative research center 637 "Autonomous cooperating Logistic Processes: A Paradigm Shift and its Limitations", the following definition of autonomous cooperating logistic systems is given: "Autonomous control describes processes of decentralized decision-making in heterarchical structures. It presumes interacting elements in non-deterministic systems, which possess the capability and possibility to render decisions independently. The objective of autonomous control is the achievement of increased robustness and positive emergence of the total system due to distributed and flexible coping with dynamics and complexity" [59]. According to this definition autonomous control is characterized by a shift of decision-making capabilities from the total system to its elements, which allows intelligent logistic objects to route themselves through a logistic network according to their own objectives [60]. In the context of this definition intelligent logistic object may be either physical objects (e.g., trucks, machines, etc.) or immaterial objects (e.g., production orders or transport orders). Modern information and communication technologies can provide an infrastructure, which enables an exchange of information about current local system states between these objects. On this basis the objects are able to generate decisions according to different autonomous decision policies. Due to these multiple decentralized decisions the local and the global behavior should be influenced in a positive manner, for example, in terms of improving the handling of dynamics caused by unforeseen events (e.g., machine breakdowns) [60].

In the past, ADPs have been developed for all areas of the logistic chain: There exist ADPs for transportation and route planning (e.g., [38]), production logistics (e.g., [43]), transport collaborations [6], or production networks (e.g., [45]). In the following, different ADPs for production systems and production networks are presented.

4.2 Autonomous Decision Policies

Generally, ADPs enable decision making of intelligent logistic objects. In the context of production systems and networks all existing ADPs facilitate decision making of parts or jobs (semi-finished products), to decide about possible routes through the system. Scholz-Reiter et al. [48] propose a classification of ADPs according to local information methods and information discovery methods. Information discovery methods, i.e., the distributed logistics routing protocol (DLRP), collect information from other objects. The DLRP is inspired by communication protocols of wireless ad hoc networks. Intelligent logistic objects using the DLRP send requests into the logistic network. By receiving replies, the object collects information about the system, which is used for local decision making. This discovery does not cover the whole system, but it is directed to information that is relevant for the actual decision. The DLRP is designed for production environments [50] as well as for transport logistic routing problems [38]. However, this contribution focuses on local information methods. Local information methods enable jobs to decide about further processing steps. Jobs using one of these methods only gather local information about states of direct succeeding buffers and machines.

According to a classification introduced by Windt and Becker [58] local information methods can be further divided into rational policies, bounded rational strategies and mixed forms. Rational strategies use solely rational measures (e.g., throughput times or due dates) for the decision-making process. In contrast, biologically inspired strategies which belong to the class of bounded rational strategies, try to transfer mechanisms from biological self-organizing systems to the decision-making in production networks. Table 1 presents different ADPs, which can be applied to production networks. It differentiates between shop-floor related and network-related strategies and presents their main characteristics as well as a short overview about the algorithmic scheme.

The QLE policy enables parts in a production system to estimate the waiting and processing times of different alternative processing resources. It uses exclusively local information to evaluate the states of the alternatives. The application of this policy leads to a better system performance regarding throughput times compared to classical scheduling algorithms in highly dynamic situations [50].

Similar to the QLE, the DUE policy estimates waiting and processing times. While the QLE uses this information for minimizing part-related throughput times, the DUE policy orientates at the tardiness of parts. A part using this policy decides for an alternative resource which offers the lowest difference between estimated due date and pre-planned due date [47].

In contrast, the PHE policy is a bio-inspired strategy. The approach is based on the idea to imitate the process of ants marking possible routes to food sources. Ants leave pheromone marks between the nest and food sources. Other ants can detect those pheromones and will follow the trail with the highest concentration of pheromones [32, 33]. This is transferred to logistic systems: During the production process, the parts leave information about their processing and waiting times at a corresponding

Table 1 Autonomous decision policies for production networks

ADP	Purpose	Type	Algorithm scheme
Queue length estimator (QLE)	Allocation decision of parts on the shop-floor	Rational	1. Parts calculate waiting times for all alternatives 2. Parts decide for the machine with the lowest waiting time
Due date policy (DUE)	Allocation decision of parts on the shop-floor	Rational	1. Parts calculate waiting times for all alternatives 2. Parts compare waiting time estimation with own due date 3. Parts choose the machine with the lowest difference between estimation and due date
Pheromone-based policy (PHE)	Allocation decision of parts on the shop-floor	Bounded rational — bio-inspired	1. Parts collect pheromone information available from all possible alternative machines 2. Parts select the machine with the highest artificial pheromone concentration 3. After processing parts leave time information as artificial pheromones
Honey bee algorithm (HBA)	Allocation decision of parts on the shop-floor	Bounded rational — bio-inspired	1. Parts advertise a particular machine according to the machine quality 2. Parts detect all actual advertisement signals of machines 3. Parts decide for the machine with the best advertisement

(continued)

Table 1 (Continued)

ADP	Purpose	Type	Algorithm scheme
Chemotaxis policy (CHE)	Allocation decision of parts on the shop-floor	Bounded rational — bio-inspired	1. Parts detect different target values of all possible alternatives as an attractant
			2. Parts start an iterative random biased process
			3. Part decides for the machine reached at the end of the iteration
Network related Queue length estimator (nQLE)	Allocation decision of parts on the network level	Rational	1. Parts estimate the processing times of all parts on the transport route to a particular plant
			2. Parts compare all estimated waiting times
			3. Part chooses the plant with the lowest waiting time
Network related Pheromone based policy (nPHE)	Allocation decision of parts on the network level	Bounded rational – bio-inspired	1. Parts collect pheromone information available from all possible alternative plants
			2. Parts select the plant with the highest concentration
			3. After processing parts leave time information about transport and processing times as artificial pheromones

machine. Following parts entering a stage of the shop-floor compare this artificial pheromone concentration by computing average value of the waiting time data of the last five parts and choose a production line. Thus, the pheromone concentration depends on waiting and processing times of previous parts. To model the evaporation process of natural pheromones a moving average of waiting time data is used [3].

The honey bee algorithm (HBA) is another bio-inspired strategy. It uses the foraging mechanisms of honey bees' colonies. In nature bees advertise possible food sources with a so-called 'waggle dance'. The duration of this dance depends on the ratio between energy consumption of the flight (between hive and food source) and available energy of the source. The probability of bees recognizing the dance of a dancing bee is proportional to the dancing duration. According to this principle parts are able to advertise different alternative production resources by means of the machining quality, which is determined by calculation of the benefit provided by a particular machine and the throughput time needed for this step [44].

The natural process, which inspires the CHE policy, differs from the PHE and the HBA policy. It is not inspired by coordination principals of social insects, but on movement processes coming from micro-biology. Natural bacteria are able to direct their movement according to the concentration of attractants (e.g., food substances) or repellants (e.g., toxic substances). Therefore, bacteria perform a random biased walk to find appropriate food sources. This basic movement principle is transferred to autonomous decision making by the CHE policy. Parts using this policy decide according to the gradient of logistic target values of different decision alternatives [49].

All ADPs described above were implemented in the past for several production logistic scenarios. In general, these policies can also be used for the decision making on the network level. Currently, the QLE and the PHE policy have already been transferred to decision making on the network level. The nQLE enables the decision making on the network level similar to the QLE policy. The network-related version enables an allocation decision of parts to plants. Therefore, the nQLE estimates, similar to the QLE policy on the shop-floor level, the transport duration from one plant to the next and estimates the processing times in the respective plant. The part chooses the plant with the lowest estimated transport and processing times. The nPHE policy is based on the same principles as the PHE policy. Intelligent parts choose one of alternative succeeding plants according to information about the processing and waiting times of previous parts. In contrast to the PHE policy this information is not limited to the waiting times at the next machine, but is based on the time spent to pass the transport system and the corresponding plant [50]. After processing in one plant the part leaves this information as an artificial pheromone at the plant, which can be detected by the following parts.

Concerning ADPs, Scholz-Reiter et al. [42] present a framework for choosing the right policy for a particular production scenario. The underlying evaluation of this framework applies evaluation methods, which will be presented in Sect. 5. This contribution presents tools for evaluation of ADPs in production networks.

5 Modeling and Analysis of ADPs in Production Networks

When dealing with production networks, it is self-evident to consider an integrated modeling of production networks that covers both job-shop scheduling and transport logistic problems from an integrated point of view. Such an integrated modeling approach for production networks is presented in Sect. 5.1. The representation of time in models of production networks varies throughout the literature: it can be distinguished between discrete event and continuous simulation models [29]. Sections 5.2 and 5.3 present different modeling and simulation approaches, i.e., a mathematical approach and a discrete event simulation (DES) in order to validate the obtained simulation results concerning ADPs in production networks against each other. Production networks usually have to deal with dynamic variations, which can be caused by internal factors or by the (external) environment. Hence, a pure static analysis of logistic performance indicators seems to be not sufficient to cover the effects and the interdependencies of these dynamics. Thus, Sect. 5.4 presents appropriate measures for analyzing production networks.

5.1 Integrated Modeling of Production Networks

For the purpose of analyzing autonomously controlled production networks a matrix-like production network scenario was introduced by Scholz-Reiter et al. [45]. This matrix-like model allows the analysis of the dynamical behavior, the stability and the logistic performance of a multi echelon production network with detailed shop-floor and transport models.

Figure 2 shows the generic structure of this model. It consists of a variable number of network stages, which comprise a variable number of production plants per stage. Furthermore, these production plants are represented as a shop-floor scenario. Each of these shop-floors is a matrix-like model (similar to Scholz-Reiter et al. 2005).

Accordingly, different production resources (buffers and machines) are located on a variable number of production stages. Figure 2 depicts this relation. Transport systems connect the production plants on the network level with each other. The network is able to process different job types. The arrival rate $u(t)$ describes the input of jobs to the network as a function of time. In order to model different demand situations this function can be modeled as an arbitrary mathematical function. For example, a sinusoidal function can be chosen for modeling seasonal demand fluctuations (similar to [43, 45]). However, stochastic inputs can be chosen as well. All transports in this model are direct deliveries, in terms of a door-to-door delivery. This means that each transport between two plants is initiated and operated separately. The model allows integrating different direct transport strategies, like a "go-when-full-policy" or a "frequency-based-policy" with pre-defined departure times as described by Crainic [10]. Trucks using the "go-when-full-policy" will depart at a particular plant, whenever their total load capacity q is reached. In a "frequency-based-policy"

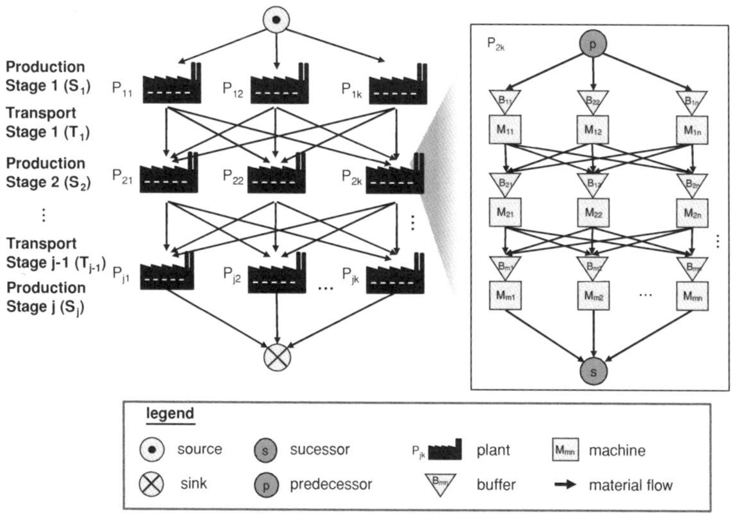

Fig. 2 Generic production network model with kxj plants and mxn machines per plant (see [45])

all departure times are predefined and scheduled. This means that the transportation starts at a certain time, which consequently may lead to variations in the trucks load quantity q. Both transportation policies are commonly used in door-to-door transports in long haul operations.

The modeling of particular scenarios can be done by different modeling tools, depending on the purpose of the analysis. This contribution introduces a mathematical approach and an engineering-orientated simulation approach. The formulation of a mathematical model, which is based on differential equations, is a suitable approach to analyze the stability of a production network [12]. Moreover, the engineering-orientated simulation approach can be used to refine and validate the results of a mathematical stability analysis. It can be used to analyze the logistic performance of ADPs in production network.

5.2 Mathematical Modeling and Stability Analysis

In order to analyze and to make statements about the dynamics of production networks, the mathematical modeling by differential equations [20] can be used, which can be called the macroscopic view. Each plant of a production network is called subsystem. General production networks consist of n subsystems and each subsystem is modeled by a differential equation that describes the incoming and outgoing material or information flows as follows:

$$\dot{x}_i(t) = \sum_{j=1, j\neq i}^{n} c_{ij}(x(t))\tilde{f}_j(x_j(t)) + u_i(t) - \tilde{c}_{ii}(x(t))\tilde{f}_i(x_i(t)), \quad i = 1, ..., n.$$

Here x_i denotes the *state* of the ith subsystem and is a positive real value. The state can be interpreted as the number of unprocessed parts of a subsystem, but one can choose any other state variable such as the number of unsatisfied orders for example. The state of the whole network is denoted by $x = (x_1, \ldots, x_n)^T$. The positive real value t denotes the time of the system. The term u_i represents an external input into the subsystem, e.g., supply of raw material.

Each plant processes the material with a production rate $\tilde{c}_{ii}(x(t))\tilde{f}_i(x_i(t))$, where $\tilde{c}_{ii} \in \mathbb{R}_+$ and $\tilde{f}_i(x_i(t))$ is a continuous, positive definite and monotone increasing function. The processed material is sent to the jth subsystem of the network with the rate $c_{ji}(x(t))\tilde{f}_i(x_i(t))$, $j \neq i$ where c_{ji} is a positive real value and represents a distribution parameter for processed material from subsystem i to j. The term $\sum_{j=1, j\neq i}^{n} c_{ij}(x(t))\tilde{f}_j(x_j(t))$ is the internal input of material from other subsystems to the subsystem i.

Denoting $c_{ii} := -\tilde{c}_{ii}$ the above equations can be rewritten as an interconnected system, which represents the whole network in a vector form

$$\dot{x}(t) = C(x(t))\tilde{f}(x(t)) + u(t), \tag{1}$$

where $\tilde{f}(x(t)) = (\tilde{f}_1(x_1(t)), ..., \tilde{f}_n(x_n(t)))^T, u = (u_1, ..., u_n)^T, C(x(t)) = (c_{ij}(x(t)))_{nxn}$.

ADPs are modeled by the production rate $\tilde{c}_{ii}(x(t))\tilde{f}_i(x_i(t))$ and the distribution coefficients c_{ij}. The production rate depends on the state of a subsystem and the ability to adapt the production speed of a plant in a network can be modeled by an appropriate choice of this rate. Namely, if there is a lot of unprocessed material, the plant increases the production and, conversely, if there is less unprocessed material the production speed goes down. For example, one can choose $\tilde{f}_i(x_i(t)) = x_i^2$ or $\tilde{f}_i(x_i(t)) = (1 - \exp(-x_i))$.

By the distribution coefficients c_{ij} a centralized or decentralized planning scenario can be modeled, where constant coefficients are identified as central planning. For example, the nQLE, nPHE or other ADPs can be implemented by the following choices of the coefficients c_{ij}:

for the nQLE

$$c_{ij} := \frac{\frac{1}{x_i+\varepsilon}}{\sum_k \frac{1}{x_k+\varepsilon}},$$

for the nPHE

$$c_{ij} := (1 - v_i)\frac{\tilde{f}_i(x_i)}{\sum_k \tilde{f}_k(x_k) + \varepsilon} + \sum_{k\neq i} v_k \frac{\tilde{f}_k(x_k)}{\sum_q \tilde{f}_q(x_q) + \varepsilon},$$

for an integrated ADP

$$c_{ij} := \frac{\frac{\tilde{f}_i(x_i)}{x_i+\varepsilon}}{\sum_k \frac{\tilde{f}_k(x_k)}{x_k+\varepsilon}},$$

where k and q are the indices of the subsystems, which get material from subsystem j, v_i are evaporation constants, $i = 1, \ldots, n$ and ε is a positive constant to assure that the term c_{ij} is well-defined.

So far, one important circumstance that occurs in production networks has been left out: transportation times of material from one plant to another. These transportation times can be modeled using time-delay systems as follows:

$$\dot{x}_i(t) = \sum_{j=1, j\neq k}^{n} c_{ij}(t)\tilde{f}_i(x_j(t-\tau_{ij})) + u_i(t) - \tilde{c}_{ii}(t)\tilde{f}_i(x_i(t)), \quad i = 1, ..., n. \quad (2)$$

Transportation times are represented as time-delays $\tau_{ij} \in \mathbb{R}_+$, which denote the time needed for transportation from subsystem j to i. In Eq. (2) the time-delays are included in the terms which represent the inflow of material from other subsystems, where c_{ij} can also depend on a time-delay. In the terms which represent the external input and the internal production rate no insertion of time-delays is necessary.

The consideration of transportation times makes the analysis of production networks more complex, but more realistic too. Due to the abstract level of this view, the model (1) or (2) is used to analyze the dynamics and make general statements about the dynamics of production networks from a macroscopic view. The results can be used to adapt the simulation model and, conversely, the results can be refined by results of the simulation view. This will be explained with examples in Sect. 5.4.

5.3 Simulation Models

Simulation approaches are often used for the analysis of stability of production networks in order to refine the mathematically found stability regions. Furthermore, they can be used to investigate different system aspects like logistic target achievement for time-varying systems parameters. For the analysis of ADPs in production networks, several simulation approaches were used in the past (e.g., [43]). These simulation models can be classified according to their general function principle concerning the representation of time in the simulation model. Continuous time simulations represent the time of the simulation model as a real variable. All system states in the simulation model change with dependence on the simulation time variable. In contrast to this, in a discrete time simulation model the time elapses in predefined equidistant time steps. A particular variant of discrete time models is discrete event simulation models [4, 62]. Here, the time elapses in non-equidistant time steps. The states of the simulation model change according to events. In the production logistic context these events describe the arrival of raw material in a source or the end of a particular production step [25], for example. Besides the representation of time, some authors discuss the purpose of analysis as a possible classification characteristic. Morecroft and Robinson [29] describe differences in the modeling representation and the interpretation between discrete event and continuous simulation models. Accordingly, discrete event simulations are applied for the representation of very detailed scenarios, which should be investigated with regard to the interaction of single system elements, while continuous simulations are used for investigations of general dynamic aspects of the system [8, 29]. The usage of different modeling and simulation approaches helps to validate the obtained simulation results concerning ADPs in production networks against each other. Possible differences or mistakes can be detected easily. A comprehensible example for this approach is the determination of stability regions of autonomously controlled production networks. In Scholz-Reiter et al. [51] a discrete event simulation approach is used for the refinement procedure.

5.4 Measures for Analysis

Production networks are steadily exposed to dynamic variations, caused by internal reasons and by the external environment. Hence, a pure static analysis of logistic performance indicators seems to be not sufficient to cover the effects and the interdependencies of these dynamics. Nevertheless, classical logistic performance indicators should not be neglected. According to Wiendahl [56] the logistic key performance measures are throughput time (TPT), delivery liability, work-in-process (WIP) and utilization. The throughput time is the time-span spent by a particular product in a production system. From a customer's point of view short TPTs are desirable, due to the shorter possible delivery times. Another aspect of this customer's perspective is the delivery reliability, which means a delivery of goods to the customer at the right time in the right quantity. The performance indicators WIP and utilization belong to the logistic costs. A high level of WIP means that the buffers of the system are filled with numerous semi-finished goods and raw material. This leads consequently to high degree of capital lock-up. From an economical point of view the WIP should be at a low level, while the system is fully utilized.

Windt et al. (2008) developed a vector-based approach which allows one to weight these targets according to the subjective preferences and to aggregate these weighted targets in one performance indicator, called logistic target achievement. This value depends on pre-defined targets, the operative target achievement and weight factors. The total logistic target achievement gives information about the performance as a percentage value. By applying this approach, different configurations of production systems can be compared easily. For the objective of the analysis of autonomously controlled production systems, this vector-based approach can be used to compare different ADPs in a defined way (e.g., [46]). Furthermore, Windt et al. [58] introduce an autonomous control application matrix (ACAM), which proposes an evaluation of different scenarios with ADPs on the basis of this vector-based approach.

Additionally, the identification of stability regions is generally crucial for planning and operating logistic networks as an aspect of the dynamic systems behavior. In this context, mathematical models are often used to determine stability regions. Typical examples of unstable behavior are unbounded growth of unsatisfied orders or unbounded growth of the queue of the workload to be processed by a plant or a machine. This causes high inventory costs and loss of customers. To avoid instability of a network it is worth to investigate its behavior in advance.

Stability means, roughly speaking, that the number of unsatisfied orders or the number of unprocessed parts remains bounded over time. More precisely, the *local input-to-state stability* (*LISS*) property from control theory is used and by the means of LISS the state of a system can be estimated. More details about this property can be found in Dashkovskiy and Rüffer [11].

A useful tool to verify the LISS property of a system is a Lyapunov function, which is positive definite and radially unbounded and can be interpreted as the energy of the systems state. A LISS Lyapunov function V_i of the ith subsystem has the property that if $V_i(x_i) \geq \max\left\{\max_{j \neq i} \gamma_{ij}(V_j(x_j)), \gamma_i(|u_i|)\right\}$ holds, where the gains γ_{ij} and γ_i are positive definite, zero at zero and strictly increasing functions, then the energy decreases. If $V_i(x_i) < \max\left\{\max_{j \neq i} \gamma_{ij}(V_j(x_j)), \gamma_i(|u_i|)\right\}$ then the energy of the system is bounded by the expression on the right side of the previous inequality. Overall, the trajectory of a system is bounded. More details can be found in [11].

By the gains, statements about the behavior of the system can be made. For example, they offer information about the upper bound of the trajectory of a system or in other words about the highest inventory level of a system. This information is helpful for plant owners, because they can plan the size of the inventory in advance and they can also design their plant in a way to assure stability.

The tool of a Lyapunov function can be used for the stability analysis in the following way:

Consider a network consisting of n subsystems and assume that each subsystem has a LISS Lyapunov function, i.e., each subsystem has the LISS property. Then, the overall network has the LISS property provided that the *small-gain condition* (*SGC*) is satisfied (see [11]).

Simply speaking, the SGC states that along every existing circle in the network the composition of the corresponding gains is less than the identity (see [11]).

Concluding this, to verify if a system is stable, one has to find LISS Lyapunov functions for the subsystems, the corresponding gains and to check the SGC, then stability is verified. Otherwise, one has to find other LISS Lyapunov functions candidates and gains. If all efforts are not successful, then no statement about stability is possible.

To assure the stability of a network by using the properties of the Lyapunov functions and the SGC one gets conditions on relevant system parameters, as the production or distribution rates and the external inputs. Using the model (1) in Sect. 5.2 describing general production networks and assuming that the distribution coefficients are bounded, the following condition for unbounded production rates can be derived:

If there exist $a \in \mathbb{R}^n$, $a_i > 0$ and $\varepsilon \in \mathbb{R}^n$, $\varepsilon_i > 0$, $\quad i = 1, ..., n$, such that

$$C(t)a < \varepsilon$$

holds, then the whole network (1) has the ISS property, which is the global variant of LISS.

For production rates, which are bounded up to a certain limit $\alpha_i := \sup_{x_i} \left\{ \tilde{f}_i(x_i) \right\}$ the condition

$$C(t)\alpha + \|u\|_\infty < \varepsilon,$$

can be derived to assure that a network has the LISS property, where $\alpha = (\alpha_1, ...\alpha_n)^T$ and $\|u\|_\infty$ denotes the essential supremum norm of the external input. Taking transportation times into account one gets similar conditions to assure stability of a production network modeled by the equations as in (2).

These conditions form a stability region: for parameter constellations (i.e., set of parameters) within this region, stability is guaranteed. For parameter constellations outside this region the tool of a Lyapunov function does not offer a statement about stability. At this stage, simulations are performed to refine the stability region.

By the analysis using Lyapunov functions a large set of parameter constellations which assure stability are identified. Only few parameter constellations have to be simulated. To identify stable or unstable parameter constellations a truncation criterion needs to be defined. Then, the simulation results refine the stability region.

This dual approach using the analytical and the simulation model has the advantage of less time consumption to identify stability regions in contrast to a pure simulation approach. This is especially relevant since the time needed for a simulation run increases exponentially as the number of plants, links and parts in a network are increased. To identify parameter constellations which assure stability and to make statements about the inventory levels of the plants of a complex network with a large number of plants is a problem which cannot be solved in an acceptable time. The dual approach presented helps to derive and refine parameter constellations in reasonable time by assuring stability (see [51]) and is presented in the following Fig. 3, where a scheme of a stability analysis is displayed.

6 Examples of ADP Implementation

This section presents the approach of modeling and analyzing autonomous decision policies in two exemplary cases of production networks. The first focuses on stability analysis and a refinement of stability regions of a relatively simple network. The second is used to demonstrate the performance evaluation of different autonomous decision strategies (for the structure see Fig. 4). The first network consists of three

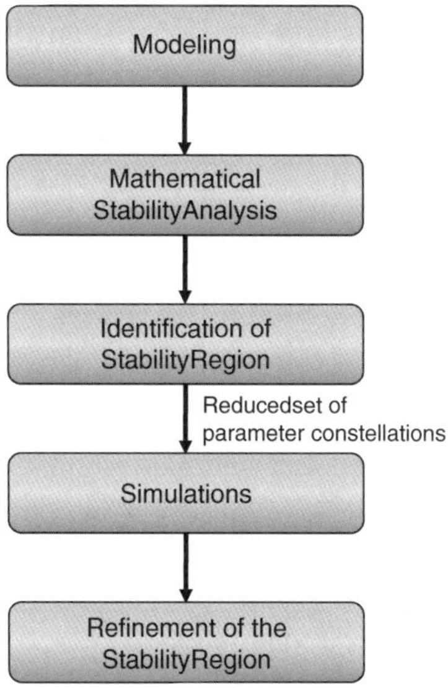

Fig. 3 Scheme of the stability analysis (similar to [51])

plants, the second has six plants. In the first example a macroscopic view is considered; the second example is investigated in detail representing the shop-floor of the plants, consisting of 3 × 3 machines (see Fig. 4).

6.1 Stability Analysis

According to Fig. 4a the material flow between the plants is defined as follows: The input of raw material arrives at plant 1 and plant 3. All material produced in plant 1 is delivered to plant 2. From here 50% of the goods are delivered to the customers and 50% for further processing to plant 3. Plant 3 sends 50% of its output to plant 1 and plant 2 each. In order to model seasonal demand fluctuations both inputs to plant 1 and plant 3 are modeled as a sinusoidal function $u_i(t)$:

$$u_i(t) := AV_i \cdot (\sin(t) + 1) + 5, \quad i = 1, 3.$$

The parameter AV_i determines the intensity of the fluctuations in terms of the amplitude. In this example this parameter is used to generate different input situations. The plants are able to decide autonomously about their current production rate $\tilde{f}_i$ at the time point t. It is assumed that this decision depends on the actual workload in the following form:

$$\tilde{f}_i(x_i(t)) := \alpha_i(1 - \exp(-x_i(t))), \quad i = 1, 2, 3., \quad \alpha_i \in \mathbb{R}_+$$

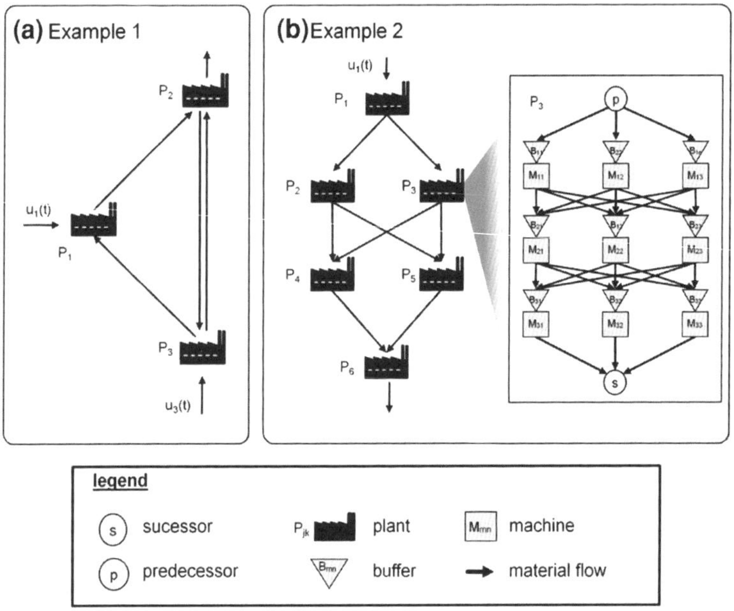

Fig. 4 **a** Production network scenario with three plants. **b** Production network scenario with six plants

This means that the production rate of plant i depends on the WIP of the plant. According to this equation, the production rate will be very low in cases of low WIP in the plant. Otherwise the production rate will be set to its maximum α_i for a high WIP level.

This example aims at determining the lowest possible values of α_i or a certain input situation, which depends on AV_i. To do so, the internal structure of the plants is neglected in a first step. Stability conditions will be derived only macroscopically on the network level. These results will be subsequently refined by a simulation model.

In this first step the network is modeled by differential equations:

$$\begin{aligned}\dot{x}_1(t) &= u_1(t) + 0.5 \cdot \tilde{f}_3(x_3(t)) - \tilde{f}_1(x_1(t)),\\ \dot{x}_2(t) &= \tilde{f}_1(x_1(t)) + 0.5 \cdot \tilde{f}_3(x_3(t)) - \tilde{f}_2(x_2(t)),\\ \dot{x}_3(t) &= u_3(t) + 0.5 \cdot \tilde{f}_2(x_2(t)) - \tilde{f}_3(x_3(t)).\end{aligned}$$

These equations describe the change of WIP in the three plants and consider the transport connections and quantities as well as the current production rate of a plant. According to the scheme in Fig. 3 the next step is to derive stability conditions using Lyapunov functions and gains. A very detailed technical description for this can be found in Scholz-Reiter et al. [51]. For this network the following stability conditions can be calculated:

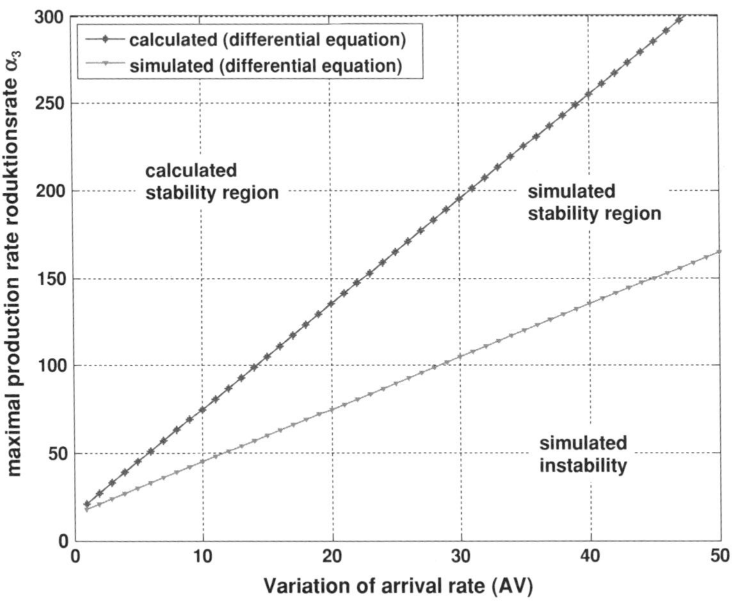

Fig. 5 Calculated and simulated stability regions for plant three

$$\alpha_1 > 0.5 \cdot \alpha_3 + \max_t\{u_1(t)\},\ \alpha_2 > 0.5 \cdot \alpha_3 + \alpha_1,\ \alpha_3 > 0.5 \cdot \alpha_2 + \max_t\{u_3(t)\}.$$

By solving this system of inequalities, the maximal production rate for that stability can be guaranteed and can be calculated for the corresponding value of AV_i. For example, choosing $AV_i \equiv 5$ leads to:

$$\alpha_1 > 37.5,\ \alpha_2 > 60,\ \alpha_3 > 40.$$

From a mathematical point of view the stability of the production network can be guaranteed for the values indicated. This does not mean that the network is unstable for values that violate these inequalities. To illustrate this, a continuous simulation of the differential equation model is conducted.

For different values of AV_i the production rate of all plants is reduced stepwise in several simulation runs. The simulation model is considered to be unstable, whenever the WIP of a plant starts to rise continuously about 10% in a time period of 30 days. Figure 5 depicts the results of the simulation model for plant 3 and compares it with the calculated results.

These results show that the simulation model is still stable in this case, even if the production rate is below the calculated stability boundary. Using simulations, the calculated stability region can thus be refined.

Figure 6 clarifies this. It presents simulation results for different values of α_i and $AV_i \equiv 5$ taken from the calculated stability region, the simulated stability region and the simulated instability region. The WIP remains bounded over time for the calculated stability region and the simulated stability region. By contrast the

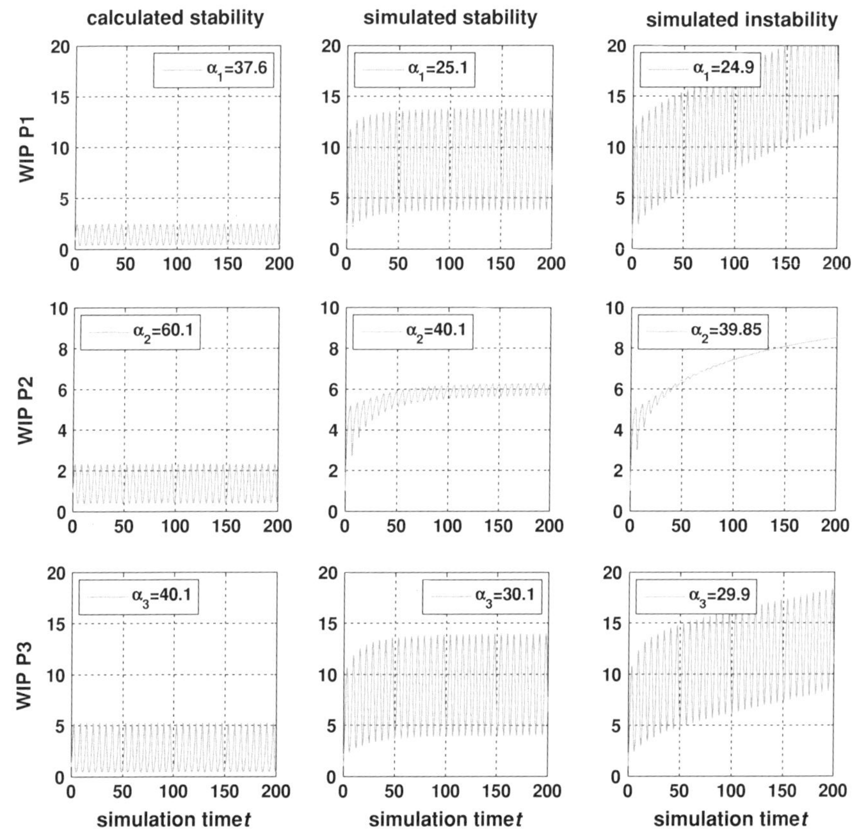

Fig. 6 WIP against time for examples of the calculated stability region, the simulated stability region and the simulated instability

WIP grows continuously in plant P_1, P_2 and plant P_3. The WIP in $P1$ increases continuously about 0.049, in P_2 about 0.02 and in P_3 about 0.029 units per time unit.

According to the results presented in Fig. 6, the mathematical determination of a stability region provides a good starting point for the refinement. Hence, this approach allows the identification of the border of the stability region with less time efforts than a pure trial and error simulation approach.

In general, different simulation approaches can be applied for the refinement. Scholz-Reiter et al. [51] successfully applied a discrete event simulation and a continuous time simulation based on differential equations for the refinement of stability regions. It was shown that the refinement results of both simulation approaches provide similar stability results.

Table 2 Weighted adjacency matrix

From plant	To plant					
	P_1	P_2	P_3	P_4	P_5	P_6
P_1	–	200	200			
P_2		–		200	200	
P_3			–	200	200	
P_4				–		200
P_5					–	200
P_6						–

6.2 Implementation and Evaluation of Different Autonomous Decision Policies

The second example presents the modeling and implementation of different ADPs on the shop-floor and on the network level. Therefore, the network depicted in Fig. 4b is considered. In order to keep this example simple, only one logistic target measure is considered. This example focuses on the total throughput time, which denotes the time spent by the parts to pass through the entire network.

This scenario has six different plants on four network stages. On stage one and on stage four there is only one plant. On stage two and three there are two parallel plants each. Additionally, every plant consists of a shop-floor with 3×3 machines. The distances between the plants are summarized in Table 2.

The transports between plants are triggered by a "frequency-based" policy. This means that a transport starts in pre-defined time intervals. The interval in this example is set to 15 h.

There are three different job types in this scenario. These job types differ in their processing times on the shop-floor level in every plant. The processing times are summarized in Table 3.

As in the first example, the arrival rate of jobs in this scenario is set to a sine function in order to model demand fluctuations:

$$u(t) = \lambda + AV \cdot \sin(t + \varphi)$$

This function has a phase shift $\varphi = 1/3$ of a period for each job type, so that the maximal arrival rates of all job types are not simultaneous. The variable λ defines the mean arrival rate and is set to 0.4 1/h in all simulation runs. The second variable *AV* determines the intensity of the arrival rate fluctuation, as in example 1, *AV* is set to 0.125 1/h.

The purpose of this example is to describe how to choose an applicable combination of different ADPs for this particular network configuration. Therefore, the ADPs QLE, PHE, nQLE and nPHE are implemented on the shop-floor and the network level to a simulation model, exemplarily. Table 4 shows the different combinations of ADPs and summarizes the results of the simulation runs in a form which is comparable to the ACAM described above.

Table 3 Processing times (h:mm)

	P_1; P_6			P_2; P_4			P_3; P_5		
Type / line	1	2	3	1	2	3	1	2	3
Type A	2:00	3:00	2:30	3:00	4:00	3:30	5:00	6:00	5:30
Type B	2:30	2:00	3:00	3:30	3:00	4:00	5:30	5:00	6:00
Type C	3:00	2:30	2:00	4:00	3:30	3:00	6:00	5:30	5:00

Table 4 Simulation results

ADP (shop-floorlevel)	ADP (network level)	Logistic performance mean total throughput time (h)	Standard deviation of mean total throughput time (h)	Rank
QLE	nQLE	86.59	5.21	2
QLE	nPHE	85.84	3.23	1
PHE	nQLE	110.85	10.67	3
PHE	nPHE	117.95	12.37	4

This relatively simple example demonstrates that a detailed analysis of different combination of network-related and shop-floor-related ADPs is necessary. In this example the combination of QLE and nPHE performs best with respect to minimizing the total throughput time (TTPT), which is the time span needed by a part to pass through the entire network. On the other hand the combination of PHE and nPHE seems not to be suitable for this particular network. This combination leads to the highest mean TTPT.

Figure 7 depicts these results in more detail. It presents the TTPT against the simulation time of each possible combination. The nPHE leads to smoother patterns in the TTPT compared to the combination of the QLE/nQLE (Fig. 7a, 7b). A comparison of the corresponding standard deviations confirms this. In combination with the shop-floor-related QLE method the best result of mean TTPT (85.84 h) is realized.

A similar curve shape is observed for the combination of the PHE/nQLE, but the absolute values differ: in this case the mean TTPT is 21.89% higher than in the first case. This can be explained by the time horizon used by the methods. The pheromone-based method uses data from past events, while the QLE is based on actual information. In the case at hand, the data on the shop-floor level used by the PHE method does not represent the current situation properly, which leads to unsuitable autonomous decisions on the shop-floor level and consequently to a longer TTPT.

The results of the network-related pheromone-based approach are different. In combination with the shop-floor-related QLE method the lowest mean TTPT (85.84 h) is realized.

Figure 7d presents the TTPT of the PHE/nPHE combination. Again, the shop-floor-related pheromone-based method leads to high throughput times in the plants, which corresponds to the effects discussed concerning the PHE/nQLE combination.

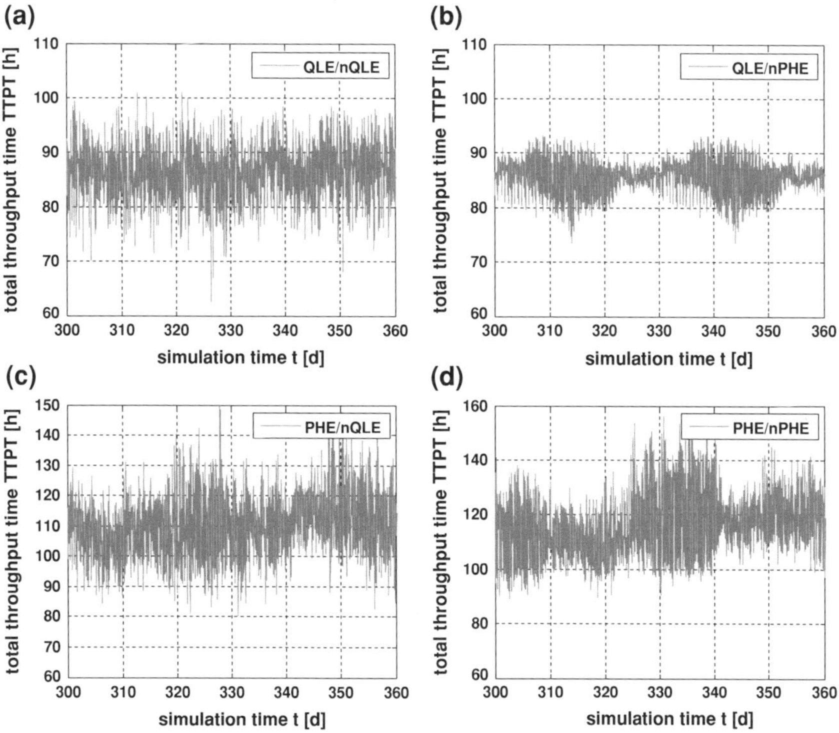

Fig. 7 Total throughput time against simulation time: **a** combination QLE/nQLE, **b** combination QLE/nPHE, **c** combination PHE/nPHE, **d** combination PHE/nPHE

Additionally to this, the nPHE method uses the information of the throughput times in the plants for the autonomous decision making on the network level. Due to the time-varying and imprecise information allocation decisions made by the nPHE method are not suitable in this situation. Consequently, this combination leads to the highest TTPT value.

This example illustrates the potentials of the application of autonomous decision policies in production networks. Combined autonomous decision policies on the network level and on the shop-floor level may lead to an acceptable logistic performance. However, the underlying dynamics and their consequences should not be neglected. In the case at hand the combination of the PHE/nPHE method leads to a dynamic interplay between network and shop-floor-related decisions which are undesirable and consequently decreases the logistic performance. Thus, the design and implementation of autonomous control strategies in production networks should be integrated in an intensive analysis of relevant system properties such as stability and systems performance.

7 Summary

This contribution described the integrated coordination between production and transport processes as an essential task of operating production networks. Different central planning and scheduling functions for shop-floor and transport operations were presented. In this context, different ADPs and the concept of autonomous cooperating processes were introduced as a novel approach to coordinated logistic processes in production networks. Several ADPs were introduced and described. Additionally, approaches for modeling and analyzing ADPs in production networks were presented and discussed in mathematical terms and via simulative approaches. Based on the mathematical modeling approach, criteria for the stability of production networks can be derived, which subsequently can be refined by simulations. Finally, two examples for analyzing the stability and the performance of autonomous decision policies in production networks were given.

Acknowledgments This research is funded by the German Research Foundation (DFG) as part of the Collaborative Research Centre 637 'Autonomous Cooperating Logistic Processes: A Paradigm Shift and its Limitations'.

References

1. Allahverdi A, Ng CT, Cheng TCE, Kovalyov M (2008) A survey of scheduling problems with setup times or costs. Eur J Oper Res 187(3):985–1032
2. Alvarez E (2007) Multi-plant production scheduling in SMEs. Robotics Comput Integr Manuf 23(6):608–613
3. Armbruster D, de Beer C, Freitag M, Jagalski T, Ringhofer Ch (2006) Autonomous control of production networks using a pheromone approach. Physica A 363(1):104–114
4. Banks J, Carson JS, Nelson BL, Nicol DM (2010) Discrete-event system simulation. Prentice Hall, Upper Saddle River
5. Bertazzi L, Speranza MG (2005) Worst-case analysis of the full load policy in the single link problem. Int J Prod Econ 93–94:217–224.
6. Bloos M, Kopfer H (2009) Efficiency of transport collaboration mechanisms. Commun SIWN 6(1):23–28
7. Camarinha-Matos L, Afsarmanesh H (2003) Elements of a base VE infrastructure. Comput Ind 51:139–163
8. Chahal K, Eldabi T (2010) A multi-perspective comparison for selection between system dynamics and discrete event simulation. Int J Bus Inf Syst Arch 6(1):4–17
9. Comelli M, Gourgand M, Lemoine D (2008) A review of tactical planning models. J Syst Sci Syst Eng 17(2):204–229
10. Crainic TG (2000) Service network design in freight transportation. Eur J Oper Res 122(2): 272–288
11. Dashkovskiy S, Rüffer B (2010) Local ISS of large-scale interconnections and estimates for stability regions. Syst Control Lett 59(3):241–247
12. Dashkovskiy S, Görges M, Naujok L (2009) Local input to state stability of production networks. In: Proceedings of 2nd international conference on dynamics in logistics (LDIC 2009). Springer, Bremen
13. Domschke W, Scholl A, Voß S (1997) Produktionsplanung. Springer, Berlin

14. Dunbar WB, Desa S (2007) Distributed MPC for dynamic supply chain management. Assessment and future directions of nonlinear model predictive control. Lect Notes Control Inf Sci 358:607–615
15. Erengünc SS, Simpson NC, Vakharia AJ (1999) Integrated production/distribution planning in supply chains. Eur J Oper Res 115(2):219–236
16. Fleischmann B, Gietz M (2008) Transport- und Tourenplanung. In: Arnold D, Isermann H, Kuhn A, Tempelmeier H (eds) Handbuch Logistik, Springer, Heidelberg pp 137–152
17. Garey MR, Johnson DS (1979) Computers and intractability: a guide to the theory of NP-completeness, Freeman, San Francisco
18. Gudehus T (2005) Logistik. Springer, Berlin
19. Guinet A (2001) Multi-site planning: a transshipment problem. Int J Prod Econ 74(3):21–32
20. Hinrichsen D, Pritchard AJ (2005) Mathematical systems theory I. series: texts in applied mathematics, vol 48. Springer, Berlin
21. Ivanov D (2009) An adaptive framework for aligning (re)planning decisions on supply chain strategy, design, tactics, and operations. Int J Prod Res 48(13):3999–4017
22. Jungwattanakit J, Reodecha M, Chaovalitwongse P, Werner F (2008) Algorithms for flexible flow shop problems with unrelated parallel machines, setup times, and dual criteria. Int J Adv Manuf Technol 37(3):354–370
23. Jungwattanakit J, Reodecha M, Chaovalitwongse P, Werner F (2009) A comparison of scheduling algorithms for flexible flow shop problems with unrelated parallel machines, setup times, and dual criteria, Comput Oper Res 36(2):358–378, Scheduling for Modern Manufacturing, Logistics, and Supply Chains
24. Kim J–H, Duffie NA (2004) Backlog control for a closed loop PPC system. Ann CIRP 53(1):357–360
25. Kuhn A., Wenzel S (2008) Simulation logisitscher systeme. In: Arnold D, Isermann H, Kuhn A, Tempelmeier H (eds) Handbuch Logistik. Springer, Berlin, pp 73–92
26. Martinez MT, Fouletier P, Park KH, Favrel J (2001) Virtual enterprise organisation, evolution and control. Int J Prod Econ 74(1–3):225–238
27. Meyr H, Wagner M, Rohde J (2005) Structure of advanced planning systems. In: Stadtler H, Kilger C (eds) Supply chain management and advanced planning. Springer, Berlin, pp 109–115
28. Min H, Zhou G (2002) Supply chain modeling: past, present and future. Comput Ind Eng 43(2):231–249
29. Morecroft J, Robinson S (2006) Comparing discrete-event simulation and system dynamics: modelling a fishery. In: Proceedings of the operational research society simulation workshop 2006. Operational research society, Birmingham, pp 137–148
30. Müller F, Otto A (2007) Anwendungsarchitekturen in supra-adaptiven Logistik-netzwerken. In: Günthner WA (eds) Neue Wege in der Automobillogistik: die Vision der Supra-Adaptivität; mit 14 Tabellen. Springer, Berlin, pp 149–166
31. Muth JF, Thompson GL (1963) Industrial scheduling. Prentice-Hall, Englewood Cliffs
32. Parunak HV (1997) Go to the ant: engineering principles from natural multi-agent systems. Ann Oper Res 15:69–101
33. Peeters P, van Brussel H, Valckenaers P, Wyns J, Bongaerts L, Kollingbaum M, Heikkilä T (2001) Pheromone based emergent shop floor control system for flexible flow shops. Artif Intell Eng 15(4):343–352
34. Philipp T, de Beer C, Windt K, Scholz-Reiter B (2007) Evaluation of autonomous logistic processes—analysis of the influence of structural complexity. In: Hülsmann M, Windt K (eds.) Understanding autonomous cooperation and control in Logistics—the impact on management, information and communication and material flow. Springer, Berlin, pp 303–324
35. Pinedo ML (2008) Scheduling—theory, algorithms, and systems. Springer, New York
36. Quadt D, Kuhn H (2007) A taxonomy of flexible flow line scheduling procedures. Eur J Oper Res 178(3):686–698

37. Rabelo L, Helal M, Lertpattarapong C, Moraga R, Sarmiento A (2008) Using system dynamics, neural nets, and eigenvalues to analyse supply chain behaviour. A case study. Int J Prod Res 46(1):51–71
38. Rekersbrink H, Makuschewitz T, Scholz-Reiter B (2009) A distributed routing concept for vehicle routing problems. Logist Res 1(1):45–52
39. Rohde J, Meyr H, Wagner M (2000) Die supply chain planning matrix. PPS Manag 5:10–15
40. Sauer J (2006) Modeling and solving multi-site scheduling problems. In: van Wezel W, Jorna R, Meystel A (eds.) Planning in intelligent systems: aspects, motivations and method. Wiley, Hoboken, pp 281–299
41. Scholz-Reiter B, Freitag M, de Beer C, Jagalski T (2005) Modelling and analysis of autonomous shop floor control. In: Proceedings of 38th CIRP International Seminar on Manufacturing Systems. Universidade Federal de Santa Catarina, Florianopolis
42. Scholz-Reiter B, Böse F, Jagalski T, Windt K (2007) Selbststeuerung in der betrieblichen Praxis - Ein Framework zur Auswahl der passenden Selbststeuerungsstrategie. Industrie Management 23(3):7–10
43. Scholz-Reiter B, de Beer C, Freitag M, Jagalski T (2008) Bio-inspired and pheromone-based shop-floor control. Int J Comput Integr Manuf 21(2):201–205
44. Scholz-Reiter B, Jagalski T, Bendul J (2008) Autonomous control of a shop floor based on bee's foraging behaviour. In: Haasis, H-D, Kreowski H-J, Scholz-Reiter B (eds.) First international conference on dynamics in logistics. LDIC 2007, Springer, Berlin, pp. 415–423
45. Scholz-Reiter B, Mehrsai A, Görges M (2009) Handling the dynamics in logistics - adoption of dynamic behavior and reduction of dynamic effects. Asian Int J Sci Technol Prod Manuf Eng (AIJSTPME) 2(3):99–110
46. Scholz-Reiter B, Görges M, Philipp T (2009) Autonomously controlled production systems—Influence of autonomous control level on logistic performance. CIRP Ann Manuf Technol 58(1):395–398
47. Scholz-Reiter B, Görges M, Jagalski T, Mehrsai A (2009) Modelling and analysis of autonomously controlled production networks. In: Proceedings of the 13th IFAC symposium on information control problems in manufacturing (INCOM 09). Moscow, Russia, pp 850–855
48. Scholz-Reiter B, Rekersbrink H, Görges M (2010) Dynamic flexible flow shop problems - scheduling heuristics vs. autonomous control. CIRP Ann Manuf Technol 59(1):465–468
49. Scholz-Reiter B, Görges M, Jagalski T, Naujok L (2010) Modelling and analysis of an autonomous control method based on bacterial chemotaxis. In: 43rd CIRP international conference on manufacturing systems (ICMS 2010). Neuer Wissenschaftlicher Verlag, Wien, pp 699–706
50. Scholz-Reiter B, Lensing T, Görges M, Dickmann, L (2010) Classification of dynamical patterns in autonomously controlled logistic simulations using echo state networks. In: International conference on Harbor, Maritime and Multimodal Logistics Modelling and Simulation (HMS 2010). DIPTEM University of Genova, Genova, pp 85–92
51. Scholz-Reiter B, Dashkovskiy S, Görges M, Naujok L (2011) Stability analysis of autonomously controlled production networks. Int J Prod Res 49(16). DOI:10.1080/00207543.2010.505215
52. Stadtler H (2005) Supply chain management and advanced planning-basics, overview and challenges. Eur J Oper Res 163(3):575–588
53. Sydow J (2006) Management von Netzwerkorganisationen—zum Stand der Forschung. In: Sydow J (ed) Management von Netzwerkorganisationen, Gabler, Wiesbaden pp 385–469
54. Toth P, Vigo D (2002) An overview of vehicle routing problems. In: Toth P, Vigo D (eds.) The vehicle routing problem, SIAM monographs on discrete mathematics and applications, Philadelphia
55. Wagner B (2006) Hub & Spoke-Netzwerke in der Logistik, Deutscher Universitäts-Verlag/GWV-Fachverlage GmbH, Wiesbaden
56. Wiendahl H-P (2008) Betriebsorganisation für Ingenieure. München, Hanser
57. Wiendahl H-P, Lutz S (2002) Production in networks. Ann CIRP Manuf Technol 51(2):1–14

58. Windt K, Becker T (2009) Applying autonomous control methods in different logistic processes—a comparison by using an autonomous control application matrix. In: Proceedings of the 17th mediterranean conference on control and automation. Thessaloniki, Greece
59. Windt K, Hülsmann M (2007) Changing paradigms in logistics—understanding the shift from conventional control to autonomous cooperation and control. In: Hülsmann M, Windt K (eds.) Understanding autonomous cooperation and control—the impact of autonomy on management, information, communication, and material flow. Springer, Berlin, pp 4–16
60. Windt K, Böse F, Philipp T (2005) Criteria and application of autonomous cooperating logistic processes. In: Gao JX, Baxter DI, Sackett PJ (eds) Proceedings of the 3rd international conference on manufacturing research. Advances in manufacturing technology and management, Cranfield
61. Windt K, Philipp T, Böse F (2008) Complexity cube for the characterization of complex production systems. Int J Comp Integr Manuf 21(2):195–200
62. Zeigler BP, Praehofer H, Kim TG (2007) Theory of modeling and simulation—integrating discrete event and continuous complex dynamic systems, second edn (reprint). Academic Press, Amsterdam

Optimal Order and Distribution Strategies in Production Networks

Simone Göttlich, Michael Herty and Christian Ringhofer

Abstract Production networks are usually defined as a set of processes utilized to efficiently integrate suppliers, manufacturers, and customers so that goods are produced and distributed in the right quantities, to the right locations, and at the right time and in order to reduce costs while satisfying delivery conditions. We focus on a network of suppliers or producers which order goods from each other, process a product according to orders, and receive payments according to a pricing strategy. Modeling manufacturing systems is characterized by many different scales and several different mathematical approaches. We follow a dynamic approach: we are interested in the time behavior of the entire system. Therefore we introduce a coupled system of ordinary differential delay equations, where time-dependent distribution and order strategies of individual manufacturers influence the flow of goods and the total revenue. We also allow manufacturers to face bankruptcy. All order and distribution strategies are degrees of freedom which can vary in time. We determine them as solution to an optimization problem where additionally economic factors such as production and inventory costs and credit limits influence the maximization of profit. Instead of using a simulation-based optimization procedure, we derive an efficient way to transform the original model into a mixed-integer programing problem.

S. Göttlich (✉)
School of Business Informatics and Mathematics, University of Mannheim, 68131 Mannheim, Germany
e-mail: goettlich@uni-mannheim.de

M. Herty (✉)
Department of Mathematics, RWTH Aachen University, Templergraben 55, 52056 Aachen, Germany
e-mail: herty@mathc.rwth-aachen.de

C. Ringhofer
School of Mathematical and Statistical Sciences, Arizona State University, Tempe, AZ 85287-1804, USA
e-mail: ringhofer@asu.edu

D. Armbruster and K. G. Kempf (eds.), *Decision Policies for Production Networks*,
DOI: 10.1007/978-0-85729-644-3_11, © Springer-Verlag London 2012

1 Introduction

A supply chain consists of suppliers, manufacturers, warehouses, and stores where parts are produced and distributed among different production facilities. Mathematical models are used to monitor cost- efficient distribution of parts and to measure current business processes. Naturally, depending on the scale, these models are characterized by several approaches which are either discrete or continuous. The main difference between these two mathematical concepts is the description of parts as individuals at discrete time instances or as a dynamic flow (measured as parts per unit time).

Simulations in general represent a powerful computing technique to analyze manufacturing systems while performing numerical experiments of the models. Most of the mathematical approaches are discrete and based on considerations of individual parts, e.g., discrete event simulations [2, 11, 13], queueing theory [16, 22, 36, 63], mixed integer models [14, 54, 60, 65, 66], and the references therein. In case of discrete event simulations the evolution of the system is viewed as a sequence of significant changes in time, also called events, for each part separately. For instance, consider a supply chain consisting of numerous consecutive facilities, where parts arrive, get processed and depart when their production is completed. The transportation of parts from one production step to another is characterized by dynamic events that can be easily evaluated using performance measures such as the number of parts in the system, the individual waiting times, and so forth. Discrete event simulations serve as one level of description of interacting part-based systems but with the drawback of exponentially increasing computational time for large-scale systems.

A well-known class of stationary models are queuing theory models. These models permit the derivation and calculation of several performance measures including the mean waiting time of parts in the system, the proportion of time the processors are busy and the probability of encountering the system in particular states.

An alternative modeling approach is differential equations. In contrast to the discrete event simulation, the evolution of averaged quantities are predicted and dynamics inside the different production steps is included. To derive accurate continuous models as many features as possible of the detailed and complex discrete model have to be transfered to the continuous level of the differential equation. This will be achieved dynamic flows, i.e., parts per time unit, instead of individual parts. Since numerical schemes for differential equations allow for fast simulation, production problems with multiple manufacturers and thousands of parts are solved in a cost-effective way. Differential equations furthermore allow to apply optimization methods as shown below. An advantage of continuous models is the possibility to easily include nonlinearities, see e.g., Sect. 2 of this contribution.

In recent years these continuous models have been intensively investigated. Continuous models such as [2, 4, 7, 20, 25, 27, 36, 40, 45, 54, 55, 59, 60, 63–65] use partial differential equations to model production flow in an aggregate way, leading to deterministic coarse-grained and fast models. The fundamental setup will be described in detail in Sect. 2.1. A high-volume multi-stage production line is considered and the time evolution of the lot-density $\rho(t, x)$ is used to describe the current work in progress (WIP) at time t at a position x. Here the position x is the degree of completion in a production process. The basic equation is a conservation law

$$\frac{\partial \rho}{\partial t} + \frac{\partial F}{\partial x} = 0. \tag{1}$$

Typical models for F are given in [5] or clearing function approaches as in [7, 20, 36, 40, 54, 55, 59, 60, 63–65] and will also be discussed in the next section.

Recently those models based on scalar conservation laws have been reformulated in the framework of network models where the dynamics on the arcs is governed by a partial differential equation (PDE); see [30, 37–39]. This approach is inspired by other recent discussions on other problems on networks; see, e.g., [9, 10, 24, 30, 43, 46, 47, 49].

We consider a problem modeled by PDEs on a network of suppliers. Each supplier receives orders for its output from the other suppliers in the network, as well as from a final customer. Each supplier orders its input from a select set of other suppliers in the network as well as from a raw material supplier. Each supplier receives payments for delivered items at a certain fixed set of prices and has to pay production costs at certain rates for each item produced. This setup is similar to the ones studied in [8, 12, 21], where it is shown that the resulting dynamics can exhibit quite complicated behavior.

The modeling of the payments allows for the definition of capital flows through the network and consequently, for the occurrence of bankruptcies if the capitalization of a node in the network falls below a certain threshold. We consider in the last section the optimization of profits, i.e., given a set of production costs, to choose ordering strategies to maximize income.

2 The Mathematical Model

In this section we introduce a macroscopic model (1) for a production network with orders. We focus on deterministic models and assume that a description similar to the gas dynamics context is reasonable [4, 26]. This assumption is reasonable if the number of parts is large and the number of machines is large, too [2, 5]. We refer to [2] for a rigorous mathematical derivation of a continuum model from production dynamics. Here we assume that the macroscopic model is given by a scalar, one-dimensional nonlinear partial differential equation. Before giving the details of the modeling of the order process we give some introductory comments in Sect. 2.1 on the relation of the solution to this differential equation and models used in production planning. The full model is then introduced in Sect. 2.2.

The use of a model based on partial differential equations is motivated by some observations. First, starting from a discrete event model description and considering a formal limit for a large number of parts and large number of machines a partial differential equation could be rigorously derived [2]. Details of this mathematical justification in various situations can be found for example in [1, 3, 5, 6, 31]. Second, assuming the validity of the partial differential equation we recover a class of proposed production models [34, 36, 54, 60, 63–65] as shown below. This is related to the fact that all production models obviously assume the conservation of parts or processed items. This fact, also present in gas dynamics, already allows to write the conservation law (1). The crucial part which is subject to modeling is then the precise definition of the flux function F, which is called clearing function in the literature on production planning [7, 40, 41, 52, 55]. From the partial differential equation's point of view many properties can be deduced without the precise knowledge on F—only monotonicity or curvature properties need to be specified. The use of a partial differential equation necessarily implies a temporal dynamic of the production line. The need of a temporal grid or epochs is not present until numerical computations need to be performed. This allows to keep nonlinearities in the clearing function and can also be useful when applying gradient-based optimization procedures (see e.g. [50, 57]). However, additional modeling features such as stochastic demands or service levels are more difficult to include into the original model. Further, the numerical solution of the partial differential equation might lead to a nonlinear formulation of the discrete problem instead of LP models. The latter aspect is discussed in Sect. 3. Here we present one possibility to reduce the complexity of the nonlinear discretization using a mixed-integer problem. However, this approach depends heavily on the specific choice of the clearing function.

2.1 Preliminary Discussion and Relation to Existing Models

As outlined previously we model a deterministic, dynamic single production line for a single product using a partial differential equation. The basic quantity in the model is the product density $\rho(t, x) \geq 0$ describing the number of parts in process per unit length at time t and position x. The position x in the production line can be seen as measure of the percentage of completion of the product (stage) within production line. We therefore normalize x such that $0 \leq x \leq 1$. Here $\rho(t, 0)$ describes the parts at stage $x = 0$ entering the production line and $\rho(t, 1)$ the leaving parts. Clearly, in many applications only these two quantities are relevant and we describe below the relation of ρ to inventories, demands, and supplies. In the production engineering context the work-in-progress (WIP) [55, 59–61] is often used to measure properties of the production line. In the context of the partial differential equation the WIP denotes the number of parts in the line. The relation to the product density is therefore

$$WIP = \int_0^1 \rho(t, x)dx. \tag{2}$$

The partial differential equation states that in the transition from $x = 0$ to $x = 1$ over time no parts are lost:

$$\partial_t \rho(t,x) + \partial_x F(t,x,\rho(t,x)) = 0 \tag{3}$$

Here F is called flux function. The flux F has the unit of parts per time and is also called the throughput. In the production engineering context a function F is called clearing function (CF) if it depends on the work-in-progress W only. If $F(t,x,\rho)$ is independent of (t,x), then the steady states of (3) satisfy $F(\rho) = const$ yielding $\rho = const$. In this case only and only in this case WIP and production density are equal due to (2). However, in the following we call the function $F(t,x,\rho)$ clearing function even if it depends on t and x. From now on we consider only CFs such that $F = F(\rho) = F(WIP)$. The CF therefore expresses the throughput of parts depending on the WIP level of the factory. Several choices for the shape of the CF have been presented in the literature [7, 20, 40, 55, 59, 60]. Without giving a complete list examples are due to Graves [40–42], Asmundsson et al. [7], Karmarkar [55], or Hopp and Spearman [52]. The constants τ and μ are lead time and the maximal capacity, respectively. All the previously mentioned functions are nonnegative, montone and concave. In our notation e.g. the CF of Graves [40] reads $F(\rho) = \rho/\tau$ or the 'best case' model of Hopp et al. [52], e.g. $F(\rho) = \min\{\rho/\tau, \mu\}$. A similar shape is also obtained by models using capacitated production and limited capacity, see Asmundsson et al. [7] and Selcuk et al. [62]. Note that in the context of gas dynamics the CF is typically written as $F(\rho) = \rho v(\rho)$ with the interpretation of v as velocity of the transported parts. Hence, in the above examples we find $\frac{1}{\tau} = v(\rho)$ or $\min \frac{1}{\tau}, \frac{\mu}{\rho} = v(\rho)$. Since the production line is parameterized by [0,1] we observe that τ corresponds to the time the parts need to travel through the production line. Typically, L is called lead time [42] in production engineering and is obtained from (3) by computing $\frac{1}{v(\rho)}$. As discussed in many references on production models, the first choice $F(\rho) = \rho/\tau$ leads to lead times independent on the WIP (or equivalenty) the density of the production factory. Hence, the parts are produced without a workload dependence. Again, using Eq. (3) and $F(\rho) = \rho/\tau$ we observe that

$$\partial_t \rho + \frac{1}{\tau}\partial_x \rho = 0 \Leftrightarrow \begin{pmatrix} 1 \\ \frac{1}{\tau} \end{pmatrix} \cdot \nabla \rho(t,x) = 0.$$

Hence, $\rho(t,x)$ is constant along $(t, x - t/\tau)$ and therefore parts entering the production line at $x = 0$ leave the production line at $x = 1$ after time $t = \tau$. This is consistent with the notion of lead time of the parts. However, this linear CF does not reflect observed behavior of production processes and therefore nonlinear CFs have been proposed and also analytically investigated. For a more detailed discussion refer for example to [55, 59, 62]. In the following we will focus the discussion on the particular choice of a combined CF [44, 62] with saturation [7, 59] at level μ

$$F(\rho) = \min\{\rho v, \mu\}. \tag{4}$$

Here μ describes a maximal throughput (also called capacity) and $v = \frac{1}{\tau}$ is the velocity or the inverse of a fixed lead time below a certain maximum throughput. The throughput F is WIP or density dependent. The function F is piecewise linear.

If we integrate (3) with respect to x we obtain

$$\partial_t WIP = \partial_t \int_0^1 \rho(t,x)dx = F(\rho(t,0)) - F(\rho(t,1)), \tag{5}$$

i.e., a first version of an inventory model: the change in the WIP is given by the inflow to the production line minus the outflow. A further discretization in time with time steps of length Δt leads to

$$WIP(t) - WIP(t-\Delta t) = \Delta t\,(F(\rho(t,0)) - F(\rho(t,1))\,.$$

Using a *linear* CF with lead time τ we further obtain

$$WIP(t) - WIP(t-\Delta t) = \Delta t(F(\rho(t,0)) - F(\rho(t-\tau/\Delta t,0)).$$

If we choose $\Delta t = 1$ and denote by $\psi(t) = F(\rho(t,0))$ (released parts to the production line or influx) and by $\phi(t) = F(\rho(1,t))$ (planned production quantity or outflux) we obtain the basic formulation of some planning models in the literature [44, 60, 62]:

$$WIP(t) = WIP(t-1) + \psi(t) - \phi(t) \,\text{and}\, \phi(t+\tau) = \psi(t). \tag{6}$$

In particular, the relation $\phi(t+\tau) = \psi(t)$ is not valid, if F is a piecewise linear CF since the lead time depends on the WIP. One possibility is to drop the equality and a so-called outer linearization representing a nonlinear CF by line segments [29, 62]. Additional inequalities are introduced as constraints on the planned production quantity $\phi(t)$. In the following we discuss an approach to treat the CF (4) using the continuous formulation (3). The function (4) is monotone increasing and therefore Eq. (3) is well posed provided boundary conditions for ρ at $x = 0$ are prescribed. Given the release rate ψ in parts per time, the desired boundary condition to determine $\rho(t,0)$ is

$$\psi(t) = F(\rho(t,0)). \tag{7}$$

However, if the influx ψ exceeds the maximal possible throughput ($\psi > \mu$), then (7) cannot be solved. In the following, we derive a mathematically well posed formulation of Eqs. (3) and (7). It turns out that this formulation can be seen as an inventory model. We proceed as follows. An Upwind [50, 58] discretization in x by the method of lines yields

$$\frac{d}{dt}\rho_j + \frac{1}{\Delta x}\left(F_j - F_{j-1}\right) = 0,\;\; F_j = \min\{v\rho_j, \mu\}, j = 1,\dots,J,\;\; F_0 = \psi. \tag{8}$$

Here $\Delta x \rho_j$ is the content of cell number j, located in (x_{j-1}, x_j) and Δx is a (uniform) spatial cell size such that $J\Delta x = 1$. Since the flux F_{j-1} into cell j is limited by μ, we have $v\rho_j \leq \mu$ for all cells, except in the first cell $j = 0$, where the release rate is given by ψ. Hence, we can drop the constraint on the maximal capacity in the interior(!) cells and the solution to (8) and (3) coincide if we define the cell fluxes F_j as

$$F_j(\rho) := \begin{pmatrix} \psi & j = 0 \\ \min\{v\rho_j, \mu\} & j = 1 \\ v\rho_j & j = 2, \ldots, J \end{pmatrix}. \tag{9}$$

Now, we rewrite (8) and obtain

$$\frac{d}{dt}(\Delta x \rho_1) = \psi - \min\{v\rho_1, \mu\} \tag{10}$$

and

$$\frac{d}{dt}\rho_j = \frac{1}{\Delta x}\left(v\rho_j - v\rho_{j-1}\right), \; j = 2, \ldots, J. \tag{11}$$

Note that $\Delta x \rho_1$ has the unit of a parts and is located at the beginning of the production process. Therefore, we call from now on $p := \Delta x \rho_1$ the input inventory. Its dynamics (9) are well defined even if ψ exceeds μ. Further, Eq. (11) is an Upwind discretization of the partial differential equation with linear CF $\bar{F}$ and lead time $\frac{1}{v}$.

$$\partial_t \rho + \partial_x \bar{F} = 0, \bar{F} = v\rho, \rho(t, 0) = \min\left\{v\frac{p}{\Delta x}, \mu\right\}. \tag{12}$$

Denoting by $\varepsilon := \frac{\Delta x}{v}$ we summarize the previous computations: using a numerical discretization of (3) and (7) and properties of the transport process we observe that formally Eqs. (10) and (17) are equivalent to

$$\frac{d}{dt}p = \psi - \min\left\{\frac{p}{\varepsilon}, \mu\right\}, \; \partial_t \rho + \partial_x \bar{F} = 0, \rho(t, 0) = \min\left\{\frac{p}{\varepsilon}, \mu\right\}. \tag{13}$$

The formulation (11) therefore can be seen as different possibility to treat the dynamics of the piecewise linear CF. Here the piecewise linearity of the CF only appears in the equation for the input inventory p. In fact, the previous equations allow for some additional interpretation: since $\bar{F}$ is a linear CF independent of the WIP ρ, it can be solved as described above and we obtain $\rho(1, t) = \rho(0, t - \frac{1}{v})$. The ingoing product density $\rho(0, t)$ is defined by the outflux $\min\{\frac{p}{\varepsilon}, \mu\}$ of the input inventory. The inventory equation itself is piecewise and reads

$$\frac{d}{dt}p = \psi - \min\left\{\frac{p}{\varepsilon}, \mu\right\} = \max\left\{\psi - \mu, \psi - \frac{p}{\varepsilon}\right\} \tag{14}$$

In the previous derivation the value of ε is supposed to be small since it is related to the formal spatial discretization in space. Hence, if there is inventory $p > 0$ the change in the inventory will be proportional to $\psi - \mu$ and the outflux of the inventory released to the factory is μ. However, if the inventory level is decreasing ($\psi - \mu < 0$), the inventory in (12) decays until $p \leq \varepsilon\mu$ holds, i.e. until the inventory is almost empty. From this point on Eq. (12) reads $\frac{dp}{dt} = \psi - \frac{p}{\varepsilon}$ and the inventory level decays exponentially to ψ on an $O(\frac{1}{\varepsilon})$ time scale. Hence, Eq. (12) can be seen as a smoothed version of the ordinary differential equation

$$\frac{d}{dt}p = \begin{pmatrix} \psi - \mu & p > 0 \\ \psi & p = 0 \end{pmatrix}. \tag{15}$$

The second line in (13) guarantess that the inventory p is positive. Hence, restating (11) on a time discrete level we obtain the following production line model for a CF given by Eq. (4). The release rate is $\psi(t)$, the planning production or outflux is $\phi(t)$, the difference in the time steps is $\Delta t = 1$, the constant lead time $\tau := \frac{1}{v}$, the maximal throughput μ, and the inventory buffer is at time t is $p(t)$:

$$p(t+1) = p(t) + \psi(t) - \min\left\{\frac{p(t)}{\varepsilon}, \mu\right\}, \ \phi(t+\tau) = \min\left\{\frac{p(t)}{\varepsilon}, \mu\right\}. \tag{16}$$

Hence, using the properties of the partial differential equation (3) and (4) lead to Eq. (13). This might be seen as alternative formulation of the effect of a nonlinear CF on the production dynamics. In contrast to other approaches based on a piecewise linear outer approximation of the CF and additional constraints on the throughput only a single evaluation of the function $\min\{\frac{p(t)}{\varepsilon}, \mu\}$ is required at every time step. The lead time τ is then constant and and the transport equation can be solved exactly. A possibility to evaluate $\min\{\frac{p(t)}{\varepsilon}, \mu\}$ within a production planning problem is presented in Sect. 3. Therein, the key idea is summarized in Lemma 2 using a reformulation of the minimum function with binary variables.

We have a final remark on the outflux or planned production. Since F in (3) is assumed to be monotone increasing the solution at $x = 1$ to the partial differential equation is uniquely defined by initial and boundary conditions. Hence, ϕ is defined by the release rate of the input inventory. This quantity is then linked toward the clearing function and the release rate by inequality relations. In the more refined model introduced below a different strategy is proposed. There, the outflux is defined through the release rate of the inventory according to the dynamics of the partial differential equation. However, the process does not necessarily deliver this outflux to another processor, but stores the outflux in an yet to be defined output inventory. Other producers then access the output inventory. Details will be given in Sect. 2.2.

Summarizing, we model a production line by assuming the validity of the partial differential equation (3). This equation allows for more general clearing functions and therefore WIP-dependent lead times. It further might be viewed as a time-continuous version of basic production models in the following sense: the differential equation describes the balance of local production densities or upon integration of local inventories (9). The presented introductory model is simple in the sense that it only states the WIP balance. No stochastic effects, no service levels, no release rate rules are present, and many further important production properties are absent. However, the basic model allows for non-constant lead times and using the above computations yields a reformulation of the WIP balance without the necessity to include piecewise linear capacity constraints as commonly used in the production engineering context.

In the remaining part of the presentation we include a few further modeling aspects into the basic model. We plan to present one possibility to extend the above model to a production network with orders. We consider a clearing function of type (4) and therefore our starting point will be the reformulated inventory equation (13).

2.2 A Production Model Based on a Partial Differential Equation

We repeat the basic setup for convenience: p denotes the number of parts at time t in the input inventory to a production process. The maximal throughput of the production process is μ. ε is a smoothing factor supposed to be small compared to the inventory level. The lead time of the production process is denoted by τ. The clearing function is WIP-dependent and is given by (4), i.e.,

$$F = \min\left\{\frac{1}{\tau}\rho, \mu\right\} \tag{17}$$

The inflow or release rate to the process is $\psi(t)$. Having the previous discussion in mind we may consider Eq. (12) for describing the input inventory, i.e.,

$$\frac{dp}{dt} = \max\left\{\Psi(t) - \mu, \Psi(t) - \frac{p(t)}{\varepsilon}\right\}, \tag{18}$$

and denote by

$$\phi(t) = \min\left\{\mu, \frac{1}{\varepsilon}p(t-\tau)\right\}$$

the outflux of the process. The previous computations have shown that these two equations describe the same process as the WIP balance for the piecewise linear clearing function. This is the continuous analog to Eq. (6) for WIP-dependent lead times.

Next, we introduce a possible modeling of an extension of production line to a network of processors. There are many possibilities to extend the dynamics of a single production line to a production network leading to possibly different formulations and models. We present an approach based on the following simplifying assumptions

- A production network is modeled as a connected graph $G(J, K)$ where the set of arcs is denoted by J and the set of nodes by K. Each node $k \in K$ is a processor and has an input inventory p_k *and* an output inventory q_k (not present before) and a possible time delay τ_k (lead time). The inventories are connected to allow for the distribution of items.
- Each process is described by a clearing function of the shape (4) with possibly different maximal processing rates μ_k and lead times τ_k.

- The output inventory delivers parts according to received orders Ω_k.
- The distribution parts between output inventory k and input inventories l is determined by the placed orders.

We introduce now the full model by discussing the implications of the previous set of assumptions and discuss thereby possible extensions and modifications. We assumed to have K nodes, i.e., processors. Items are taken from the input inventory, processed by the given constant lead time τ_k, put into the output inventory and instantaneously delivered according to received orders Ω_k. We assume that each node $k \in K$ therefore has two inventories: a front-end (or input) inventory with an inventory level $p_k(t)$ (number of parts), and a back-end (or output) inventory with inventory level $q_k(t)$. The dynamics are given by the reformulated partial differential equation in order to fulfill the dynamics induced by the piecewise linear clearing function.

The input inventory p_k of node number $k = 1 : K$ receives an influx ψ_k, and has an outflux determined by the processing rate μ_k and hence (12) reads at processor k :

$$\frac{dp_k}{dt} = \max\left\{\psi_k - \mu_k, \psi_k - \frac{p_k}{\varepsilon}\right\} = \psi_k - \min\left\{\mu_k, \frac{p_k}{\varepsilon}\right\} = \psi_k - \phi_k. \qquad (19)$$

As discussed before, the output inventory is given by $\phi_k(t - \tau_k)$, i.e. the time-delayed outflux of the input inventory. The evolution of the output inventory level q_k is then modeled in the same way by

$$\frac{dq_k}{dt} = \max\left\{\phi_k(t-\tau_k) - \Omega_k, \phi_k(t-\tau_k) - \frac{q_k}{\varepsilon}\right\} = \phi_k(t-\tau_k) - \min\left\{\Omega_k, \frac{q_k}{\varepsilon}\right\} = \phi_k - f_k \qquad (20)$$

with Ω_k the rate of orders received by processor k, and f_k the total outflux of node number k. Again, f_k cannot exceed Ω_k, and node number k cannot deliver at a faster rate than orders are received. Once again, we stretch the fact that in this simple model the inventory releases parts at constant rate to the factory. The only control is due to the orders with the imposed policies. Summarizing, Eqs. (16) and (17) provide the underlying order and distribution model.

Given the evolution of the input and output inventories, defined previously, we have to define the interaction of the different nodes in the production network. We define the influx ψ_k of node number k in (14) in terms of the outfluxes f_k of the other nodes, given by (15), and we have to decide on a rule for the order rates Ω_k in (15). One of the key mechanisms governing the dynamics of the system is obviously the policy of placing orders. In general, we will denote with Ω_{jk} the rate at which node number j places orders to node number k, and the total rate of orders received by node number k in (15) is given by $\Omega_k = \sum_j \Omega_{jk}$. Concerning the rules governing the node policies we also refer to recent literature on the theory of traffic flow and internet traffic. Therein, similar problems appear and a suitable definition of the distribution of parts or vehicles among input and output inventories has to be modeled. Obviously, a wide variety of choices exist. We refer to [15, 17, 18, 19, 23, 28, 35, 37, 49] for more details. Here we model a very simple distribution policy. We define by F_{jk} the flux from node k into node j. Consequently, $f_k = \sum_j F_{jk}$ is the total outflux of node number k. Because of (15) we already have that $f_k \leq \Omega_k$ holds. The need for a distribution policy arises when f_k is actually strictly less than Ω_k, i.e. node number k cannot satisfy all its orders and has to make a decision how to distribute its limited resources.

We write the flux from node number k to node number j as $F_{jk} = A_{jk}f_k$, where the matrix $\mathbf{A} = \{A_{jk}\}$ is a Markov matrix, i.e. a matrix with non-negative entries whose column sums equal unity. For any admissible distribution policy, the matrix $\mathbf{A}$ should satisfy the following two criteria:

- $F_{jk} = A_{jk}f_k \leq \Omega_{jk}$, i.e. node k cannot deliver more to node j than node j is ordering from node k.
- If $f_k = \Omega_k$ holds in (15), then $F_{jk} = \Omega_{jk}$ should hold, i.e. if node number k can satisfy all its orders, then it will do so.

This assumption is also found in the traffic and internet traffic flow literature, see e.g. [23, 49, 51]. It is also well known that, in addition, we have to define an external supplier of raw materials and a final customer. For simplicity, we assume a single raw material supplier, defined as node number $k = 0$, and a single customer, defined as node number $k = K + 1$. So Ω_{j0} denote the rates at which raw materials are ordered, $\Omega_{K+1,k}$ denote the rates at which the final customer orders, and $F_{K+1,k}$ are the rates at which product is delivered to the customer.

There are many different ways to define a distribution policy, i.e. a matrix $\mathbf{A}$, satisfying the two criteria above. The precise choice of the distribution policy is a modeling choice and depends on the production facility at hand. An example is choosing the flux distribution proportional to the place orders. This policy implies that, if not all orders can be satisfied, node number k distributes the product proportionally according to the orders received. Mathematically, the policy reads

$$A_{jk} := \frac{\Omega_{jk}}{\Omega_k}, j = 1 : K + 1, k = 1 : K, \quad \Omega_k := \sum_{j=1}^{K+1} \Omega_{jk}, k = 1 : K. \qquad (21)$$

Equation (18) satisfies the previous admissibility criteria since, by definition, $F_{jk} = \frac{\Omega_{jk}f_k}{\Omega_k}$ holds. Since $f_k \leq \Omega_k$ the inequality.

In summary, the dynamics of the flow of the network is given by Eqs. (16) and (17) for all processor nodes $k = 1 : K$, where the influx functions ψ_k in (16) are determined by the outflux rates f_k in (17) through the connectivity matrix $\mathbf{A} = \{A_{jk}\}$. So

$$\psi_j = \sum_{k=1}^{K} A_{jk}f_k + \Omega_{j0}, j = 1 : K \qquad (22)$$

holds. Ω_{j0} denote the external inputs into the system, from the raw material supplier which, assuming an unlimited supply, equal the orders placed to the raw material supplier. After choosing an order matrix $\mathbf{Z} = \{\Omega_{jk}, j = 1 : K+1, k = 0 : K\}$ the dynamics of the system are therefore completely defined. However, the system

(16)–(19) represents an open system, i.e. mass is not conserved, due to the external influx and outflux at the supplier and the customer node. For analytical purposes, it will be convenient to replace the open system by a closed system by introducing an artificial 'recycling step'. That is we artificially identify the raw material supplier with the customer, and feed the delivered product back into the system as raw material. So, Ω_{j0} still denote the order rates from the raw material supplier and $\Omega_{0k} = \Omega_{K+1,k}$ denote now the rates at which the customer orders. With this change in notation, (18) and (19) become

$$(a)\quad A_{jk} = \frac{\Omega_{jk}}{\Omega_k}, j = 0 : K, k = 0 : K, \quad \Omega_k = \sum_{j=0}^{K} \Omega_{jk}, k = 0 : K, \tag{23}$$

$$(b)\quad \psi_j = \sum_{k=0}^{K} A_{jk} f_k, j = 0 : K.$$

ψ_0, the influx into the raw material supplier/customer is now the rate at which final product is delivered. The advantage of this notational trick is that it allows for a uniform treatment of all the nodes and yields a mass conserving system. In order not to change the dynamics of the system we have to design node number $k = 0$ in such a way, that there is a limitless supply. This is easily done by giving the raw material supplier/customer formally an infinite production capacity and a zero processing time, and by making its output inventory large enough at the beginning such, that it never runs dry, i.e. we set formally $\mu_0 = \infty$, $\tau_0 = 0$ in (16) and make $q_0(0)$ sufficiently large. This has the effect that all the product delivered to the final customer immediately goes to the output inventory, and the system is fed from this (sufficiently large) output inventory q_0, i.e. we have $\phi_0 = \psi_0, f_0 = \Omega_0$ in (16). Alternatively, we could simply change the definition of the fluxes in (16) and (17) for the node $k = 0$ to

$$\phi_0 = \psi_0, \quad f_0 = \Omega_0,$$

allowing the output inventory q_0 to become negative. The system is now closed and the total product is conserved, since the columns of the square matrix $\mathbf{A} = \{A_{jk}, j, k = 0 : K\}$ all add up to unity. If we define the contents of the processor number k at time t by $r_k(t)$ its evolution is given according to (16)–(17) by $\frac{dr_k}{dt} = \phi_k(t) - \phi_k(t - \tau_k)$, and the evolution of the total mass in the system is given by

$$\frac{d}{dt} \sum_{k=0}^{K} (p_k + q_k + r_k) = \sum_{k=0}^{K} \psi_k - f_k = 0.$$

We note that this modification of the system is done only for technical convenience, and that, if the initial supply inventory $q_0(0)$ is chosen large enough, the dynamics of the network remain unchanged.

We refer [39, 48] to have for more details concerning the mathematical analysis of the proposed model. We summarize the results briefly. In [48] it has been shown that the model describing a network of suppliers and a nonlinear, monotone, concave clearing function describing the inventory is well posed. Therefore, there exists a solution for L^1–initial data. In [39] some conclusions concerning the steady-state solutions to the previous model are obtained: in the steady-state case the influx into each node equals its outflux. The condition $\psi_k = f_k$ implies, that the fluxes f_k are eigenvectors of the connectivity matrix $\mathbf{A}$, i.e., $\mathbf{A}f = f$. Due to eqution (20) all entries of the matrix $\mathbf{A}$ are non-negative entries and its column sums equal unity. The matrix $\mathbf{A}$ furthermore possess an eigenvalue $\bar{\lambda}$ equal to unity and the eigenvector z to that eigenvalue has only non-negative elements. For given matrix $\mathbf{A}$ this eigenvector z defines all possible steady-states of the system. However, in the previous model we only obtain steady-states provided that the production capacities are sufficiently large, which in general, might not be the case. We therefore continue in the following with the numerical treatment of the dynamical process.

3 A Mixed-Integer Programing Approach

An important question in the context of production networks are optimal order and distribution strategies. Depending on the model, many aspects will be of interest: inventory and production costs, distribution of goods, and supply and demand. Here we introduce a special cost functional and choose the distribution matrix $\mathbf{A}$ and the order matrix $\mathbf{Z}$ dynamically to optimize the functional. The solution to this problem can of course only be given numerically. We start, by discretizing the dynamical system defined in Sect. 2. On the discrete time level, the dynamics will actually appear as a large set of piecewise linear constraints for a given cost functional. There are essentially two ways to treat the the resulting constrained optimization problem: either the constraints are formulated on a continuous level using an adjoint calculus to compute a restricted gradient direction as in [50, 56], or a nonlinear programming approach is used, introducing additional binary variables. The latter leads to a mixed integer program MIP [33].

3.1 Numerical Discretization

We start with a proper discretization of the differential equations (16) and (17) for the inventories p_k and q_k. We discretize on a uniform mesh in time setting $p_k^n = p_k(n\Delta t), n = 0 : N$ with $T = N\Delta t$ the final time of the simulation. As is almost always the case in relaxation models, we have introduced an artificial $O(\frac{1}{\varepsilon})$ time scale in the system, and thus artificially created a stiff system. In order not to impose too severe a restriction on the time step Δt we discretize equation (14) implicitly, giving

$$p_k^n = p_k^{n-1} + \Delta t \max\left\{\psi_k^{n-1} - \mu_k, -\frac{p_k^n}{\varepsilon}\right\}, n = 1 : N, \tag{24}$$

with ψ_k^{n-1} the influx at the previous time step. Equation (21) can be inverted explicitly for p_k^n, by using the following

Lemma 1 *The function $u(x) = \min\{ax + b, cx + d\}$ is invertible for $a > 0, c > 0$. This inverse is given by $u^{-1}(y) = \max\{\frac{y-b}{a}, \frac{y-d}{c}\}$.*

Reordering Eq. (21), we have

$$\min\left\{p_k^n - \Delta t(\psi_k^{n-1} - \mu_k), \left(1 + \frac{\Delta t}{\varepsilon}\right)p_k^n\right\} = p_k^{n-1}$$

and therefore, setting $a=1, b = -\Delta t(\psi_k^{n-1} - \mu_k), c=1 + \frac{\Delta t}{\varepsilon}, d = 0$ in Lemma 1, we obtain

$$p_k^n = \max\left\{p_k^{n-1}+\Delta t(\psi_k^{n-1}-\mu_k), \frac{p_k^{n-1}}{1+\frac{\Delta t}{\varepsilon}}\right\} = p_k^{n-1}+\Delta t \max\left\{\psi_k^{n-1}-\mu_k, -\frac{1}{\varepsilon + \Delta t}p_k^{n-1}\right\}$$

which we write as

$$p_k^n = p_k^{n-1} + \Delta t(\psi_k^{n-1} - \phi_k^{n-1})$$

with

$$\phi_k^{n-1} = \min\left\{\mu_k, \psi_k^{n-1} + \frac{1}{\varepsilon + \Delta t}p_k^{n-1}\right\} \tag{25}$$

Therefore the implicit discretization of Eq. (16) can be written in explicit form as

$$p_k^n = p_k^{n-1} + \Delta t(\psi_k^{n-1} - \phi_k^{n-1})$$

with the numerical outflux ϕ_k^{n-1} given by (22). Note, that this eliminates any restriction on the time step, since the flux function ϕ_k^{n-1}, as defined in (22), is well defined in the limit $\varepsilon \to 0$. We employ the same implicit discretization strategy for the evolution of the output inventory q_k in (17). To avoid any additional interpolation procedure, we assume that all the processing times $\tau_k, k = 0 : K$ are integer multiples of the time step Δt. So $\frac{\tau_k}{\Delta t} \in \mathbb{N}$ holds. Thus, we obtain, setting

$$\psi_k^n \to \phi_k^{n-\tau_k/\Delta t}, \quad \mu_k \to \Omega_k^{n-1},$$

$$(a) \quad q_k^n = q_k^{n-1} + \Delta t(\phi_k^{n-\tau_k/\Delta t} - f_k^{n-1}), \tag{26}$$

$$(b) \quad f_k^{n-1} = \min\left\{\Omega_k^{n-1}, \phi_k^{n-\tau_k/\Delta t} + \frac{1}{\varepsilon + \Delta t}q_k^{n-1}\right\}$$

Any optimization problem, involving the discretization of the dynamical system, as formulated above, will still be piecewise linear due to Eqs. (22) and (23). The goal of this section is to formulate an optimization problem for the dynamics defined in Sect. 2 in a framework close to a linear programing problem, i.e. as a mixed-integer programing problem. A mixed-integer program is a linear program which includes binary switches, i.e. variables taking only values in $\{0, 1\}$. The advantage of linear- and mixed- integer programing approaches is that they are capable of dealing with an enormous amount of free variables. The basic tool to convert an optimization problem involving the piecewise linear flux functions (22, 23) into a MIP is given by the following Lemma (see [33]):

Lemma 2 *Let $\xi \in \{0, 1\}$ be a binary variable. Let $M > |a - b|$ be a sufficiently large constant. Then, for a given constant M, the solution of the inequalities*

$$a - M\xi \le \phi \le a, \quad b - M(1 - \xi) \le \phi \le b \tag{27}$$

is given by

$$\phi = \min\{a, b\}.$$

Using Lemma 2, we replace the definition (22), (23) of the fluxes ϕ_k^n, f_k^n by the constraints

$$(a) \quad \mu_k - M\xi_k^n \le \phi_k^n \le \mu_k, \tag{28}$$

$$(b) \quad \psi_k^n + \frac{p_k^n}{\varepsilon + \Delta t} - M(1 - \xi_k^n) \le \phi_k^n \le \psi_k^n + \frac{p_k^n}{\varepsilon + \Delta t}$$

$$(c) \quad \Omega_k^n - M\eta_k^n \le f_k^n \le \Omega_k^n,$$

$$(d) \phi_k^{n-\tau_k/\Delta t} + \frac{q_k^n}{\varepsilon + \Delta t} - M(1 - \eta_k^n) \le f_k^n \le \phi_k^{n-\tau_k/\Delta t} + \frac{q_k^n}{\varepsilon + \Delta t}$$

with the binary variables $\xi_k^n, \eta_k^n \in \{0, 1\}, k = 0 : K, n = 0 : N$, and a constant M, chosen a priori sufficiently large.

Remark 1 For some special models it is even possible to derive a linear programing model (LP) instead of mixed-integer one. In [32], a complete proof can be found.

The MIP approach allows us to optimize the order strategies, given by the matrix $\mathbf{Z}$. In order to encode the topology of the network, we define the elements of the order matrix $\mathbf{Z}$ as $\Omega_{jk}^n = \theta_{jk}\tilde{\Omega}_{jk}^n$, where the matrix $\Theta = \{\theta_{jk}, j, k = 0 : K\}$ denotes the adjacency matrix of the graph defining the network topology, i.e. $\theta_{jk} = 1$ if node number j can order from node number k, and $\theta_{jk} = 0$ otherwise. Similarly, we define $F_{jk}^n = \theta_{jk}\tilde{F}_{jk}^n$ for the fluxes. In the context of the MIP approach, the orders and fluxes $\tilde{\Omega}_{jk}^n, \tilde{F}_{jk}^n$ at each time step are treated as free variables to be optimized. In order to guarantee conservation of product, we have to add the constraint

$$f_k^n = \sum_{j=0}^{K} \theta_{jk} \tilde{F}_{jk}^n, \quad \boldsymbol{\psi}_k^n = \sum_{j=0}^{K} \theta_{kj} \tilde{F}_{kj}^n, \quad k = 0 : K, n = 0 : N, \tag{29}$$

and in order to guarantee that fluxes cannot exceed orders, we enforce the constraints

$$0 \le \tilde{F}_{jk}^n \le \tilde{\Omega}_{jk}^n, \quad k = 0 : K, n = 0 : N. \tag{30}$$

Of course, only the orders and fluxes for adjacent nodes, i.e. for nodes j and k for which $\theta_{jk} = 1$ holds, have to be used in the actual program. The solution of the MIP implicitly defines an, adaptive and time dependent, distribution policy matrix **A**, given by

$$A_{jk}^n = \frac{\theta_{jk} \tilde{F}_{jk}^n}{f_k^n},$$

and the amount of orders received by node number k, which is used in the constraint (25b), is given by

$$\Omega_k^n = \sum_j \theta_{kj} \tilde{\Omega}_{kj}^n. \tag{31}$$

So, altogether, the external variables, which have to be supplied to the MIP, are

- μ_k the processor capacities.
- θ_{jk} the adjacency matrix of the graph.
- $\tilde{\Omega}_{Kj}^n$ the time-dependent orders of the final customer.
- p_k^0, q_k^0, the initial inventory levels.
- $\phi_k^n, n = -\tau_k/\Delta t : 0$, the past influx of the processors, defining the processor contents at time $t = 0$.

The free variables to be optimized consist of

- $\tilde{\Omega}_{jk}^n, j = 1 : K-1, k = 1 : K$: the internal orders
- $\tilde{F}_{jk}^n, f_k^n, \boldsymbol{\psi}_k^n, \phi_k^n, \Omega_k^n$: the partial and total fluxes and total orders received for each node, given in terms of the $\tilde{\Omega}_{kj}^n$ by the constraints (25)–(28)
- ξ_k^n, η_k^n : the auxiliary binary variables used in the MIP formulation.

There are various possible goals to be followed when defining the cost functional to be optimized. A general cost functional might be of the form

$$\sum_{k=1}^{K} \sum_{n=1}^{N} \mathscr{J}(p_k^n, q_k^n), \tag{32}$$

where $\mathcal{J}$ is a function depending on the queue-loads p_k^n and q_k^n. For example, think of the minimization of storing costs or the maximization of fluxes. Combining all discretization we observe that the optimization problem is in a fact mixed-integer programing problem and given by

$$\max(29) \text{ subject to } (21)\text{–}(28). \tag{33}$$

Remark 2 The optimization problem (30) can be extended introducing an additional equation for the flow of capital. We assume that each node in the network charges a production cost β_k, $k = 0 : K$ per item delivered, and has a specific price b_k per unit delivered. The evolution of the amount of capital κ_k^n of node number k is therefore given by the discretized equation

$$\kappa_k^n = \kappa_k^{n-1} + \Delta t\Big(\beta_k f_k^{n-1} - \sum_{j=0}^{K} A_{kj}\beta_j f_j^{n-1} - b_k\phi_k^{n-1}\Big), \quad n = 1 : N. \tag{34}$$

In order to study the actual profitability of a given policy, it is necessary to adapt the objective. The simplest choice is then to optimize the capitalization of all the interior nodes at the final time:

$$\mathcal{J} = \sum_{k=1}^{K} \kappa_k^N. \tag{35}$$

In order to further incorporate bankruptcies, we make the possible orders dependent of the capitalization rates of the individual nodes, and force each node to cease ordering as soon as its capital falls below a certain threshold $\underline{\kappa}_k$. In the context of the MIP approach, we implement this by introducing another binary variable $v_k^n \in \{0, 1\}$ and by defining the total rate at which node k orders as

$$\sigma_k^n = \sum_{j=0}^{K} \theta_{kj}\tilde{\Omega}_{kj}^n. \tag{36}$$

To force the node into bankruptcy as soon as its capital falls below the threshold, we add the constraints

$$\sigma_k^n \le M v_k^n, \quad M(v_k^n - 1) \le \kappa_k^n - \underline{\kappa}_k \le M v_k^n, \tag{37}$$

where M again denotes a sufficiently large a priori constant. Again, as in the Lemma 2, there are two ways to satisfy the constraint (34). For $v_k^n = 1$ we have $0 \le \kappa_k^n - \underline{\kappa}_k$ and essentially no constraint for σ_k^n (provided that M is sufficiently large). For $v_k^n = 0$, the bankruptcy case, we have $\kappa_k^n - \underline{\kappa}_k \le 0$ and $\sigma_k^n = 0$. Adding the variables κ_k^n and σ_k^n to the the system together with the constraints (31–34) allows now for the optimization of the cost functional (32) together with the possibilities of bankruptcies. The optimization problem

$$\max(32) \text{ subject to } (21) - (28), (31), (33), (34). \tag{38}$$

gives rise to a case study where three different order and distribution rates are compared:

I. As 'benchmark' scenario we denote the problem as stated in Eq. (35). The solution to this problem should give the maximal possible profit since there are no additional constraints on the distribution and order rates.

II. We require the order policies to be time independent. This amounts to add the constraints

$$\Omega_{jk}^{n+1} = \Omega_{jk}^{n}. \tag{39}$$

This choice is reasonable if the suppliers do not want to change their policy dynamically. Clearly, this additional constraint restricts the set of possible solutions and we expect a lower profit. The optimization problem hence reads

$$\max(32) \text{ subject to } (21) - (28), (31), (33), (34) \text{ and } (36) \tag{40}$$

and this scenario will be called 'time–independent orders' in the numerical results.

III. We impose the following rule: whenever the supplier S_k is *not* bankrupt, the supplier has to order up to his capacity μ_k :

$$\sum_j \Omega_{jk}^{n} \leq v_k^n \mu_k. \tag{41}$$

where v_k^n is the binary variable introduced in Eq. (34). This rule is motivated by the fact that the complete production line should have the highest possible utilization.The optimization problem hence reads

$$\max(32) \text{subject to} \quad (21) - (28), (31), (33), (34) \text{ and } (38) \tag{42}$$

and this scenario will be called 'order–up to capacity' for short.

3.2 Computational Experiments

The optimization problem is solved using the mixed-integer programing framework. We use the commercial software ILOG CPLEX V11.0 [53] with default parameters. Here we study the behavior of the optimal controls Ω_{jk} and A_{jk} on two different networks and the three different cases **I** – **III** and study the complexity of the mixed-integer problem. If not stated otherwise we use default parameters when running the commercial solver with a maximum computation time of $24h$. All computations are done on a AMD 2 Ghz personal computer.

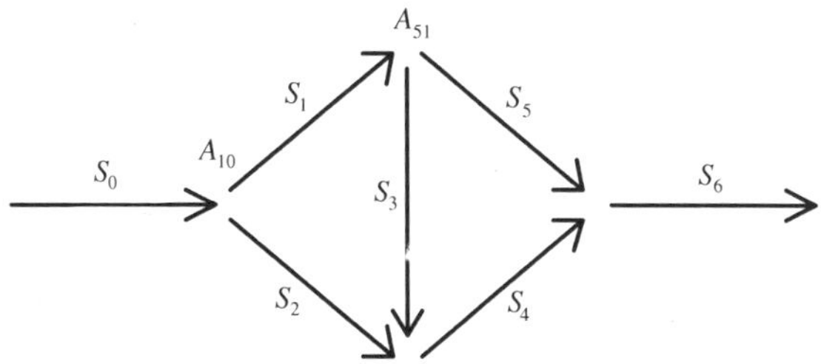

Fig. 1 Sample network of six suppliers and a customer

Table 1 Specification of suppliers S_k present in the diamond network

Processor k	b_k	μ_k	β_k
1	1	2	0
2	2	1	1
3	1	1	1
4	1	1	10
5	1	1	10

In the model we set $\tau_k \equiv \Delta t \equiv \epsilon \equiv 1$ for all examples. The network we are interested in consists of seven suppliers and six vertices, see Fig. 1.
Each supplier (except for customer $k = 6$ and raw material supplier $k = 0$) has specific prices, production capacities, and production costs as given in Table 1.
For each scenario we consider a constant inflow $\psi_0^n = 2$ and a threshold for going bankrupt of $\underline{\kappa}_k = -5$. The total simulation time is $T = 40$. Note that, for the data given the processing chains 'raw material supplier $\rightarrow$ supplier 1 $\rightarrow$ supplier 5 $\rightarrow$ customer' is the preferred one. However, the inflow of two parts per time cannot be passed through supplier 5 and there is the possibility to either store the goods in the input inventory of 5 or redistribute along suppliers 3 and 4.

We give computational results for the scenarios **I–III** by solving (35), (37), and (39), respectively. In Table 2 we report on the size of the optimization problem (non–zero variables), the computational time used by CPLEX (CPU time), the optimal profit $\sum_k \kappa_k(T)$, and the amount of delivered parts over time at customer, i.e., $\int_0^T f_6(t)dt$.
On a simple factory of six suppliers rule (**I**) yields the highest profit, but rule (**III**) yields the most delivered parts since the policy requires to order as many parts as possible. The highest profit is only obtained at the expense of the bankruptcy of some internal suppliers for a very long time. The optimal choice in the benchmark case is to accept the bankruptcy of most of suppliers in order to maximize the total profit. In the case of the time-dependent policy most of the suppliers do not go bankrupt, however, there is nearly no part delivered and the overall profit is the least of all cases.

We have seen that a modeling approach based on nonlinear(!) differential equations can be transformed and interpreted as a discrete optimization problem. The underlying dynamic is still conserved in this context. Therefore it is possible to determine time-dependent order strategies for several highly complex problems.

Table 2 Comparison of computation results for the diamond network

	Benchmark (**I**)	Time–independent orders (**II**)	Order up to capacity (**III**)
# nonzero vars	1497	1907	1600
CPU-times [sec]	3480	406	140
Bankruptcy percentage	24%	1%	14.5%
Optimal profit	260,4	16,72	31,00
Delivered parts	3,6	1,54	4,00

4 Conclusion

The presented production network is suitable for a wide range of applications including order and distribution policies and money flow as well. The model is an essential extension to recently proposed continuous production network models. The distribution and order rates are determined by an optimization problem for maximizing the money flow, where the discretized maximization problem is solved by mixed-integer programing techniques. We added a case study for a sample network where we studied a priori determined order and distribution strategies, namely, each supplier can decide at every time on his own how to order and deliver (**I**), each supplier has to fix an order and delivery rule for the full production process (**II**), or each supplier has to order up to his capacity as long as he is not bankrupt (**III**). As indicated by the case study, the presented model might be used to compare different a priori given rules (e.g., **I–III**) or to detect costly suppliers. Concluding, future work should include two essential issues: the speed-up of the black-box solver CPLEX and the extension to stochastic processing times and demands. The former needs the derivation of suitable heuristics to provide good starting solutions. For simulation purposes only, the latter can be achieved by introducing time-dependent randomness and the performance of Monte Carlo simulations. However, stochastic optimization problems require more sophisticated methods.

Acknowledgments This work has been supported by DFG grant HE5386/6-1, DAAD 50756459 and 50727872.

References

1. Armbruster D, de Beer C, Freitag M, Jagalski T, Ringhofer C (2006) Autonomous control of production networks using a pheromone approach. Physica A 363:104–114
2. Armbruster D, Degond P, Ringhofer C (2006) A model for the dynamics of large queuing networks and supply chains. SIAM J Appl Math 66:896–920
3. Armbruster D, Degond P, Ringhofer C (2007) Kinetic and fluid models for supply chains supporting policy attributes. Bull Inst Math Acad Sin (NS) 2:433–460
4. Armbruster D, Marthaler D, Ringhofer C (2004) Kinetic and fluid model hierarchies for supply chains. Multiscale Model Simul 2:43–61

5. Armbruster D, Marthaler D, Ringhofer C, Kempf K, Tae-Chang Jo (2006) A continuum model for a re-entrant factory. Oper Res 54:933–950
6. Armbruster D, Ringhofer C (2005) Thermalized kinetic and fluid models for reentrant supply chains. Multiscale Model Simul 3:782–800
7. Asmundsson JM, Rardin RL, Turkseven CH, Uzsoy R (2009) Production planning models with ressources subject to congestion. Naval Res Logist 56:142–157
8. Bak P, Chen K, Scheinkman J, Woodford M (1993) Aggregate fluctuations from independent sectoral shocks: Self- organized criticality in a model of production and inventory dynamics. Ricerche Economiche 3:3–30
9. Banda MK, Herty M, Klar A (2006) Coupling conditions for gas networks governed by the isothermal Euler equations. Netw Heterog Media 1:295–314
10. Banda MK, Herty M, Klar A (2006) Gas flow in pipeline networks. Netw Heterog Media 1:41–56
11. Banks J, Carson JS (1984) Discrete-event system simulation. Prentice-hall international series in industrial and systems engineering. Prentice-Hall Inc, Englewood Cliffs
12. Battiston S, Delli Gatti D, Gallegati M, Greenwald B, Stiglitz JE (2007) Credit chains and bankruptcy propagation in production networks. J Econ Dyn Control 31:2061–2084
13. Baumol W-J (1972) Economic theory and operations analysis, Prentice-Hall Inc. Prentice–Hall International Series in Management, Englewood Cliffs
14. Bixby R, Simchi-Levi D, Martin A, Zimmermann U (2004) Mathematics in the supply chain. Oberwolfach Rep 1:963–1036
15. Blandin S, Bretti G, Cutolo A, Piccoli B (2009) Numerical simulations of traffic data via fluid dynamic approach. Appl Math Comput 210:441–454
16. Bolch G, Greiner S, de Meer H, Trivedi KS (2006) Queueing networks and Markov chains. Modeling and performance evaluation with computer science applications. 2nd edn. Wiley, Hoboken
17. Bretti G, D'Apice C, Manzo R, Piccoli B (2007) A continuum-discrete model for supply chains dynamics. Netw Heterog Media 2:661–694
18. Bretti G, Natalini R, Piccoli B (2006) Fast algorithms for the approximation of a traffic flow model on networks. Discrete Contin Dyn Syst Ser B 6:427–448
19. Bretti G, Natalini R, Piccoli B (2006) Numerical approximations of a traffic flow model on networks. Netw Heterog Media 1:57–84
20. Buzacott JA, Shanthikumar JG (1993) Stochastic models of manufacturing systems. Prentice-Hall, Englewood Cliffs
21. Caldarelli G, Battiston S, Garlaschelli D, Catanzaro M (2004) Emergence of complexity in financial networks. In: Ben-Naim E, Frauenfelder H, Toroczkai Z (eds) Lecture notes in physics "Complex Networks". Springer 650:399–423
22. Chen H, Yao DD (2001) Fundamentals of queueing networks. Springer, New York
23. Coclite GM, Garavello M, Piccoli B (2005) Traffic flow on a road network. SIAM J Math Anal 36:1862–1886
24. Colombo RM, Guerra G, Herty M, Schleper V (2009) Optimal control in networks of pipes and canals. SIAM J Control Optim 48:2032–2050
25. Daganzo CF (2003) A theory of supply chains. Lecture notes in economics and mathematical systems, vol 526. Springer, Berlin
26. Apice DC, Göttlich S, Herty M, Piccoli B (2010) Modeling, simulation, and optimization of supply chains. Society for industrial and applied mathematics (SIAM), Philadelphia
27. D'Apice C, Manzo R (2006) A fluid dynamic model for supply chains. Netw Heterog Media 1:379–389
28. D'Apice C, Manzo R, Piccoli B (2006) Packet flow on telecommunication networks. SIAM J Math Anal 38:717–740
29. de Kok AG (1990) Computationally efficient approximations for balanced flowlines with finite intermediate buffers. Int J Prod Res 28:410–419

30. Degond P, Göttlich S, Herty M, Klar A (2007) A network model for supply chains with multiple policies. Multiscale Model Simul 6:820–837
31. Degond P, Ringhofer C (2007) Stochastic dynamics of long supply chains with random breakdowns. SIAM J Appl Math 68:59–79
32. Fügenschuh A, Göttlich S, Herty M, Kirchner C, Martin A (2009) Efficient reformulation and solution of a nonlinear PDE-controlled flow network model. Computing 85:245–265
33. Fügenschuh A, Göttlich S, Herty M, Klar A, Martin A (2008) A discrete optimization approach to large scale supply networks based on partial differential equations. SIAM J Sci Comput 30:1490–1507
34. Garavello M, Piccoli B (2006) Traffic flow on a road network using the aw-rascle model, Comm. Partial Differ Equ 31:243–275
35. Garavello M, Piccoli B (2006) Traffic flow on networks vol.1 of AIMS series on applied mathematics, American institute of mathematical sciences (AIMS), Springfield, MO
36. Gordon WJ, Newell FG (1967) Closed queuing systems with exponential servers. Oper Res 15:254–265
37. Göttlich S, Herty M, Klar A (2005) Network models for supply chains. Commun Math Sci 3:545–559
38. Göttlich S, Herty M, Klar A (2006) Modelling and optimization of supply chains on complex networks. Commun Math Sci 4:315–350
39. Göttlich S, Herty M, Ringhofer C (2010) Optimization of order policies in supply networks. Eur J Oper Res 202:456–465
40. Graves SC (1986) A tactical planning model for a job shop. Oper Res 34:552–533
41. Graves SC (1998) A dynamic model for requirements planning with application to supply chain optimization. Oper Res 46(3):S35–S49
42. Graves SC (1999) A single-item inventory model for a nonstationary demand process. Manuf Serv Oper Manag 1:50–61
43. Gugat M, Leugering G, Schittkowski K, Schmidt EJPG (2001) Modelling stabilization and control of flow in networks of open channels in online optimization of large scale systems. Springer, Berlin, pp 251–270
44. Hackman ST, Leachman RC (1989) An aggregate model of project oriented production. IEEE Trans Syst Man Cybern 19:220–231
45. Helbing D (1997) Verkehrsdynamik. Springer, Berlin
46. Herty M, Kirchner C, Moutari S (2006) Multi-class traffic models for road networks. Commun Math Sci 4:591–608
47. Herty M, Klar A (2004) Simplified dynamics and optimization of large scale traffic networks. Math Model Methods Appl Sci 14:579–601
48. Herty M, Klar A, Piccoli B (2007) Existence of solutions for supply chain models based on partial differential equations. SIAM J Math Anal 39:160–173
49. Herty M, Rascle M (2006) Coupling conditions for a class of second order models for traffic flow. SIAM J Math Anal 38:592–616
50. Herty M, Ringhofer Ch (2007) Optimization for supply chain models with policies. Physica A 380:651–664
51. Holden H, Risebro NH (1995) A mathematical model of traffic flow on a network of unidirectional roads. SIAM J Math Analysis 26:999–1017
52. Hopp WJ, Spearman ML (2001) Factory physics foundations of manufacturing management. McGraw-Hill, Boston
53. IBM ILOG CPLEX. (2010) IBM Deutschland GmbH
54. Johnson LA, Montgomery DC (1974) Operations research in production, planning scheduling and inventory control. Wiley, New York
55. Karmarkar US (1989) Capacity loading and release planning with work-in-progress (wip) and leadtimes. J Manuf oper manag 2:105–123
56. Kelley CT (1999) Iterative methods for optimization. Frontiers in applied mathematics, xv, 180 p. Society for Industrial and Applied Mathematics, Philadelphia

57. La Marca M, Armbruster D, Herty M, Ringhofer C (2010) Control of continuum models of production systems. IEEE Trans autom control 55(11):2511–2526
58. LeVeque RJ (2002) Finite volume methods for hyperbolic problems. Cambridge texts in applied mathematics. Cambridge Unviversity Press, Cambridge
59. Missbauer H (2002) Aggregate order release planning for time-varying demand. Int J Prod Res 40:688–718
60. Missbauer H, Uzsoy R (2010) Optimization models for production planning. Planning production and inventories in the extended enterprise. In: Kempf KG, Keskinocak P, Uzsoy R (eds) A state of the art handbook. New York, Springer, pp 437–508
61. Pahl J, Voss S, Woodruff D L (2005) Production planning with load dependent lead times. 4OR Q J Oper Res 3:257–302
62. Selcuk B, Fransoo JC, De Kok AG (2007) Work in process clearing in supply chain operations planning. IEEE Trans 40:206–220
63. Solberg JJ (1981) Capacity planning with stoachstic workflow models. AIIE Trans 13(2): 116–122
64. Sterman J D (2000) Buisness dynamics. Systems thinking and modeling for a complex world. McGraw-Hill, New York
65. Voss S, Woodruff D (2003) Introduction to computational optimization models for production planning in a supply chain. Springer, Berlin
66. Wolsey L, Pochet Y (2006) Production planning by mixed integer programming. Springer, New York

The Production Planning Problem: Clearing Functions, Variable Lead Times, Delay Equations and Partial Differential Equations

D. Armbruster

Abstract Determining the production rate of a factory as a function of current and previous states is at the heart of the production planning problem. Different approaches to this problem presented in this book are reviewed and their relationship is discussed. Necessary conditions for the success of a clearing function as a quasi steady approximation are presented and more sophisticated approaches allowing the prediction of outflow in transient situations are discussed. Open loop solutions to the deterministic production problem are introduced and promising new research directions are outlined.

Keywords Supply chains · Production planning problem · Conservation laws · Clearing functions

1 Introduction

The production planning problem, the starts into a factory that generate a desired production profile in the future, is either explicitly or implicitly a major theme of almost half of the chapters in this book. Aouam and Uzsoy [1] have production planning in the title, Lefeber [15] deals with the issue in the context of the reference tracking problem using Model Predictive Control, Göttlich et al. [11] change the control variable from production starts to outing probabilities and then try to match a particular output pattern, Braun and Schwartz [7] assume a model for a production planning problem and deal with the nervousness of the scheduling algorithm, Perdaen et al. [20] measured the success of controlling a reentrant manufacturing line through the Push-Pull-Point via its missed production targets and Ringhofer [21] uses traffic

D. Armbruster (✉)
School of Mathematical and Statistical Sciences,
Arizona State University, Tempe, AZ 85287-1804, USA
e-mail: armbruster@asu.edu

D. Armbruster and K. G. Kempf (eds.), *Decision Policies for Production Networks*,
DOI: 10.1007/978-0-85729-644-3_12,

type models based on hyperbolic partial differential equations (PDEs) to study the production planning problem with priority rules.
The common problem of determining the output of a production unit (machine, factory, supply chain node) is dealt with at very different levels of sophistication. This paper will connect some of these approaches and determine their applicability and their approximation errors. Finally we discuss some of the practical questions of choosing an appropriate clearing function mode for a production planning problem and identify the open questions associated with those problems in general and the clearing function approach in particular.

2 Clearing Function Models

Typically, all production units are stochastic and hence the production process is a stochastic process. As a result, the mathematical model that comes closest to reality is a discrete event simulation model. However, even if every production detail is modeled, the characterization of the stochastic processes involved is non-trivial generating another *model* of reality. Given the fact that the stochastic processes are not well understood and given that they are very time consuming to simulate, the need for aggregate models is generally accepted and this need drives the discussion. Hence we will discuss deterministic models that, one hopes, represent *average* behavior in some sense.
Depending on the perspective of the author the number of items a production unit produces is either characterized by a flux (or outflux), typically denoted by $F(t)$ and defined as the rate of production as a function of time, or by the number of units produced in a time interval n (shift, day, etc.) often called X_n. Given that the production unit has a finite production rate μ or capacity C the simplest constraint is to require that

$$\begin{aligned} X_n &= C_n, \quad \text{or} \\ F(t) &= \mu. \end{aligned} \tag{1}$$

This constraint is only true for a system that is overloaded and hence produces at constant average production rate μ. At the same time, due to the stochastic nature of the production process, a truly overloaded system leads to increasing work in progress (wip), making this an unrealistic assumption for the characterization of the production process for any significant length of time.

If arrival rates are less than maximal production rates, mass-balance equations model the time evolution of an inventory I:

$$\begin{aligned} I_{n+1} &= I_n + R_n - X_n, \quad \text{or} \\ \frac{\mathrm{d}I}{\mathrm{d}t}(t) &= \lambda(t) - F(t), \end{aligned} \tag{2}$$

where R_n are the starts in the time interval n (Aouam and Uzsoy [1] use the term release rates) and $\lambda(t)$ and $F(t)$ are the start rate and the outflux, respectively. Notice that $R_n = \int_{t_{n-1}}^{t_n} \lambda(s)\mathrm{d}s$. As only $\lambda(t)$, (or R_n) is controlled, the system is not defined without a model for the outflux $F(t)$ (or X_n). This is where the clearing function first introduced by Karmarkar [14] comes in. The clearing function is a state equation that defines the outflux F as a function of the wip W in steady state, i.e.

$$F = \Phi(W). \tag{3}$$

The functional form of the clearing function Φ has been determined in many different ways: Measured in real factories, modeled via an $M/M/1$ queue, modeled after the fundamental diagram of a traffic model [17] etc. (see e.g. [1, 11]),

$$\begin{aligned} \Phi &= \frac{\mu W}{1+W} && M/M/1 \\ \Phi &= \mu W - W^2 && \text{fundamental diagram of traffic.} \end{aligned} \tag{4}$$

Lefeber [15] defines the clearing function by its functional inverse, i.e. the wip as a function of the flux, approximated as an $M/G/1$ queue:

$$W = \frac{c_B^2 + c_E^2}{2} \frac{r^2}{1-r} + r \tag{5}$$

where r is the utilization of the machine $r = \frac{\lambda}{\mu}$ and c_B and c_E are the coefficient of variation of the arrival and machine departure processes. Both Aouam [1] and Lefeber [15] notice that the clearing function can be approximated by piecewise linear functions, making the production planning problem an Integer-LP optimization problem.

Even at this low level of approximation there is a basic inconsistency: the clearing function is supposed to describe the outflux *in steady state* as a function of wip level. However, the clearing function is used with a wip level that is a function of time and is updated constantly to determine the outflux as a function of time. Hence, by making the outflux follow instantaneously any change in the wip level, the fundamental assumption is that the wip level changes slowly relative to the damping time of the underlying stochastic process. Therefore the fundamental assumption that justifies the use of a clearing function is that by the time the wip-level has reached a new state, the stochastic process determining the outflux is back in steady state. As a result the outflux is never in transient and always characterized by its steady state behavior. This is known as the *quasi-steady assumption* or the *adiabatic model*.

The quasi-steady assumption poses a major problem for the applicability of any type of clearing function approach. Since almost no research in production planning is concerned with the specific nature of the stochastic process, there are no good estimates to my knowledge about the damping time of the stochastic processes. In fact, even the concept is ill-defined without discussing the timescales and magnitudes of the stochastic disturbances. One way presumably would be analogous to Aouam and Uzsoy's [1] formulation of their ZOIP algorithm in assuming that there is the everyday stochasticity (in their case restricted to the demand variability) which would be captured by the clearing function and there are extraordinary events that require extraordinary measures for which the clearing function approach is not well suited. I would consider operator overload, operator negligence and scheduled machine maintenance to be part of the everyday stochasticity. For semiconductor production lines the time that e.g. a scheduled machine shutdown would be felt could be described as of an order less than the cycle time. The stochastic damping time would therefore be of the order of a day or two. Hence, to stay with the semiconductor production model, ramping start-ups by 20% over a weekly schedule would not violate the quasi-steady assumption of a clearing function model but ramping up within a day would.

3 Dealing with Delays

All clearing function approaches so far have considered the wip at the current time interval as the independent variable determining the outflux at the current time. As most production is not started and completed within a day and no production process is instantaneous this is in general not a good model. This is especially true for cycle times that are long relative to the planning period since parts that have just entered the production process will not be involved in determining the current outflux, unless the factory is reentrant or some other special circumstances apply. There are two approaches that cover the delayed response of a production unit in this book (but see also Hackman and Leachman [12] who have a detailed discussion of delayed timing issues for linear models of production systems): Effective processing times (EPT) and partial differential equation models.

3.1 The Effective Processing Time Approach

The effective processing time t_e is the mean time that a part needs to get through a stochastic processing unit, without considering the waiting time [13, 15]. Hence for a single machine it can be considered as $\frac{1}{\mu}$, with μ the average machine processing rate. By focussing on the start rates of the machines the delay experienced by a part will be fixed, independent of the buffer length. We define u_{up} to be the uptake rate of the machine immediately upstream and t_e its effective processing time. Calling the u_d the uptake rate of the machine immediately downstream of an inventory $I(t)$ (buffer), its time evolution can be written as

$$\frac{\mathrm{d}I}{\mathrm{d}t} = u_{\text{up}}(t - t_e) - u_d(t). \qquad (6)$$

By using an ordinary differential equation for the time evolution of the inventory, parts are losing their identity and, without additional modeling, the cycle time through a factory cannot be recovered from this model unless queuing is minimal. However, by using the functional inverse of the clearing function (Eq. 5) to bound the uptake of a machine, we can approximate the overall production rate of a production line rather accurately. Notice though that using the clearing function (or its functional inverse) still makes the effective process time model a quasi-steady state model with all the problems discussed before. In particular, the production rates of a machine are based on the average behavior of the machine and arrival processes and hence fast changing transients may not be resolved properly.

The relationship between the effective processing time approach and clearing functions in fact is complicated and not completely understood. In particular, a constant effective processing time is not equivalent to a linear clearing function. A clearing function describes the interaction between the stochastic processes that describe the machine availability and the stochastic processes that describe the product availability whereas the effective processing time focusses on the machine availability alone, making it necessary to develop a model for the uptake of the machine again. Lefeber [15] uses the clearing function as a bound for this uptake model but one could imagine more sophisticated approaches.

3.2 Transport Equations

Considering a factory as a pipe and parts flowing through the factory as a fluid, we can describe the transport through the factory via standard transport equations studied extensively in fluid mechanics. In contrast to fluid mechanics, the spatial variable defining the transport direction is not given by physical space but rather by the degree of completion of the part, or the stage of the production. Calling $x \in [0, 1]$ the degree of completion, $\rho(x, t)$ describes the density of parts at stage x at time t. If the fluid moves with a velocity field $v(x, t)$ then the flux is described as $F(x, t) = v(x, t)\rho(x, t)$. Mass conservation then is given by the partial differential equation

$$\frac{\partial \rho}{\partial t} + \frac{\partial F}{\partial x} = 0. \tag{7}$$

Since $v(x, t) \geq 0$ the fluid moves from left to right, allowing a boundary condition to be imposed at $x = 0$. Typically the boundary condition is $F(0, t) = \lambda(t)$, i.e. the local flux at zero is the arrival rate of the parts into the factory. Together with an initial wip profile $\rho(x, 0) = \rho_0(x)$ this sets up a well defined hyperbolic problem. Notice that we are describing a flow that is continuous in its parts and continuous in its spatial direction. This should be distinguished from the so-called fluid equation models of queueing theory [6] which are continuous in its parts but describe a flow through a finite and distinct number of queues, leading to a set of Ordinary Differential Equations (ODEs).

To clarify issues, let us examine the solutions to the mass conservation Eq. (7) for a constant velocity $v(x,t) = c$. In this case the transport equation becomes a linear first-order wave equation with the solution

$$\rho(x,t) = \begin{cases} \rho_0(x - ct) & \text{for}\, t < \frac{x}{c} \\ \frac{\lambda(t - \frac{x}{c})}{c} & \text{for}\, t > \frac{x}{c}. \end{cases} \tag{8}$$

Integrating Eq. (7) over x and defining the total wip $W(t) = \int_0^1 \rho(x,t)\mathrm{d}x$ we get

$$\frac{\mathrm{d}W}{\mathrm{d}t} = F(0,t) - F(1,t) = \lambda(t) - \lambda\left(t - \frac{1}{c}\right). \tag{9}$$

Comparing the inventory $I(t)$ in the EPT approach and the wip $W(t)$ in the PDE approach, we see that the two delay Eqs. (6) and (9) only differ in their accounting of the parts—the EPT approach counts them *after* the machine whereas the PDE approach counts it *in* the machine.

When we integrate the transport equation over the completion space all wip is aggregated into one variable, the total wip, and any uneven distribution of the wip is lost. Hence again, cycle time of an individual part cannot be recovered in the delay equation model nor are we resolving short term fluctuations of the outflux.

In contrast, in the PDE model, we can clearly follow the transport of any local wip portion given by $\rho(x,t)\mathrm{d}x$ over time through the factory. Hence, if the observation time interval Δt and the cycle time τ satisfy $\tau \gg \Delta t$, a PDE model (or its discretization) is the only one that allows us to follow the flow of parts through the production unit. For cycle time of the order of the observation times, the clearing functions based on the total wip are appropriate. This observation is independent of the velocity model that is used to describe the flow through the factory, i.e. independent of the type of clearing function that is used.

3.2.1 Clearing Functions for PDEs

Since the Karmarkar clearing function $F = \frac{\mu W}{1+W}$ [14] is only a good approximation if the cycle time is of the order of the observation time, a clearing function describing the flux in a spatially extended partial differential equation should depend on the local variable x. In particular $F(1,t)$ should depend on the density $\rho(1,t)$. We have argued in [3] that for a strongly re-entrant flow with FIFO dispatching rules the velocity should be uniform over the total completion space and hence

$$F(x,t) = v(W(t))\rho(x,t) = \frac{\mu}{1+W}\rho(x,t) \tag{10}$$

would be a good flux function.

For acyclic flows (linear production lines) heuristic discussions lead to space dependent clearing functions given the local production rate at stage x either just as a function of the local density e.g.

$$F(x,t) = \frac{\mu}{1+\rho(x,t)}\rho(x,t) \tag{11}$$

or, as in Ringhofer's chapter [21] as a linear interpolation between the flux expected for the whole factory and the flux expected at the very last machine.

$$F(x,t) = \frac{1-x}{\tau\left(\int_0^1 \rho(z,t)\mathrm{d}z - x\tau_0\right)}\rho(x,t), \tag{12}$$

where $\tau(x,t)$ is the time to completion of a production sitting at stage x at time t.

Another heuristic model that develops a clearing function for a linear production line with finite buffers has recently been developed by Armbruster et al. [4],

$$F(x,t) := \begin{cases} \frac{\mu\rho}{1+\rho+k\rho(1-x)} & \text{for } \rho < M \\ 0 & \text{for } \rho \geq M, \end{cases} \tag{13}$$

where k is an adjustable constant and M is the maximal buffer space. Experiments (Goossens [10]) that shut down the last machine in the factory and subsequently restart the whole factory with full buffers show a cascading collapse of production traveling upstream in the factory and, once the last machine has been repaired, a slower recovery to steady state. PDE simulations using the flux (Eq. 13) show good, though not perfect, agreement with the discrete event simulations.

4 Transient Clearing Functions

We have seen that, using any type of clearing function model, whether in a discrete mass balance equation describing inventories or in a continuous flow model which is characterized as a hyperbolic PDE, the model assumes that the local production rate instantaneously adjusts to the one given by the equilibrium relationship between flux and wip described by the clearing function. Recently Missbauer [19] has studied the issue of clearing functions for systems that are not in steady state. He considers a simple M/M/1 queue with a production rate of $\mu - 1$ and studies the expected output $E[X]$ over five time units as a function of the expected load $E[L]$ at the end of the five time units, depending on the initial wip w_0 and the arrival rate $\lambda(t)$, i.e.

$$E[L] = w_0 + \int_0^5 \lambda(s)\mathrm{d}s. \tag{14}$$

He argues that clearing functions that describe such transient behavior should not just depend on the total load of the system but on three variables:

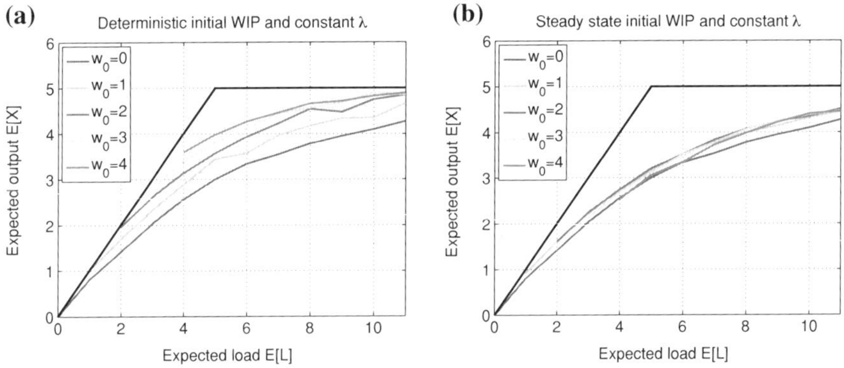

Fig. 1 Simulation of an $M/M/1$ system for different initial wip and load, 10^3 simulations per data point. **a** Deterministic initial wip, **b** Initial wips sampled from steady state distributions

- the expected initial wip level,
- the expected input during the period,
- the probability distribution of the wip level at the beginning of the period.

Fonteijn [9] extended Missbauer's study. Figure 1a shows that with a deterministic intial wip and a constant rate of influx $\lambda = \frac{E[L]-w_0}{5}$, the output over five time periods depends crucially on the initial wip as shown before by Missbauer.

However, if the initial wip is randomly distributed corresponding to an expected value of the initial wip of w_0, the curves describing the mean outflux behavior (averaged over the initial wip distribution) as a function of the expected load all pretty much collapse into each other. Figure 1b shows that the differences between different mean initial wips become very small.

The assertion that the *total* expected input during the observation period determines the outflux can be shown to be wrong by looking at the outflux for the same total input, distributed differently over time: Fig. 2a shows the clearing function for an experiment where the necessary influx to generate the expected load over the time period of five time units is generated at the *beginning* of the time period. As a result, the initial wip is instantaneously increased and hence the outflux is higher than in the case of a constant influx in time as shown in Fig. 1b. Figure 2b shows the clearing function for an experiment where the necessary influx to generate the expected load over the time period of five time units is generated at the *end* of the time period. As a result, none of the influx will come out of the factory within the time period and the outflux is only determined by the initial wip.

We can conclude from these four figures that a steady state-based clearing function will be a reasonably good description of the outflux, if the system is increased from an average of 20% of production capacity to an average of 90% of production capacity with a constant ramp within five cycle times. If ramping is done much faster or if the system is prepared in a particular initial state, the initial condition matters and so does the timing of the ramp.

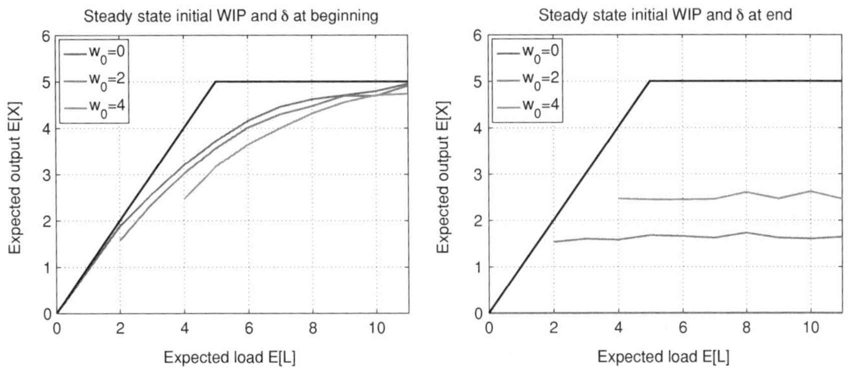

(a) Total influx at $t = 0$ (b) Total influx at $t = 5$

Fig. 2 Clearing functions for an $M/M/1$ queue with different average initial wips and (**a**) all influx arriving at the beginning of the observation interval and (**b**) at the end of the observation interval

Another approach to go beyond the quasi-static models based on regular clearing functions follows from Armbruster et al. [2]. They have developed a hierarchical set of moment equations that models the time dependent behavior of the flux and higher order moments like the variance, etc. The hierarchy of moments generate an infinite set of hyperbolic equations which by itself is not of practical use. The standard approach to such hierarchies is to truncate them at some level via a *moment closure* that defines a higher-order moment whose time dependence is not resolved any more by a relationship to lower-order moments. The simplest such closure is the clearing function, defining the flux in terms of the density. The next more sophisticated approach leads to a system of two PDEs where the flux becomes a dynamic variable. In [2] the following system of two partial differential equations is derived:

$$\begin{aligned} \frac{\partial \rho}{\partial t} + \frac{\partial F}{\partial x} &= 0 \\ \frac{\partial F}{\partial t} + \frac{\partial v(x,t)F}{\partial x} &= 0, \end{aligned} \tag{15}$$

where again $F(x,t) = v(x,t)\rho(x,t)$. Intuitively, the second equation describes the fact that a perturbation in the production rate, e.g. a region of high production rate, travels downstream with a velocity v . Treating again a factory as a single $M/M/1$ queue, the flux at the beginning of the factory is given as

$$F(0,t) = \rho(0,t)\frac{v_0}{1+W}. \tag{16}$$

This model performs better than any other model for Missbauer's test cases [19] but the errors are still significant for highly variable inputs [9].

5 Solving the Production Planning Problem Using PDE Models

In [18] La Marca et al. determine the start rate as a function of time that minimizes the mismatch between a desired production rate over a given time interval and the actual production rate according to the solution of a PDE model of a production flow. Specifically they define the cost function

$$J(\rho, \lambda) := \frac{1}{2} \int_0^{\tau} (\mathrm{d}(t) - F(1,t))^2 \, \mathrm{d}t, \tag{17}$$

where $\mathrm{d}(t)$ is the instantaneous demand rate and $F(1,t) = \rho(1,t)v(1,t)$ is the instantaneous outflux. Minimizing the cost functional $J(\rho, \lambda)$ over all possible influx functions $\lambda(t)$, subject to the PDE–dynamics introduced previously, i.e.

$$\begin{aligned} &\min_{\lambda(t)} J(\rho, \lambda) \text{ subject to} \\ &\frac{\partial \rho(x,t)}{\partial t} + \frac{\partial}{\partial x}(F(x,t)) = 0 \\ &\lambda(t) = v(\rho)\rho(0,t) \\ &\rho_0(x) = \rho(x,0) \\ &F(x,t) = \frac{\mu \rho(x,t)}{1 + \int_0^1 \rho(s,t)\mathrm{d}s} \end{aligned} \tag{18}$$

solves the production planning problem over the time horizon τ. The method is based on the formal adjoint method for constrained optimization, incorporating the hyperbolic PDE as a constraint of a nonlinear optimization problem.

Figure 3 shows the optimal influx for a sinusoidally varying demand. Notice that the nonlinear dependence of the lead time on the wip in the system generates a sawtooth-like form of the optimal influx. In contrast, a constant lead time would have lead to an input function that has the exact same functional form as the demand, just phase-shifted.

6 Conclusion and Open Problems

6.1 Practical Considerations

The production planning problem can be considered a prototype problem of multiscale modeling. We have seen that, depending on the modeling context, different time scales become relevant. At the same time different approximations are appropriate for those different time scales. It is my contention that the current literature on production planning and clearing function does not pay enough attention to the time scale issue and how the purpose and usage of the model chooses its time scale and model sophistication.

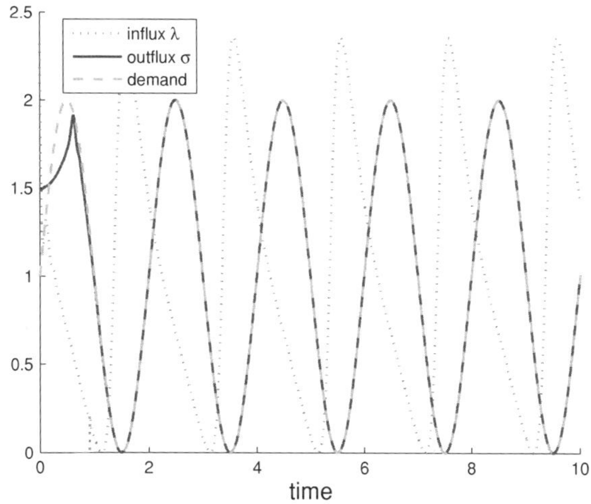

Fig. 3 A sinusoidally oscillating demand function (from [18])

A clearing function approach (and all derivatives discussed here) is based on a model that assumes that the production rate is described by a steady state relationship to the wip. Since the processes that are described here are stochastic processes, the definition of the stochastic process becomes very important. In particular, a steady state that allows us to determine a clearing function is defined by a stationary process, where the average quantities (e.g. wip and outflux) over the the relevant time interval do not differ in a relevant way from the long term averages. The concept of ordinary (captured in the average description) versus extraordinary events (not captured) becomes crucial. No extraordinary event can be incorporated into a clearing function approach although extraordinary events may define initial conditions and recovery from the initial conditions may be modeled by a clearing function approach. Specifically, a disaster like the Japan earthquake in 2011 cannot be modeled in an aggregate description of a supply chain but modeling the recovery from the disaster could be attempted.

Hence any situation that allows the production system to adjust to its steady state *before* the outflux is measured can in principle be well approximated by a clearing function approach. Any situation that requests a higher temporal resolution for the outflux will be badly modeled by a clearing function and will need models that describe the time evolution of the wip in the factory and the flux together.

When a clearing function model is appropriate there are still other timescale considerations that influence the choice of a model: If the cycle time is much larger than the observation time interval a PDE model (Eq. 7) or an effective processing time model (Eq. 6) with a suitable delay are the only ones that properly model the flow of parts through the production unit. If the cycle time is of the order of the observation times, an instantaneous clearing function model (Eqs. 2 and 3) without additional delays based on the total wip is appropriate. If the cycle time is much shorter than the observation time, an instantaneous clearing function based on the wip at the moment of the change in influx is appropriate. For instance, if a change in influx is done at the beginning of the time interval, the outflux should be based on the wip at the end of the time interval. If the influx is changed at the end of the time interval, the outflux is based on the wip at the beginning, etc.

6.2 Continuous Versus Discrete Models

There are two ways to consider the relationships between the different approaches discussed here: Once could start with time intervals based on production time units (shifts, days, weekly schedules) and on production intervals based on machines and come to the iterative model shown in Eq. (2). Going to a continuum in time leads to ODE models which are called fluid models in the queuing theory context and are the bases of the effective processing time models (Eq. 6). Assuming a large number of machines allows us to go to the continuum in production space leading to the PDE models (Eq. 7). Alternatively one could consider the ODE model a discretization of the PDE in space and the discrete time model a subsequent discretization of the ODE model in time. Usually discretization of a PDE in space and time can be done on any scale and with many different schemes while the discretizations in Eqs. (2) and (6) are based on the granularity of the actual production process. However, for simulations of hyperbolic PDEs, the space and time discretization are not independent. In order to have a stable algorithm they have to satisfy a necessary condition known as the CFL condition [16]. Daganzo [8] has shown that the CFL condition is not just a numerical analysis issue but that it is equivalent in choosing order policies that prevent the bullwhip effect.

6.3 Further Work

The current understanding of the production planning problem and the promising approaches presented in this book generate a multitude of research problems, at least some of which are well within reach of current mathematical modeling and optimization techniques.

- As Aouam [1] noted, the production planning problem has really two parts, production planning under nonlinear lead times, which has been the topic for most of this chapter, and production planning under stochasticity leading to stochastic optimization. While La Marca et al. [18] conceptually solved the tracking problem for nonlinear lead times, their approach falls short of a usable and robust algorithm since it deals only with the open loop problem. Hence, any perturbation that disturbs demand or production during the planning time horizon will typically invalidate the optimal production plan calculated with La Marca's algorithm and hence will require a complete replanning. Model predictive control linking the tracking algorithm to a discrete event simulation that provides the reality against which the re-planning will have to occur is a promising direction. However other stochastic optimization approaches should also be tried for the PDE-based clearing function models.

At this point we also have a connection to the control theoretical approaches discussed by Braun and Schwartz [7]. Any closed loop feedback control system will have to deal with schedule nervousness and limit the amount of variations that are allowed for an optimal schedule. Inherently, the approaches in [7] do not care where the errors in the model come from—they could be coming from variations in the demand but they could also be coming from the linearization of a fundamentally nonlinear clearing function. As the modeling errors increase and the request for scheduling stability stays the same, at some time there will not be a feasible solution that is at the same time smooth enough and accurate enough. Whether real industrial problems can satisfy these constraints and whether general rules for the success of this approach can be developed are open problems.

Alternatively, limiting schedule changes in a closed loop version of La Marca's model could be done in much the same way as in [7] through frozen horizons, move suppressions and schedule change suppression. This would have the advantage of a much better—nonlinear—model but the disadvantage that the LP-based optimization tools would not suffice any more.

- Deriving clearing function models from first principles is an extremely hard problem as it adds another layer to the already very hard problem of the relationship between queuing systems and their fluid models. A fluid model as it is used in approximation theory for queuing theory treats the products arriving at a queue as a continuum flux leading to an ODE description for the average behavior of a queuing system. That problem is still not completely solved for multi-class queueing networks and arbitrary priority rules at the machines. However for single class queueing networks the equivalence between the fluid model and the long term average behavior is well understood (see [6] for a introduction into this subject).

  To derive a clearing function for a supply chain or a factory with a large number of machines requires the additional limit of infinitely many production steps modeled through a continuum motion along the completion line. No first principle theory dealing with the interplay of a large number of products going through a large number of production steps exists to my knowledge.
- It will be much easier to determine more sophisticated approaches for highly transient systems. While the studies of Fonteijn suggest that using two PDEs (Eq. 15) based on the multi-moment expansion is better than one, a complete study of the approximation errors associated with these equations has not been done. In particular, it is unclear for which acyclic production systems the closure (Eq. 16) is the correct one and what other closure options are available.
- Asmundsson [5] has discussed the production planning problem for production of more than one product type. The Asmundsson ACF approach is a rough way of doing this with big time buckets and potentially restrictive assumptions, but seems to work well in many cases as a practical approach. There is an obvious

relationship to Ringhofer's [21] service rule discussion for PDE models that has not been explored.

It is hoped that this book serves as an incentive for many researchers in applied mathematics, industrial engineering and operations research to study some of these fascinating problems.

Acknowledgements This research was supported in parts by a grant from the Stiftung Volkswagenwerk, by NSF grant DMS 1023101 and by the INTEL Research Council.

References

1. Aouam T, Uzsoy R, An exploratory analysis of production planning in the face of stochastic demand and workload-dependent lead times, this volume
2. Armbruster D, Marthaler D, Ringhofer C (2004) Kinetic and fluid model hierarchies for supply chains. SIAM Multiscale Model Simul 2(1):43–61
3. Armbruster D, Marthaler D, Ringhofer C, Kempf K, T-C Jo (2006) A continuum model for a re-entrant factory. Oper Res 54(5):933–950
4. Armbruster D, Göttlich S, Herty M (2011) A scalar conservation law with discontinuous flux for supply chains with finite buffers. SIAM J App Math 71(4):1070–1087
5. Asmundsson JM, Rardin RL et al (2009) Production planning models with resources subject to congestion. Naval Res Logist 56:142–157
6. Bramson M (2008) Stability of queueing networks. Lecture notes in mathematics, 1950. Springer, Berlin
7. Braun MW, Schwartz JD, A control theoretic evaluation of schedule nervousness suppression techniques for Master Production Scheduling, this volume
8. Daganzo CA (2003) Theory of supply chains. Springer, New York
9. Fonteijn J (2009) Analysis of clearing functions and transient PDE models.Technical report TU Eindhoven, SE 420613
10. Goossens P (2007) Modeling of manufacturing systems with finite buffer sizes using PDEs. Masters thesis, TU Eindhoven, Department of Mechanical Engineering, SE 420523
11. Göttlich S, Herty M, Ringhofer C, Optimal order and distribution strategies in production networks, this volume
12. Hackman T, Leachman RC (1989) A General framework for modeling production. Manag Sci 35(4):478–495
13. Hopp WJ, Spearman ML (2000) Factory physics, 2nd edn. Irwin/McGraw-Hill, New York
14. Karmarkar US (1989) Capacity loading and release planning with work-in-progress (wip) and lead-times. J Manuf Oper Manag 2:105–123
15. Lefeber E, Modeling and control of manufacturing systems, this volume
16. LeVeque RJ (2002) Finite volume methods for hyperbolic problems. Cambridge University Press, Cambridge
17. Lighthill MJ, Whitham GB (1955) On kinematic waves. II. A theory of traffic flow on long crowded roads. Proc Royal Soc Lond A(229):317–345
18. La Marca. M, Armbruster D, Herty M, Ringhofer C (2010) Control of continuum models of production systems. IEEE Trans Autom Control 55(11):2511–2526
19. Missbauer H (2010) Order release planning with clearing functions. A queueing-theoretical analysis of the clearing function concept. Int J Prod Econ. doi:10.1016/j.ijpe.2009.09.003
20. Perdaen D, Armbruster D, Kempf K, Lefeber E, Controlling a re-entrant manufacturing line via the pushpull point, this volume
21. Ringhofer C, Traffic flow models and service rules for complex production systems, this volume